GENERAL EQUILIBRIUM ECONOMICS

Other books by Robert E. Kuenne

THE THEORY OF GENERAL ECONOMIC EQUILIBRIUM
THE ATTACK SUBMARINE: A Study in Strategy
THE POLARIS MISSILE STRIKE: A General Economic Systems Analysis
MONOPOLISTIC COMPETITION THEORY: Studies in Impact (*editor*)
MICROECONOMIC THEORY OF THE MARKET MECHANISM: A General Equilibrium Approach
EUGEN von BÖHM-BAWERK
RIVALROUS CONSONANCE: A Theory of General Oligopolistic Equilibrium
WARRANTIES IN WEAPON SYSTEMS PROCUREMENT (*with P. H. Richanbach, F. Ridell and R. Kaganoff*)
MICROECONOMIC THEORY: Theoretical and Applied, Vols I, II, and III (*editor*)
DYNAMICS AND CONFLICT IN REGIONAL STRUCTURAL CHANGE (*co-editor with Manas Chatterji*)
NEW FRONTIERS IN REGIONAL SCIENCE (*co-editor with Manas Chatterji*)
THE ECONOMICS OF OLIGOPOLISTIC COMPETITION: Price and Non-Price Rivalry

General Equilibrium Economics

Space, Time and Money

Robert E. Kuenne
Professor of Economics
Princeton University

NEW YORK UNIVERSITY PRESS
Washington Square, New York

First published in the U.S.A. in 1992 by
NEW YORK UNIVERSITY PRESS
Washington Square
New York, N.Y. 10003

Printed in Hong Kong

Library of Congress Cataloging-in-Publication Data
Kuenne, Robert E.
General equilibrium economics : space, time, and money / Robert E. Kuenne.
p. cm.
Includes bibliographical references and index.
ISBN 0–8147–4636–5 (cloth)
1. Equilibrium (Economics) 2. Space in economics. 3. Time.
4. Money. I. Title.
HB145.K84 1992
339.5—dc20 92–1414
CIP

For my children

Christopher Brian Kuenne
Carolyn Kuenne Jeppsen

with loving memories of their childhoods,
great pride in their adulthoods
and thankful joy in their companionship

Contents

Preface

General equilibrium economics has made great strides since Hicks and Samuelson began to move Anglo-American economic theory beyond the confines of Marshallian statics into the wider vistas of the Walrasian-Paretian legacy. In the 1950s Arrow and Debreu applied the techniques of real analysis, set theory and topology within a game-theoretic context to investigate the necessary and sufficient conditions for the existence of a general equilibrium. Lange and Patinkin initiated an examination into the integration of monetary and real sectors in such systems. During the same period Leontief's input–output derivative began to be applied directly to realistic problems and to be embedded in more ambitious econometric models to the same end. And during the 1960s and 1970s operational general equilibrium frameworks were developed with greater sophistication for applied work.

But much labor remains to be done to improve the field. Some progress has been made in moving general equilibrium theory away from its pure competition moorings, but that task has just begun. Welfare economics – which general equilibrium theory must motivate – must be reworked to incorporate imperfect competition. Uncertainty and imperfect information must be intruded into the theory, and space still beckons as a largely excluded dimension. And better methods of teaching general equilibrium systems analysis and instilling its 'vision' of economic reality would be welcomed.

In the present volume I have selected twenty-two of my papers in the field that are grouped into four areas of interest. The first treats the general theoretical and empirical problems and successes in general equilibrium economics, explaining its general structure and uses, and culminating in my attempt to build a model of general oligopolistic equilibrium. A second area concerns some investigations into spatial interdependence, including development of exact and approximate solution algorithms as well as investigation of a Poisson process in space. The third topic is concerned with the attempt of Austrian theorists, especially Böhm-Bawerk, to build a general equilibrium approach to capital and interest theory. And the last area concerns the integration of money and real sectors in Walrasian and Keynesian models, as well as the conditions for the existence of money in a stationary state and the role of money in an intertemporal general system.

The topics are, I believe, enduring themes in general equilibrium analysis, embracing concerns about the foundations of economic theory, and therefore likely to remain of importance to theorists. They will continue to be researched and will stimulate new and better insights into the market economy, I am sure. Today's young economist, much better equipped with the mathematical apparatus to probe into variate interdependence than were most economists of my generation, has a great potential for improvement of the understanding of the system. But there is a danger that methodology will become an end rather than a means, so that much talent will be wasted in seeking ingenious solutions to questions that are amenable to attack by such techniques but which have no issue of discernible relevance to the phenomena of this earth. If this eventuates, general equilibrium economics will be cordoned off from the body of economic analysis by a wall of antagonistic – and much-merited – disinterest. Care must be taken to choose realistic problems for research and to make the results of that research accessible to the economic practitioner.

My hope is that the present volume will encourage efforts in these directions. I have been indebted over my career to a large number of individuals who inspired my interest in economics in general and economic theory in particular. Professor Werner Hochwald of Washington University in St. Louis was especially influential in these respects. Professors Edward H. Chamberlin, William Fellner, Walter Isard, Guy Orcutt and Arthur Smithies at Harvard were exceptionally inspiring.

I am also grateful for the financial support of the Ford Foundation, Rockefeller Foundation, National Endowment for the Humanities, the National Science Foundation and the Fulbright Commission during various periods of my research. And, finally, I have enjoyed the gracious hospitality of the University of Vienna during my stay in Austria, and am especially indebted to Professors E. Streissler and E. Weissel of the *Institut für Wirtschaftswissenschaften* at that institution for their support and encouragement.

ROBERT E. KUENNE

Acknowledgements

The author and publishers wish to thank the following for permission to reproduce the chapters indicated.

John Wiley & Sons, publishers of the *Quarterly Journal of Economics*, for permission to reprint Chapters 1, 3, 15 and 21.

Helbing & Lichtenhahn Verlag AG, Basel, and Frankfurt/M, publishers of *Kyklos*, for permission to reprint Chapter 2.

Elsevier Science Publishers, publishers of A. Andersson, W. Isard and T. Puu (eds), *Regional and Industrial Development Theories, Models and Empirical Evidence*; Robert E. Kuenne, *Rivalrous Consonance: A Theory of General Oligopolistic Equilibrium*; *Review of Economics and Statistics*; and *Mathematical Programming*, for permission to reprint Chapters 4, 8, 10 and 13.

Chapman & Hall, publishers of *Applied Economics*, for permission to publish Chapter 5.

Princeton University Press, publishers of Robert E. Kuenne, *The Theory of General Economic Equilibrium*, for permission to reprint Chapter 6.

Editrice Cedam, publishers of *Rivista Internazionale di Scienze Economiche e Commerciali*, for permission to reprint Chapter 7.

The *Journal of Regional Science* for permission to reprint Chapters 9, 11 and 12.

Columbia University Press, publishers of Robert E. Kuenne, *Eugen von Böhm-Bawerk*, for permission to reprint Chapter 14.

Springer-Verlag Wien, publishers of *Zeitschrift für Nationalökonomie*, for permission to reprint Chapter 16.

The *Southern Economic Journal* for permission to reprint Chapters 17 and 18.

The Canadian Journal of Economics for permission to reprint Chapter 19.

Metroeconomica for permission to reprint Chapter 20.

Librairie Droz SA, Geneva, publishers of *Economie Appliquée*, for permission to reprint Chapter 22.

Note: the reprinted material appears as in the original publications except that minor alterations have been made to the style and presentation for the purpose of publishing this collection of readings in book form.

Introduction

The opportunity to review one's writing and research efforts over forty years permits one to become reacquainted with one's younger self, to assess what one interprets – perhaps too self-indulgently – as growth, and to judge the worthwhileness of what one has been about for a lifetime. Like all self-education that is a satisfying experience, but not without the pain of wishing that one had seen things more clearly more quickly, that one could have avoided mistakes and had not entered so many blind alleys. Like all self-judgment it is at once too forgiving and too harsh, and to be viewed with (hopefully generous) suspicion by others. All in all, however, it has been enjoyable in the present case, and I am grateful for the opportunity to do so. My hope is that the reader will find the volume valuable for his or her purposes.

General equilibrium economics has retained a dominant interest for me from the time of my research for my doctoral dissertation at Harvard to the present. It has provided the frame within which my other major effort – oligopoly theory – was inspired and given shape and dimension. Its vision of economic process as a complex mesh of interdependent variation, with multidirectional flows of causation and simultaneous determination of values, is both humbling and challenging. It is also vital, I believe, to the early education of the journeyman economist, at which time he forms the outlooks that will shape his professional *anschauung* and provide both the source and temper of his research.

It has another contribution to make to the professional economist: it defines the boundaries of his or her professional concerns, providing him or her an overview of that terrain and an ordered means of integrating his or her interests into that terrain. Few other social sciences have achieved the consensus concerning the variables within their remit that ours has, or the assurance that, however abstract, a framework exists for their joint and simultaneous determination, given information that derives wholly from other fields. Schumpeter has termed general equilibrium theory the 'Magna Charta' of the economist, the very constitution of his profession granting him independence and sovereignty within the specified turf.

My dedication to the field has taught me to be lovingly critical of its content and development, however, and the reader will find the results of that dissatisfaction throughout this volume. A first concern of mine – engendered in part by Edward H. Chamberlin's classes and work – was its adoption of pure competition as its universal market structure. The reasons for this are clear, of course, but its unreality in face of the dominance of differentiated oligopoly recommended that some effort be given to the incorporation of this more realistic and qualitatively different form of interdependence. My effort to accomplish this goal is contained in another volume, and will be discussed in only one of the papers reprinted.[1]

Another concern of mine was developing more fully the economic foundations of some of the economic constructs in the Walrasian theory. Some of them were presented in rather cryptic, almost mechanical, formulations which were not sufficiently buttressed with economic interpretations, and therefore suspect or unfinished. Walras's theory of investment and saving was unclear at points to the degree of incomprehensibility, for example, and needed further explanation. His theory of money, while making extremely important and innovatory contributions which are still not sufficiently appreciated, was unclear in its failure to identify associated markets and was imperfectly integrated with the investment sector in particular and the real sectors in general. Indeed, on close inspection it seems to incorporate Say's Identity, which denies any possibility for integration with the system as a whole and a determinate price level.

My work in Part III contains my investigation of the contributions that Austrian capital theory could make to the provision of a deeper understanding of the straightforward productivity interpretation Walras gave to capital goods and his lack of probing into the utility motivation of saving. Although I conclude that it can make no *formal* contribution to model structure, it has much to offer in the way of fleshing out with economic substance the barebones mathematical treatment in the early general equilibrium theorists. Much progress has been made in recent years in improving the structure dealing with investment and capital, particularly in an intertemporal framework, but even those analyses need the deeper motivation that Austrian theory provides.

Part IV contains my more extensive concerns with the justification, roles and integration of money in general equilibrium theory. There the reader will find my interpretation of Walras's monetary theory, my reconstruction of it to avoid Say's Law, and my defense of it against Patinkin's charge of the presence of the Invalid Dichotomy. Also, viewing Keynes's *General Theory* as another form of general equilibrium theory, I examine the manners in which it differs from 'classical' models to discover the source of his major theorem: the inability of wage declines to reduce unemployment. I find another dichotomization of the monetary and security sectors from the real sector necessary to achieve these results and term it 'Keynes's Identity'. And, in later work, I employ a set of simple dynamic models of the von Neumann type to gain insights into the role of money in an intertemporal framework and its integration with the real capital sector.

A third area of concern to me was the non-inclusion of space in the classical general equilibrium frameworks. This interest in the spatial dimension was acquired in Walter Isard's classes, extended in his direction of my dissertation, and enhanced by forty years of friendship with him and acquaintance with his work. My efforts to repair this deficiency and to cope operationally with the very difficult forms of interdependence that arise in this effort are contained in Part II. Five algorithms – three of them to obtain globally optimal solutions and two to get local solutions – are presented for some complicated spatial models, and another paper applies a spatial Poisson process to the exploration of the relation of price and market areas. Much work remains to be done in this sector, but regional science now exists as a field to provide active research for such purposes.

A last concern of mine is the sterility of general equilibrium frameworks in terms of giving insights into economic variation with theorems derived from comparative statics. It is 'nonoperational' in Samuelson's terminology, being unable to evaluate unambiguously the signs of changes in variable values brought about by parameter displacements, using information derived by second-order extremum conditions or from the assumption of stable equilibria. The alternatives were to specify such systems numerically, accepting the necessary structural simplifications, or to adopt partial analyses and sacrifice interdependence. Of course, the former route seemed preferable, and so Leontief's input–output model held a great deal of attraction for me. Early and later applications of that framework to empirical problems are contained in Part I, and presented in detail in two co-authored papers.

The remaining papers in Part I contain my thoughts about the major contributions of general equilibrium theory to the field of economics. I am encouraged by the progress that has been made in the last twenty years in applied general equilibrium analysis, especially in its level of economic sophistication. The computer has been a major enabler, of course, as have newer forms of optimization theory, but an able younger corps of economists has been the prime mover. But nonetheless I continue to think that the primary functions of the subject will be the pedagogical ones discussed above: shaping economists' broader awareness of the boundaries of their subject, the net of interdependence that makes it cohere, the placement of its narrower specialties, and the faults in its deeper foundations that prevent better integration of those specialties. That is an indirect contribution to analysis, to be sure, but it is 'roundabout' indirectness that assures better analysis in the long run.

In these respects I must record my skepticism about the usefulness of much of general equilibrium theory in the last thirty years, despite its mathematical virtuosity. Indeed, the desire to exploit set theory, topology and real analysis has steered much of it into paths that contribute little to the functions listed above. Proofs of the existence, uniqueness or stability of equilibria, or stability of systems, revolve about unrealistic assumptions concerning the characteristics of functions or sets or other exotica which cast little light on anything but the analyst's ability to adapt mathematics to problems of little relevance to reality. We are in the peculiar position of witnessing methodology driving the problems addressed rather than seeking answers raised by the empiricist. There is a place, no doubt, for the 'aesthetics' of modelling – proving theorems with fewer hypotheses or deriving them from alternative assumptions – but a great deal of effort from some of our potentially most imaginative talent has been devoted to it, and one must wonder about the opportunity cost.

Each of the chapters is discussed in the introduction to its relevant part, and so I will not repeat those comments. My hope must be that the reader in perusing them will be at once interested, instructed, inspired – and forgiving.

R. E. K.
1 March 1991
Vienna, Austria

Note

1. See *The Economics of Oligopolistic Competition: Price and Non-Price Rivalry, Collected Papers of Robert E. Kuenne*, Basil Blackwell, Oxford, 1992.

Part I

General Equilibrium Theory: The Economics of Interdependence

Introduction to Part I

For me the major challenge in economic theory has been to cope with large-scale interdependence, and my major research effort has been to attempt to do so at the conceptual and the empirical level. Large-scale modelling had been a major interest, therefore, in both endeavours. The chapters in Part I, I believe, indicate the persistence of that interest in the theoretical field over almost forty years of effort, and Parts III and IV will display its application in the fields of capital and of money. Finally, Part II will feature my empirical explorations of large-scale operational modelling, especially in the area of spatial theory.

Interdependent economic variation is the hallmark of differentiated oligopoly, and therefore my desire to integrate it into general equilibrium theory is also explicable in terms of my underlying motivation. I have devoted a major effort to accomplishing this aim, touched upon in this volume, but explored in depth in another collection of my writings revolving about that interest.[1]

Chapter 1 I view as a point of departure for a career that was to devote the overwhelming amount of time and energy to these two areas of investigation, united by my belief that realistic theory and empirical methodology had to encompass large amounts of interdependence. It is an extensive presentation, interpretation and extension of the Walrasian system, as well as the derivation of the Leontief input–output model from that system.

The paper was published before William Jaffé's critical translation into English made Walras's work accessible to the Anglo-Saxon economic community. At the time, although Walras was quoted or referenced by such economists as Hicks and Samuelson, his work in full detail was not well known. The Marshallian frame dominated the mainstream in its Chamberlinian, Robinsonian, Pigovian and Keynesian versions well into the 1950s, and, indeed, continues today to have its exclusionary devotees. We know that Marshall was antagonistic toward general equilibrium, urging Edgeworth at one point to close the *Economic Journal* to Walras.

Also, Leontief's input–output model was just beginning to penetrate the empirical field. The Bureau of Labor Statistics had just published the 50 × 50 sector model for the United States and, although many references to the framework as an 'operational form of general equilibrium analysis' were made, no attempt had been made to my knowledge to reveal the relation to the Walrasian structure in any formal and detailed manner. These conditions, therefore, motivated the article, the research for it having been performed for my doctoral dissertation under Walter Isard which featured an empirical application of input–output analysis. If the reader feels, therefore, that I am dealing with matters well known to him or her, I am pleased that progress has been made in extending professional awareness of both systems, and plead that it was not always so.

The problem of incorporating money into an economy without uncertainty was an early concern of Walras's critics, having been broached as early as 1931 by Hicks and Marget. Patinkin in the 1950s began his attack upon classical economic theory in general and Walrasian monetary theory in particular for having failed to integrate money into the real economy, making absolute price levels indeterminate and creating dichotomized real and monetary sectors. Schumpeter, untypically, seems to misinterpret Walras in these issues as well as the nature of the criticisms. I felt at the time, and continue to do so, that the critics were refusing to allow minimal amounts of friction into the Walrasian exchange system in the case of the earlier charges, and that Patinkin failed (along with a large number of other critics) to understand the Walrasian monetary system.

Walras's theory of money and money services is a quite complicated one, as the reader will discern from this chapter, and is not clearly stated in the *Eléments*. I give my interpretation in this chapter – an interpretation which I could not gain agreement upon with either Jaffé or Patinkin – but which I continue to believe is the only construction consistent with the inner logic of Walras's model. I present it in this paper and modify it in ways I believe constitute an improvement, and I employ it more extensively in Chapters 17 through 21 of Part IV in developing a rationale for money in the stationary state and in defending Walras from the charge of incorporating Patinkin's Invalid Dichotomy. As will be seen in the chapters to follow in Part I and those mentioned in Part IV, Walras's monetary theory was that part of his model which was most criticized, and yet, to my way of thinking, the area in which his achievement was most notable if least appreciated.

Some space is devoted to interpreting Walras's theory of saving and investment, which is most ingenious, in that it creates a homogeneous unit of investment and unites all capital goods and money in a single theory. That portion of his accomplishment I also feel is underappreciated.

I do criticize Walras's model on the basis of its failure to be linked intertemporally to periods over a multiperiod time horizon, its dependence upon purely competitive markets, and the omission of international trade and government sectors. Hicks's work was to go far in correcting the first, as was that of other economists who linked consumer and investment behavior to expectations over time. Keynes's work was linked to this effort in that within a macro framework he introduced the type of uncertainty that made a modern theory of money – a form of investment in Walras – integrable, although of course he did not develop the theory within a microeconomic context.

Triffin provided an early attempt to move away from pure competition by introducing Chamberlin's large-group case into general equilibrium, or, more accurately, providing a basis for doing so. In this chapter, as elsewhere, I have questioned the adequacy of this market structure as a suitable one in terms of realism. As can be seen from this article, since early in my career I have been concerned that the large-group case was an insufficient response to the challenge, and that oligopoly had to be incorporated.

The dependence upon pure competition has remained a fault in general equilibrium theory, reinforced by the introduction by Arrow and Debreu of topological and real analytical methods to establish proofs of existence and uniqueness of equilibrium. Over my career I have attempted to develop operational techniques to permit the introduction of differentiated oligopoly into the theory, and in 1986 I published a work which derives a general equilibrium model with pervasive oligopoly, parts of which are presented in Chapters 7 and 8 below.

Mosak has extended the model into international trade theory as I myself have attempted to do in my 1963 book *The Theory of General Economic Equilibrium*. Government, of course, continues to elude analysis within traditional economic motivational frameworks.

Finally, Section IV of the paper derives the Leontief system from the Walrasian model by indicating which endogenous variables become exogenous and the types of simplified functional forms adopted to make the system operational. In retrospect, I believe I was unnecessarily harsh upon the failure of the input–output model to incorporate substitution and marginal productivity: forty years of searching for operationalism has given me a great deal more sympathy for simplification. Moreover, the relation of the quantity and price models in input–output (equations I and II in the paper) is better referred to as a dual relation, indicating that one set implies the other, rather than giving the impression they are rigidly dichotomized. Although I retain a good deal of the disappointment that much interdependence was sacrificed in moving to Leontief's system, and I continue to believe that it is not in any effective way a 'general equilibrium system', I am much more sympathetic to the necessities with which it compromised, and see the system as an important step forward in rendering interdependence models operational.

In this chapter the reader is made aware of the basic model that shaped my 'vision' of economics, defining its bounds and integrating its sectors – consumer, producer, markets, investment and money – into a whole. I continue to believe that the young economist is offered no better opportunity to obtain an overview of his or her professional terrain, the manner of its unification, and the placement of his or her own specialty within the whole than by becoming familiar with the Walrasian system, suitably modified by modern advances. Even in the absence of hopes of making it capable of yielding operational theorems, the gifts of perspective and dimension are important ones in any field, and economics is fortunate in its possession of such a tool. I believe the reader will be pleasantly surprised to see the ease with which modern theoretical departures fit into the Walrasian system to give insights into their relations with subsystems other than those for which they were designed.

The reader can also see in the chapter the discontents and challenges I perceived in the model that were to provide a basic unity to the further efforts I was to devote my career to in the future: the integration of oligopoly into the model; a closer examination of the role of money in a real economy and the manner of its integration; improved methods of dealing with capital, interest and savings in a general equilibrium framework, with a strong interest in Austrian theory as a means of doing so; the search for methods of making the model operational; and the means of introducing space into the system, especially in an oligopolistic context. Those subjects will be explored in the chapter to follow.

Chapter 2 was an invited review article of William Jaffé's monumental English translation and critical annotation and commentary of Walras's *Eléments*. More than half a century after their completion, the Anglo-American economic community was given access to Walras's models, and in the paper I wonder at the time lapse and its causes. I also speculate about the speedier understanding that might have greeted Keynes's system had economists been more at home with general interdependence models. In my experience it was the Continental economists, by and large, with some acquaintance with general equilibrium systems, who greeted that 'new dawn' with a great deal more equipoise and understanding.

One comforting reflection one had on reading the translation is that the confusing theory of saving and investment remains so after the impediments of a foreign tongue are removed. Jaffé, too, found parts incomprehensible. Also, Walras's comparative statics techniques were quite informal, but he saw clearly that the purpose of the model was to derive insights into reality by parametric displacements of equilibrium. Moreover, he expressed the need for a dynamic analysis to prove that the markets' *tâtonnements* would lead the economy to an equilibrium that his system determined. And he was insistent upon identifying real world markets in which prices and quantities were determined with his notional markets. In these senses he was striving for an operational theory, and most hopeful about its achievement. The limited potential for such results would await Samuelson's *Foundations*.

But the wonder is not that Walras's reach exceeded his grasp, or that his perceptions and methodology were deficient in some areas, but rather that his imagination and analysis accomplished so much. He saw, and insisted upon others seeing, the economy as a largely self-contained decision-making mechanism whose vital and interesting functional characteristic was an interdependence among markets brought about by flows of price information. Not content with a simple exchange economy, he sought to incorporate money, investment and growth into the model, with varying degrees of success, granted, but with great perception as to the needs of the field. The work stands as an architectural masterpiece, albeit with some rooms unfinished and some windows askew.

After the publication of *Value and Capital* Hicks's concepts of stability, and especially what he called perfect stability, in an exchange economy received a great deal of attention, particularly concerning their relation to stability of equilibrium conditions derived from explicitly dynamic models. Samuelson's 1944 paper brought the issue to light and his analysis and conclusions were repeated in the *Foundations*. Chapter 3 focuses upon another aspect of the Hicksian conditions for perfect stability: it does not relate to the earlier discussion.

The argument begins with an explanation of the conditions for Hicksian perfect stability as motivated by Hicks's description. They hinge upon the readjustment of a market in disequilibrium after a succession of subsets of other markets are readjusted to equilibrium. Suppose, for example we start in an n good economy where the nth good is the *numéraire* and the system is in equilibrium, with all markets exhibiting zero excess demand. Suppose now we raise good 1's price, holding all other prices constant. Then a necessary condition for perfect stability is that excess demand in good 1's market be negative. Next, let us adjust the price of good 2 to re-equilibrate its market. We then adjust good 2 and good 3 prices to re-equilibrate them, and so forth in a sequence of subsets until all $n-2$ markets (other than good 1 and the *numéraire*) have reattained zero excess demand. As Hicks asserts, if, after the price adjustments in every subset in the sequence of markets good 1 experiences negative excess demand, its equilibrium is perfectly stable. If at one or more stages it experiences positive demand but reattains negative excess demand when the other $n-2$ markets are in equilibrium, the initial equilibrium is imperfectly stable.

The question posed in Chapter 3 is whether this process of readjustment is independent of the order in which the $n-2$ markets are chosen in sequence. The paper shows by counter-example that it is not, and hence Hicks's motivation of the mathematical conditions is faulty. Hence it is Hicks's description of the mathematical conditions that is faulty, not these conditions themselves. He should have asserted that his restrictions on excess demand must hold for *all* adjustment permutations. The meaningfulness of the Hicksian static stability conditions in addressing the concept of stability of equilibrium, which is by its nature a dynamic concept, is another issue.

Chapter 4 is a pioneering work in regional input–output analysis which seeks to derive an operational 'general equilibrium' model for a region and to employ it in projecting regional growth. At the time of its application no regional input–output tables were available for the US economy, and, indeed, the 50 × 50 sector national tables had just been published for data relating to 1947. Hence the level of aggregation in effectively available input–output tables was quite large.

In the early 1950s the US Steel Company announced its intention to build a large, integrated steel mill above Philadelphia on the Delaware River near Trenton, New Jersey. This posed a challenging research project to forecast the impacts upon output and employment that would occur directly and indirectly in every sector of the regional economy centering on the river, from Philadelphia through New Jersey, southern New York state and into Western Connecticut. These impacts were to be projected for a ten-year period to permit existing industry in the area to expand capacity and to allow 'agglomeration' among metalworking industries to occur to exploit the joint availability of markets and raw materials.

The problem, therefore, was one of operational general interdependence modelling, and input–output analysis was at hand and newly specified in numerical terms for applications. One major problem we faced was to derive the regional model from the national model. There were no data available to project from an *interregional* model, or one which would permit feedbacks from regional imports into the Delaware Valley from other regions in the form of induced exports to those regions. A *regional* model, which treated the rest of the world including the excluded portions of the United States as exogenous and leading to income leakages with no feedbacks upon our region, had to be constructed. Our rather *ad hoc* and crude techniques for doing so are explained in the chapter. Also, in an important modification of conventional input–output analyses, we made the regional consumption sector endogenous, capturing thereby the regional consumption multiplier effects.

We estimated conservatively that a minimum of 3 million tons of steel annually would be produced in the region, on the basis of past demand data and expected growth. We expressed the desire to bracket such estimates with upper and lower bounds, and to employ an intermediate estimate as our most likely projection. Interestingly, perhaps, for the younger researcher, that which is routinely possible today was impossible then. No computer existed to permit us to invert the regional Leontief matrix or to employ it in solving our problems iteratively with the input matrix. The latter had to be done by hand with a mechanical calculator, and computational effort was so burdensome that one set of estimates had to suffice. To perform those computations, for each iteration, each of 45 rows was multiplied by a 45-element vector and the result summed, and the operation was repeated to check for accuracy. This was repeated 45 times, once for each row, for each iteration. It required 6 iterations to achieve convergence of each row result to a value sufficiently close to zero to permit reliable extrapolation. The results of these iterations are recorded in Table 4.5 in this chapter. To put it mildly, the work was tedious and time-consuming, and sensitivity analysis could not be pondered.

To estimate agglomeration employment we employed statistical data on production workers in customer/supplier two- and three-digit SIC code industries related to the steel industry, recent growth data in such industries, and interviews with knowledgeable persons in the industries to arrive at estimates for 26 such sectors. The techniques are illustrated in the chapter for one such industry. In total, we estimated new production workers in these agglomeration industries within the region over the following decade would total 61 326, or about 6.7 times the new production employment associated with the new steel capacity. That ratio was about 8.9 nationally in 1947.

From these estimates as well as estimates of the demands for regionally produced intermediate products to be used directly as inputs by the new steel mills and agglomeration industries, we constructed a bill of goods for application to the regional input–output table. From the sum of the iterations discussed above and reproduced in Table 4.5 we obtained estimates of induced regional employment in 45 sectors, including both production and nonproduction workers. Our projection was a directly and indirectly induced employment of 70 089 which, with the employment by the steel and agglomeration industries, brought the total to 158 769. With independently estimated induced construction and government employees, the grand total of new employment amounted to 180 228.

Finally, the chapter briefly considers the potential impact of this growth in the region upon the Puerto Rican economy, which was of interest to the senior author at the time.

The article was a seminal effort at devising new techniques for making microeconomic estimates incorporating the interdependence of economic sectors. The field of regional science has moved well beyond it, given the availability of interregional input–output models, the computational power of the computer, and a better regional database. But the paper generated much interest over the forty years of its availability, and, to the authors' gratified surprise, continues to generate interest in the literature.

Chapter 5, written about thirty years after Chapter 4, in conjunction with research sponsored by the US Department of Energy, provides a compelling illustration of the progress over that period in improving the methodology available to the empirical analyst using interdependence modelling. It employs Clopper Almon's forecasting model, INFORUM, which integrates a 185-sector input–output model with a set of econometrically fitted regression equations that endogenize such variables as consumption, construction, investment and government spending. Such variables are normally exogenous in input–output modelling, forming the parametric 'bill of final goods'. The INFORUM input–output model itself incorporates annual changes in its input coefficients reflecting foreseeable technological advances.

The specific problem attacked in the paper was an important and topical one, and retains its policy relevance today. In 1975 the US Congress mandated a set of phased miles-per-gallon standards for the automobile industry for 1978 to 1985. Projecting the weight reductions and technological changes that could be expected over this period, we sought to determine which industries would gain or lose output and employment owing to the adaptations necessary, and by how much. The procedure of determining expected weight changes in a 'notional' or average automobile over the relevant years, of projecting the alterations in the automobile industry's input coefficients because of technological changes motivated by the law, and constructing incremental bills of goods to be impacted upon the model are explained stepwise in the paper.

Throughout the study lower and upper bounds on strategic parameters were determined to hedge against the uncertainty in such estimates. As noted in Chapter 4, such bracket estimation of parameters and resultant solution bounds were not feasible when the steel study was done. Its present feasibility constitutes, in my judgment, the best manner of coping with uncertainty in empirical modelling.

Our basic results were that the solutions implied that the most optimistic and most pessimistic parameter sets projected a slight *positive* impact upon net employment, and that even with directly affected sectors (iron and steel, petroleum and gas, wholesale and retail trade, and plastics) variations of only 0.7 to 4.0 per cent over the period would occur. Hence, policies designed to protect the environment and reduce petroleum consumption and import could proceed with minimal concern about their macro- or microeconomic employment and output effects.

The study is reproduced in juxtaposition to the steel study of Chapter 4 to indicate the increasing availability of sophisticated interdependence models to the empirical researcher, as indicated above. Such large-scale models have their structural problems, and from our experience with its use in this project INFORUM is no exception. But with the revolution in computational power over the last forty years, sensitivity analysis can reduce the implications of these defects significantly. In my own experience with employing large-scale interdependence modelling in such policy areas and with teaching their usage and interpretation to policy-makers, the greatest obstacle is getting such decision-makers to accept bracket estimates rather than demand point estimates of variables. Ever ready to appeal to uncertainty in defending their programs, they are extremely reluctant to accept its consequences when receiving advice.

Chapter 6 is extracted from my book, *The Theory of General Economic Equilibrium*, published in 1963, which contained my models of purely competitive general equilibrium, attempting to incorporate space, time and international trade into the punctiform, static, closed classical models. It is basically Walrasian in structure but contains important modifications of the Walrasian framework, especially in the monetary and investment sectors. Many of these will be discussed in fuller detail in Parts III and IV of this volume. The book drew heavily upon Samuelson's *Foundations* in its approaches to comparative statics and dynamics and upon Arrow and Debreu in the explication of proofs of existence of equilibrium. The lessons I pointed to in the work – emphasized in this chapter – were those that led me to follow two paths away from these formal models in my future work: (*1*) the discontent with the assumption of pure competition and my belief that differentiated oligopoly had to be incorporated into such models, and (*2*) a disappointment with the meager capacity of such models to yield operational qualitative theorems, and the search for quantitatively specified models with which to obtain insights into oligopolistic general models. This latter path seemed preferable to the alternatives of aggregating microeconomic sectors – which seems especially difficult with oligopolistic industries – and of 'partializing' the models with *ceteris paribus* assumptions with the necessary sacrifice of interdependence.

This chapter, which is the introductory chapter of the book, is designed to give the reader an overview of general equilibrium theory, its uses to derive theorems, distinctions among the structures of such models, and characteristics of general static and dynamic equilibrium. All of these matters are developed in fuller detail in succeeding chapters, so presentations here are concise. The degree of interdependence of models is analyzed in terms of decomposable and indecomposable matrices. The uses of general equilibrium models are seen here as those of 'framing' the field of economic variation, or 'closing' the circle of variation of concern to economists to the variables whose determination by models is the worry of other fields. These are the 'data' (or, in more current terminology, the 'parameters') of economics. Dynamic models are distinguished from static, optimization models from non-optimization, and equilibrium from non-equilibrium models. Existence, uniqueness, economic meaningfulness, and stability of equilibrium solutions are briefly explained as concepts. A great deal of time is spent in interpreting the methods and goals of comparative statics for a simple demand–supply market equilibrium, and the rapid increase in complexity of such methods as the number of endogenous variables is increased is emphasized. Ambition in a model is the enemy of usefulness.

The primary goal of the chapter, and the main reason for reprinting here, is to present an integrated view of the goals of general equilibrium theory, its important characteristics, the uses to which it can be put, the methods by which those uses are implemented, and the limitations of those methods with their implied restrictions on usefulness. Secondarily, and for the extended purposes of this volume, it presents the major lessons classical general equilibrium theory taught me about its limitations, and the paths of research which attracted me in my further work in the field.

I began to search in earnest for a realistic, operational theory of oligopoly that could be fitted into general equilibrium frameworks in the late 1960s. I became convinced that no such universal theory was obtainable, given the distinctive structural and functional characteristics of each oligopolistic industry. The search also led me to other conclusions about the nature of the desired model. Existing analyses were almost wholly game-theoretic in inspiration, and I believed that although this approach might be applicable in certain cases, in general it overstressed the competitive relations among oligopolistic firms. In general, empirical research both in Europe and in the United States seemed to point toward a mixture of the competitive and the tacitly co-operative. Mature oligopolies – the ones of central interest in a developed economy – became communities of firms which came to accept a 'rivalrous consonance of interests', leading them to a mixture of rivalry and tacit collusion in their pricing decisions for individually selfish desires to stabilize prices.

Further, empirical research seemed also to prove rather clearly the fact that oligopolistic firms – especially large ones – devised strategies that reflected multiple objectives. A primary goal might be profit maximization, limited by the co-operative urges discussed above, but secondary goals existed to constrain this search in addition to rivalrous consonance. Maintenance of or seeking target growth in market share, limits upon price changes for public relations purposes, or minimum or maximum levels for advertising budgets are examples of possible secondary goals. Indeed, the whole set of such secondary goals, defined firm by firm, was one of the important features defining the singularity of each oligopolistic industry.

Flexibility, operationalism, capability of incorporating mixtures of co-operation and rivalry, multi-objective decision-making, preservation of each firm as an analytical entity, and ability to be incorporated into a general equilibrium system – these were demanding desiderata. My solution was the *rivalrous consonance* theory of oligopoly, in which firms strive to maximize their own profits plus their rivals' profits, when those rival profits are discounted by multipliers reflecting the power structure of the industry as perceived by each firm. Hence the firm's optimization is 'crippled' by the need to incorporate its rivals' profits in its objective function, as well as restrained by its own secondary goals expressed as constraint functions. The framework, therefore, from a computational standpoint, becomes one of nonlinear programming.

Chapter 7 deals with a question that has always interested me: if agents follow certain rules of conduct that have an intuitive appeal to them, is it possible to treat such rules as first-order conditions for maximizing some 'phantom' objective function and integrate to obtain that function? Further, can such an objective function be derived for an industry from which the industry solution can be obtained without having to resort to firms' iterative and interactive decision-making leading to that solution?

These questions are investigated in this chapter within the context of rivalrous consonance with admittedly rather discouraging results with respect to the general ability to perform the integration. In performing this exercise within the rivalrous consonance framework, however, an introduction is given to that theory which will be more fully developed in Chapter 8. Also, in the course of seeking a phantom objective function for the rivalrous consonance industry, an interpretation of Lagrangean multipliers or dual variables is given which might possess independent interest to economists. It should be noted that because the goal of the chapter is to examine the feasibility of deriving an industry phantom objective function, the rivalrous consonance model used is not designed to be realistic but rather was chosen for convenience of exposition. Finally, I show that for the specific cases defined in the paper it is possible to derive extremely complicated objective functions for the industry, but that in general it will not be possible. This road, therefore, for an algorithmic approach to the solution for a rivalrous consonance industry turned out to be of restricted usefulness.

To complete Part I of this volume, Chapter 8 presents the theory of rivalrous consonance and its genesis in some detail. The selection is the introduction to my book, *Rivalrous Consonance: A Theory of General Oligopolistic Competition*, published in 1986, which is the culmination of my long-standing desire to produce a general equilibrium framework with oligopoly as its dominant market structure. Various models are presented here and elaborated upon in the book, but this chapter gives the reader a fairly good sense of the structure of the theory and the motivating factors that led to that structure. Because I have discussed both aspects of the model in the foregoing introduction, I need not repeat it.

The eight chapters of Part I constitute an overview of the work that I have done for over forty years in general equilibrium theory as a total system, both in abstract and empirical terms. Moving from a Walrasian framework with modifications to the theory of rivalrous consonance, I have sought to correct the major flaws I believe the classical theory contained. The corrective work I realize is only a beginning, and my hope is that others will improve upon it and reconstruct it where desirable. At the least I would hope the reader agrees with me about the goals of the effort, even if he or she does not accept my methodology.

In applied work, through presentation of interdependence modelling separated by a gap of thirty years, I have featured the advances made over that period. Static input–output frameworks remain the hard core of such efforts, but much improved models exist for the analyst to employ. Operational general equilibrium analysis has generated a whole new literature which I have not examined in this volume, and deserves the scrutiny of the economist convinced of the importance of operational interdependence analysis in policy formation.

In the remaining parts of the volume I narrow my focus to feature my efforts in three different sectors of general equilibrium theory: the spatial dimension, the time dimension, and the phenomenon of that peculiar good, money. In each field I have attempted to address fundamental issues as I see them and to link those efforts to general equilibrium modelling.

Notes

1. See Robert E. Kuenne, *The Economics of Oligopolistic Competition: Price and Non-Price Rivalry, Collected Papers of Robert E. Kuenne*, Basil Blackwell, Oxford 1992.

1 Walras, Leontief and the Interdependence of Economic Activities*

I INTRODUCTION

The theoretical groundwork for detailed interdependence models on a national basis was erected by Léon Walras in his *Éléments d'Économie Politique Pure*, first published in two parts: the first, dealing with the theory of exchange, in 1874, the second, concerned with the theory of production, in 1877.[1] In this work, and four subsequent editions, Walras developed his general equilibrium theory, all but totally neglected in the Anglo-American pursuit of Marshallian partial analysis.[2]

In its essence, his structure is a demonstration that (*1*) given simple premises about consumer and entrepreneurial motivation, (*2*) given consumer preferences for each individual in the economy, technological conditions of production for each product, and the amounts and distribution among individuals of economic resources, then (*3*) the logicality of behavior constrained by the above non-economic data leads to certain equilibrium conditions, which, when enforced upon the system, allow the unique determination of all economic magnitudes in the system.

Opinions of economists differ as to the significance of Walras's achievement – from that of Schumpeter who believed it to be the

* Published originally in the *Quarterly Journal of Economics*, 68(1954), pp. 323–54, and reproduced with permission. The writer is indebted to Professors E. H. Chamberlin, W. Isard, W. W. Leontief, and A. Smithies who read earlier drafts of the manuscript and gave many valuable criticisms. Some major changes have been made subsequent to their readings, and of course all errors must be laid at the writer's door.

The manuscript was in third draft when the opportunity came to read Schumpeter's critique of Walras in the monumental *History of Economic Analysis*. This brilliant analysis helped immensely, and while the writer hopes the specific debts have been acknowledged in the footnotes, a danger always exists that points at issue will obscure such debts, or that aid which cannot be documented will go unnoted. Such debts are acknowledged gratefully here.

'Magna Charta of economics' as a demonstrably separable body of human behavior, to that of those who, though impressed by its breadth and imaginative construction, think it essentially sterile of practical significance.[3]

Granted that as a device to describe to a close approximation the processes in an advanced economy it is seriously deficient, the Walrasian type of system remains, nevertheless, the best approach to a study of the nature of the complex interrelationships on a detailed basis in such an economy. Both as a pedagogical and analytical tool, the Walrasian 'vision' of the mutual interdependence of variables within the economic universe as a whole, as well as the breadth of his structure without sacrifice of detail, remain valuable.

The work of Professor W. W. Leontief in building operational interdependence models and that of the Interindustry Division of the Bureau of Labor Statistics in quantifying them, lends Walras's work a more immediate importance.[4]

The purpose of this article is to synthesize a complex Walrasian model from his work, to alter it or indicate changes which seem desirable in the light of more modern theory, to point out some of its defects, and, lastly, to indicate the simplifications which have been made in its conversion to Leontief's input–output schema.

II THE WALRASIAN MODEL

The Household Sector

Walras begins with the assumption that a group of individuals, x_i, ($i = 1, 2, \ldots m$), is located in a spaceless economy, or, less drastically, in an economy with zero transportation costs. Each has title to an endowment, q_{iz}, of z_j factor services ($j = 1, 2, \ldots n$) used in the production of goods y_k ($k = 1, 2, \ldots s$), these latter serving both as consumer goods and intermediate industrial materials.[5]

Corresponding to each factor service, z_j, there exists a capital good which is capable of reproduction and sale, Z_j. A stock of money, Q_u, which has no *direct* utility and is for use merely in facilitating exchange and as a temporary store of value, is distributed among individuals in a predetermined way, q_{iu}. Good $y_{k=1}$ is used throughout as the *numéraire*, i.e., for reckoning prices, but not as a medium of exchange.

All markets – consumer goods, factor services, and capital goods –

are assumed to operate under the familiar regime of perfect competition.[6]

If it be assumed that the satisfaction derived by consumers from a given amount of good y_k is independent of the amount of every other good consumed, and that utilities are cardinally measurable,[7] then *ms* independent utility functions are given, from which an equal number of marginal utility functions can be derived by differentiation:

$$r_{iy_k} = R_{iy_k}\,(q_{iy_k})$$

where r_{iy_k} is the marginal utility of good y_k to consumer x_i as a function of the quantity consumed.

Similarly, if it be assumed that title to all factors is in the hands of consumers, that these factor services have alternative uses for direct consumption by their possessors, and that the satisfaction obtained is independent of the quantity of all other services and consumer goods, then *mn* utility functions are defined from which an equal number of marginal utility functions can be derived:[8]

$$r_{iz_j} = R_{iz_j}\,(q_{iz_j}).$$

But such goods and services do not exhaust the objects within the consumer's scope for decision-making. Walras adopts a novel and ingenious method of introducing savings into his model by creating a commodity, E, or perpetual net revenue, whose price is the reciprocal of the net yield on capital goods, i.[9] He assumes that the economy is a 'progressive' one, so that savings become additions to a stock, q_{ie}, or revenue for each individual.[10]

The marginal utility functions for E are then given:

$$r_{ie} = R_{ie}\,(q_{ie} + X_{ie})$$

where X_{ie} is the consumer's demand for net revenue (securities).

The explicit recognition of time lags between the receipt of income and expenditures on goods and services introduces two more elements into the consumer's decision-making. If it be assumed that a period is begun with a brief lapse of time during which general equilibrium is established quickly in some manner,[11] so that at the expiration of this short space of time contracts have been made for the delivery of goods and services at equilibrium prices, Walras's schema requires that consumers have on hand (*1*) inventories of finished consumer goods yielding inventory services, y'_k, and (2) a stock of money, u, with which to replenish stocks of inventories before receipt of income.

A fundamental problem arises, however: in an economy in which no uncertainty exists about the future,[12] and all delivery dates are contract-certain, is there any need for stocks of real inventories or of cash balances? Accepting the conditions of Walras's model, and assuming for the moment the absence of cash balances, then consumers begin the period with holdings of real inventories. Clearly, if a lag occurs between inflows of current services and outputs of current goods, consumers must rely upon their stocks.

But can the same be said of cash balances? Hicks has argued that under the conditions of Walras's model, each rational person would desire to hold assets and earn interest at every point in time, since he could see perfectly the dates at which he would require money and would seek to lend between such dates. The demand for money to hold would be zero. In order to circumvent these logical difficulties, Hicks introduces frictions into the economy in the form of the costs (monetary and psychological) of lending small amounts of money for short periods of time. Such lending thus becomes irrational and money balances are held, so to speak, by default.[13]

Other solutions to this existence-of-money problem suggest themselves. Marget has stated one succinctly: 'Even in a world in which everything were perfectly foreseen, a lack of synchronization between the receipt of income and its outlay would give rise to a need for cash-balances so long as there were not perfect facilities for the borrowing of money in anticipation of receipts and the investment of money during the period elapsing between receipt and outlay.'[14] However, nothing in the nature of Walras's equilibrium period or in his specific assumptions about foresight, etc., can be appealed to in the search for imperfect facilities of the type Marget suggests. Marget must thus assume such frictions (Hicks's are of another type) or assume institutional constraints of some sort. In any even, the introduction of some such imperfection in order to provide a *raison d'être* for cash balances does no great violence to the model (and does little to achieve a realistic theory of money).

Recently, however, Patinkin has revived the controversy in a new form. His position is that the Walrasian system finds itself in error in that it does not ascribe a utility to money: 'The basic postulate of the classical monetary theory is that people do not derive any utility from holding money, and, consequently, it does not enter the utility function. . . .'[15] An immediate implication of this is 'that since the individual derives no satisfaction from money, he will not plan to hold money, which yields no satisfaction, when he can use the same resources to obtain goods from which he does derive enjoyment.'[16] Thus, the individual offers his entire stock of money, but, since demand to hold it is zero, the classical (Walrasian) system renders money a free good.

But this raises the question of how Walras determines the individual's demand for money. Walras assumes that consumers divide their inventory requirements into two accounts, those held in kind and those over which potential purchasing power in the form of cash balances is held (postponable inventories), each with separately calculable utility functions, from which the marginal utility functions below are derived:

$$r_{iy'_k} = R_{iy'_k}(q_{iy'_k})$$

where $r_{iy'_k}$ is the marginal utility derived by individual x_i from the holding of goods as inventories, and

$$r_{iy''_k} = R_{iy''_k}(q_{iy''_k})$$

$$r_{ie'} = R_{ie'}(q_{ie'})$$

where y''_k represents the amounts of inventory services of real goods and e' the 'inventory services of perpetual net revenue' which the consumer wishes to hold control over *in money form*, i.e., postponable inventory services.[17] Thus, money is endowed with an indirect utility and it does enter indirectly into the consumer's utility map.[18]

The condition is imposed that the consumer spend all his income (and also his dissaving, in the event the *individual's* demand for securities is negative) by the following balance (budget) equation,

$$\sum \bar{X}_{iz_j} p_{z_j} + \sum \bar{X}_{iy'_k} p_{y'_k} + \bar{X}_{iu} p_{u'} = \sum X_{iy_k} p_{y_k} + X_{ie} p_e \, ,$$

where $\bar{X}$ and X represent supplies and demands respectively, p represents prices and $p_{u'}$ the price of the *service* of money as opposed to p_u, the price of money itself. Combining this with the condition that marginal utilities of goods demanded and supplied be proportional to prices,

$$\frac{r_{iy_{k\neq 1}}}{r_{iy_{k=1}}} = p_{y_{k\neq 1}}\ , \quad \frac{r_{iz_j}}{r_{iy_{k=1}}} = p_{z_j}\ , \quad \frac{r_{iy'_k}}{r_{iy_{k=1}}} = p_{y'_k}\ ,$$

$$\frac{r_{iy''_k}}{r_{iy_{k=1}}} = p_{y'_k}\ , \quad \frac{r_{ie}}{r_{iy_{k=1}}} = p_e\ , \quad \frac{r_{ie'}}{r_{iy_{k=1}}} = p_{y'_{k=1}}$$

then demand and supply functions may be derived as below

$$X_{iy_{k\neq 1}} = F_{iy_k}(p_{y_{k\neq 1}}, p_{y'_k}, p_{z_j}, p_e, p_{u'})$$

and letting P denote the price vector above,

$$X_{ie} = F_{ie}(P),\ X_{iy''_k} = F_{iy''_k}(P),\ X_{ie'} = F_{ie'}(P)$$

$$\bar{X}_{iz_j} = F_{iz_j}(P),\ \bar{X}_{iy'_k} = F_{iy'_k}(P)$$

$$\bar{X}_{iu} = q_{iu} - \frac{\sum X_{iy''_k}\, p_{y'_k} + X_{ie'}\, p_{y'_{k=1}}}{p_{u'}}$$

$$X_{iy_{k=1}} = \sum \bar{X}_{iz_j}\, p_{z_j} + \sum \bar{X}_{iy'_k}\, p_{y'_k} + \bar{X}_{iu}\, p_{u'} - \sum X_{iy_{k\neq 1}}\, p_{y_{k\neq 1}} - X_{ie}p_e$$

Walras' performance in the monetary sphere has been neglected in the literature, perhaps because of its eclipse by his real analysis: (*1*) his theory of bimetallism has lost its interest with its topicality, although Schumpeter terms it the outstanding analytic performance in the field;[19] (*2*) his development of a cash-balance approach to the value of money before Marshall has been overlooked;[20] and (*3*) the 'postponable inventory' approach to the demand for money has been ignored as a possible bridge between value and monetary theory.

The latter two achievements are relevant to this study. The most striking characteristic of the analysis is the seeming 'realness' of it: that is, Walras has a system of equations which includes the consumer's demand for cash balances without explicitly introducing the *numéraire* value of money (p_u), as distinguished from the value of money services ($p_{u'}$). Can it be that consumers find it possible to decide upon the physical quantity of money they wish to hold without considering the money prices of goods?

The explanation of this phenomenon involves the entire theory of capital good or asset pricing in Walras. This process is a derivative of the pricing of the services yielded by such assets, for the purchase of capital goods or money is merely the purchase of perpetual net revenue. Let there be a market in which securities (riskless) which guarantee the payment of one unit of *numéraire* in perpetuity are exchanged against money. Now let a price be established, p'_e, for such a security, and from it a *numéraire* rate of perpetual revenue, i'. The prices of factor services, p_{z_j}, $p_{y'_k}$, and $p_{u'}$, being determined in separate asset service markets, and the supply of securities being merely a demand for perpetual asset services, such goods yielding p_{z_j}, $p_{y'_k}$, and $p_{u'}$ per unit of service will be sold for $p_{z_j} p'_e$, $p_{y'_k} p'_e$, and and $p_{u'} p'_e$, or for $\frac{p_{z_j}}{i'}$, $\frac{p_{y'_k}}{i'}$, and $\frac{p_{u'}}{i'}$. In Walras, assets *always* exchange in proportion to their net revenues.

Lying behind the demand for money services is the demand for real services. If these services were of the same nature as (for example) capital good services, the division of the *numéraire* value of the inventory services desired, by the *numéraire* value of money, p_u, would yield a cash balance magnitude. But Walras does not postulate that consumers' inventories are ordinarily acquired by renting their services. Rather, he assumes that such inventories are acquired by purchase of goods at their asset values. Thus, dividing the value of inventory services by $p_{u'}$ $(= i\, p_u)$ yields a cash balance large enough to command the cash value of such assets, providing the sales prices of assets are always equal to the capitalized value of their services, whether the system is in or out of equilibrium. Also, in the demand and supply equations, either $p_{u'}$ or p_u may be included, since the use of the first implies the second.

But is not the postponable inventory approach to the demand for cash balances a bit too ingenious for its surroundings? Has not Walras's desire for mathematical symmetry led to a rather overwrought design for money? Does the concept of 'postponable inventory services of perpetual net revenue' $(X_{ie'})$ have any but mathematical meaning? Abstracting from the Keynesian type of uncertainty which gives money a true asset function, cannot money be looked upon more realistically as a flywheel converting discontinuous income receipts into a smooth path of consumption through time? That is, assuming that a consumption budget has been decided upon for a period of finite length, money serves a passive role in Walras' economy and cannot properly be envisaged as a vehicle for services which are conceptually capable of a continuous utility function.[21]

In view of these objections and the unnecessary complications of the model which they indicate, a disaggregated Marshallian k approach seems preferable to the writer. As Walras himself tacitly recognizes on the entrepreneurial demand side for cash balances, in so far as he does not bring it under the regime of marginal productivity, it is not necessary to adopt a 'nicely calculated less or more' philosophical buttressing for supply and demand functions in order to get a theory of money of the type he adopts. Although it might well traumatize Walras's mathematical psyche, an appeal to institutional determinants need not constitute a mere refuge in vagueness; in this case, the writer believes, it improves the model.[22]

Under these assumptions, the individual's supply curve of cash balances becomes,

$$\bar{X}_{iu} = q_{iu} - \frac{\sum k_{iy_k} X_{iy_k} p_{yk} + k_{ie} X_{ie} p_e}{p_u},$$

where k_{iy_k} and k_{ie} are institutionally determined data. In this way the desire to hold money is related to the flow expenditure directly. The demand and supply of cash balances serves to 'determine' (in so far as any one equation in a system determines any one variable) the price of money as an asset instead of the interest rate. The postponable inventory utility functions may be deleted. If it be assumed that these deletions have been effected, and the remaining demand–supply equations have been summed over all individuals, the first system of equations in this modified Walrasian model is derived and denotes the market supply and demand functions of consumers:

$$X_{y_k \neq 1} = F_{y_k}(P) \tag{1.1.1}$$

$$X_e = F_e(P) \tag{1.1.2}$$

$$\bar{X}_{y'_k} = F_{y'_k}(P) \tag{1.1.3}$$

$$\bar{X}_{z_j} = F_{z_j}(P) \tag{1.1.4}$$

$$\bar{X}_u = Q_u - \frac{\sum K_{y_k} X_{y_k} p_{y_k} + K_e X_e p_e}{p_v} \tag{1.1.5}$$

$$X_{y_{k=1}} = \sum \bar{X}_{z_j} p_{z_j} + \sum \bar{X}_{y'_k} p_{y'_k} + \bar{X}_u p_{u'} - \sum X_{y_{k\neq 1}} p_{y_{k\neq 1}} - X_e p_e \qquad (1.1.6)$$

The question arises: how much of Walras's achievement has been lost in thus altering his monetary analysis? Most important, the equation which was integrated in the body of 'real equations' to the extent that a change in the interest rate, *ceteris paribus*, induced a change in the desire to hold command over inventories and thereby a change in the level of 'real' money desired, now becomes a mere quantity equation tacked on to the real system. Yet Walras, who was never averse to vaunting his achievements or parading his priorities, does not dwell on this advantage. Rather he reveals some doubts as to its true integration, for on page 311 he asserts that the effects of changes in the interest rate upon the consumer's choice will be 'very weak and very indirect'; earlier, he seems to revert to a 'Cambridge equation' analysis when he says that in the real world each consumer and producer knows approximately what needs of cash he will have for a forthcoming period subject to the uncertainty of the future (pp. 300–1); and, finally, in his chapters on monetary analysis, he reverts to a pure form of the nonintegrated cash-balance approach. Thus, Walras himself seems to have doubts whether, in the absence of the uncertainty which could convert money into a good with a direct utility, capable of introduction into the utility system, the desire to hold inventories in the form of money would not indeed lead to a perfectly inelastic demand function for money in terms of the interest rate (all other prices constant).

The route of the causation between interest rate changes and cash balances is more direct and of radically different character in the real world than the indirect one in Walras, and it is incapable of inclusion in his system of certain future. Whether one emphasizes the speculative motive of Keynes or the precautionary motive of Fellner and Hart, it remains true that only the germ of liquidity preference is found in Walras.[23]

But this last criticism leads us to the most important of Patinkin's criticisms of the Walrasian system. Briefly, the argument is this: in the exchange economy of the classical system, if we neglect borrowing and lending, demand and supply equations for real goods and for money are homogeneous of degree zero for all prices (including money, and measured in some arbitrary *numéraire* outside the system). Moreover, if money does not enter the utility function, demands and supplies of real goods are homogeneous of degree zero in the subset which excludes the price of money, and such equations for money are homogeneous of order one in the same subset. If all real goods prices double, the price of money remaining constant, demands and supplies for real goods will remain the same for the individual, and no excess demand for money will arise; therefore, demand for money will always equal its supply, and writing such equations in the Cambridge k fashion, for example, is misleading in that it implies that an inequality can arise. If, lastly, we may not enforce the equality of demand and supply in the money sector as an equilibrium condition, then the absolute price level is indeterminate. The dichotomization of the economy into a real and money sector is complete and effective.[24]

Patinkin's contribution on this point is a real one, but in fairness to Walras it should be pointed out that the latter never introduced money as a circulating medium into his exchange models. Because of its asset nature and the similarity of its services to capital assets services, money is introduced into a production economy with borrowing and lending.

To correct this fault of the classical system, Patinkin at first introduced the nominal amount of money into the consumers' demand functions,[25] but then in an extensive presentation of his 'new' system he reverts to Samuelson's introduction of real balances.[26] Lack of space prevents an extensive discussion of Patinkin's system, but in its major characteristics it is the Walrasian system modified for an exchange economy. The demand for money is written $\frac{p_n Z_n}{p}$, where Z_n is the individual's demand for physical units of cash, p_n is the price of money in a standard unit of account outside the system, and p is an index number representing the price level of real goods in the same unit of account. It is, then, the demand for control over a desired number of 'market basket units' of real goods. But this is exactly what Walras's demand for money is, except that Walras substitutes inventory services for 'market basket units': in similar terminology, Walras's equation is

$$Z_n = \frac{\sum Z_{y''_k} p_{y'_k} + Z_{e'} p_{y'_{k=1}}}{p_{u'}}$$

which is equivalent to

$$Z_n = \frac{\sum Z_{y''_k} p_{y_k} + Z_{e'} p_{y_{k=1}}}{p_u}$$

where prices are measured in terms of a real good, $y_{k=1}$. Had Patinkin adopted this method of price measurement and Walras's concept of money services, he could have done away with the clumsy expedient of a price index. The real values of assets enter the Walrasian utility functions just as they enter Patinkin's total function, and Walras, just as Patinkin, introduces the return on some assets (money in Walras's case, securities in Patinkin's) defined in *money* terms into the budget equation. The absolute price level, and with it the income effect nonhomogeneity which meets Patinkin's early criticisms, is introduced via the budget equation into both systems, and has nothing to do with the introduction of money into the utility function, as Patinkin recognizes.[27] In short, Patinkin follows step by step the methods of the very system he is criticizing, and ends with it.

Moreover, given this nonhomogeneity of the budget equation, the nonnecessity for introducing money into the utility function means that a Cambridge *k* alteration is perfectly legitimate. Indeed, Walras's analysis is nothing more than a marginalistic determination of the Marshallian *k*. Money need not be effectively introduced into the utility functions as would be necessary if money were a true asset in an uncertain economy. Although our conclusion is that the price of money is determinate in the Walrasian system, by virtue of the very analysis Patinkin thought he was using to attack it, its determinacy is independent of the conversion of money into a good with a direct utility, and thus does not meet the conditions for a realistic money discussed above.

The Entrepreneurial Sector

Among the data pertinent to this sector are the technological production functions for the y_k consumer goods and Z_j capital goods (numbered so as to indicate their relation to 1.4.1 and 1.4.2 below):

$$\bar{X}_{y_k} = f_{y_k}(a_{z_j y_k}, a_{y_k y_k}, a_{y'_k y_k}, a_{u y_k}, \bar{X}_{y_k}) \quad (1.4.3)$$

$$\bar{X}_{z_j} = f_{z_j}(a_{z_j z_j}, a_{y_k z_j}, a_{y'_k z_j}, a_{u z_j}, \bar{X}_{z_j})^{28} \quad (1.4.4)$$

where the a's are *average* coefficients of production, indicating the amount of each factor service required per unit of product, and $\bar{X}_{y_k}$ and $\bar{X}_{z_j}$ are gross outputs of goods y_k and capital goods Z_j respectively.

The inclusion of intermediate products having been made by assuming them to have end-uses as consumers goods, another component of demand for goods y_k, entrepreneurial demand, must be added to consumers' demand, and $X_{y'_k}$, or investment in inventories, must be anticipated at this point:

$$ {}_tX_{y_k} = X_{y_k} + X_{y'_k} + \sum_{K=1}^{s} a_{y_k y_K} \bar{X}_{y_K} + \sum_{j=1}^{n} a_{y_k z_j} \bar{X}_{z_j} \qquad (1.2) $$

where the term on the left-hand side denotes total (gross) demand for goods y_k, the last two terms on the right-hand side entrepreneurial demands for current flows of y_k, and the capital K has been introduced merely to distinguish the summation index.

Entrepreneurial demand for factor services, inventory services, and money services must be equal in equilibrium to the total supply of each:

$$ \sum_{k=1}^{s} a_{z_j y_k} \bar{X}_{y_k} + \sum_{j=1}^{n} a_{z_j z_J} \bar{X}_{z_J} = \bar{X}_{z_j} \qquad (1.3.1) $$

$$ \sum_{K=1}^{s} a_{y'_k y_K} \bar{X}_{y_K} + \sum_{j=1}^{n} a_{y'_k z_j} \bar{X}_{z_j} = \bar{X}_{y'_k} \qquad (1.3.2) $$

$$ \sum_{K=1}^{s} a_{u y_k} \bar{X}_{y_K} + \sum_{j=1}^{n} a_{u z_j} \bar{X}_{z_j} = \bar{X}_u \qquad (1.3.3) $$

In the commodities markets, conditions are imposed that demands for and supplies of goods be equal in equilibrium:

$$ {}_tX_{y_k} \ \bar{X}_{y_k} \qquad (1.4.1) $$

$$ X_{z_j} = \bar{X}_{z_j} \qquad (1.4.2) $$

where X_{z_j} is demand for (gross investment in) capital goods Z_j.

The determination of the optimum coefficients of production is more complex. They may be divided into two types: (*1*) those which are best handled by considering them as essentially independent of the relative amounts of other factors used, in the sense that they are not directly substitutable for those other factors, and (*2*) those which are for factors capable of substitution for one another.

The treatment of the demand for inventories and the services of money – the first type of factor service – is essentially one of institutional determination modified by the indirect influence of prices.

In the case of inventories of capital goods proper, Walras assumes that the holding of such stocks by the producing industry is the provision of a service to all users of the producer's good. For example, the inventories of finished lathes in the shop of a producer of lathes are considered as furnishing inventory services to those industries using lathes. Under stationary conditions this solution is a reasonable one, for such inventories will be in largest part a function of the existing machines and their rates of depreciation. Instead of assuming, as Walras does, that the $a_{z_j y_k}$ and $a_{z_j z_j}$ factor services which are produced from the Z_j reproducible capital goods include this inventory service, thus introducing such inventories into the production functions of the users, we assume such inventories to be nonexistent for the sake of simplicity.[29] Moreover, the logic of the model suggests this, for all capital goods proper are deliverable only at the end of the period during which they are bought, in order that the quantities of resources remain fixed. As a consequence, no inventories of them are needed at the start of the next period, and, if we assume that all capital goods require the whole of the period to produce, no violence is done to the model.

The holding of finished goods by their industries of origin and use, given the lags of the economy, is assumed to be a linear and homogeneous function of output, given by the equations,

$$a_{y'_k y_k} = f_{y'_k}(a_{y_k y_k}) \quad (1.5.1)$$

$$a_{y'_k z_j} = g_{y'_k}(a_{y_k z_j}) \quad (1.5.2)$$

Similarly, the amounts of money balances needed are assumed to be a linear and homogeneous function of inputs, and vary with average expenditures on factor services and intermediate products.[30]

$$a_{u y_k} = f_u(a_{z_j y_k}, a_{y'_k y_k}, a_{y_k y_k}, p_{z_j}, p_{y'_k}, p_{y_k}, p_u) \quad (1.5.3)$$

$$a_{u z_j} = g_u(a_{z_j z_j}, a_{y'_k z_j}, a_{y_k z_j}, p_{z_j}, p_{y'_k}, p_{y_k}, p_u) \quad (1.5.4)$$

Under the implicit conditions of Walras's industry production functions, the determination of the coefficients of the second type is a problem in optimum factor combinations and output, the familiar equilibrium conditions being the equality of marginal value products and the prices of factor services and intermediate products. One complication does arise, however: since the coefficients for money and inventory services are functions of the amounts of the latter used, they will vary with the relative amounts of these substitutional factor services used. Consequently, the conditions which impose the maximum profit and least cost conditions become the equality of value of marginal products to the prices of factor services plus the marginal expenditures on money balances and inventories entailed by the marginal unit of the substitutional factor.[31]

The equilibrium conditions become:[32]

$$\frac{\partial \bar{X}_{y_{k=1}}}{\partial (a_{z_{j=1}y_{k=1}} \bar{X}_{y_{k=1}})} p_{y_{k=1}} = p_{z_{j=1}} + \sum_{k=1}^{8} {}_{j=1}Y'_k + {}_{j=1}U \qquad (1.5.5)$$

and so forth, with $(s + n - 1)$ equations for each industry, where

$${}_jY'_k = \frac{\partial f_{y'_k}}{\partial (a_{z_jy_k} \bar{X}_{y_k})} p_{y'_k} \;, \quad {}_jU = \frac{\partial f_u}{\partial (a_{z_jy_k} \bar{X}_{y_k})} p_{u'}.$$

In equilibrium, competition will have ironed out economic surpluses and all goods be sold at their average costs of production. Walras's famous entrepreneur who makes '*ni bénéfices ni pertes*' receives his income from the sale of factor services (p. 195). It should be noted that these surpluses are ironed out without the intervention of new capital goods or other factors, for these latter are held fixed during this period. Rather, entrepreneurs and factors are assumed to be perfectly mobile between all employments (pp. xv, 194, 221) and temporary surpluses are absorbed into the prices of factors as well as reduced by increased production. Tacitly assumed, therefore, is a supply of entrepreneurial ability sufficient to render it a free good. The equilibrium, then, is not a Marshallian long-run normal. The no-profit conditions constitute system 1.6.[33]

$$\sum_{j=1}^{n} a_{z_jy_k} p_{z_j} + \sum_{K=1}^{s} a_{y_Ky_k} p_{y_K} + \sum_{K=1}^{s} a_{y'_Ky_k} p_{y'_K} + a_{uy_k}p_{u'} = p_{y_{k \neq 1}}, \qquad (1.6.1)$$

$$\sum_{J=1}^{n} a_{z_Jz_j}p_{z_j} + \sum_{k=1}^{s} a_{y_kz_j} p_{y_k} + \sum_{k=1}^{s} a_{y'_kz_j} p_{y'_k} + a_{uz_j}p_{u'} = p_{z_j}. \qquad (1.6.2)$$

The Investment Sector

The conditions for equilibrium as imposed by Walras in this sector are of four types:

(*1*) The prices of capital goods must equal their average costs of production (equations 1.6).

(*2*) Savings must equal the value of capital goods and new inventories *priced at average cost*:

$$X_e\, p_e = X_{z_j}\, p_{z_j} = X_{y'_k}\, p_{y_k}. \tag{1.7}$$

(*3*) Demand and supply for each capital good must be equal, and, given the equality of demand for and supply of y_k other than for new inventory purposes, the supply and demand for these goods as inventories must be equal (equations 1.2.1 and 1.4.1).

(*4*) The average cost of each capital good and inventory must equal the price of its service capitalized at the net rate of return on assets; the price of money, as an asset, must equal the value of its service capitalized at the same net rate of return; also, the definitional identity of p_e to the reciprocal of the net rate of return may be included here:

$$p_{z_j} = \frac{p_{z'_j}}{i}, \quad p_{y_k} = \frac{p_{y'_k}}{i}, \quad p_u = \frac{p_{u'}}{i}, \quad p_e \equiv \frac{1}{i}. \tag{1.8}$$

Following Stigler's suggestion, the demand for capital goods proper may be brought under explicit marginal productivity reasoning.[34] To do so, however, would introduced all the complications of consumer and entrepreneurial expectations and intertemporal uncertainty, the problems of which are discussed in Section III.[35] A simpler solution exists, and it is the one adopted by Walras: expectations of future prices are merely projections of present prices. Given this assumption, and given U-shaped cost functions, systems 1.5, 1.6, and 1.8 implicitly define demands for capital goods in terms of marginal productivity. With fixed consumer tastes and abstracting from depreciation, $X_e p_e$ becomes zero, and Walras's model collapses into a stationary state. In Wicksell's words, '. . . Walras's formula for interest . . . reduces itself, on the assumption of stationary conditions, simply to the equation $F(i) = 0$, in which $F(i)$ is the amount of annual savings conceived as a function of the rate of interest (i). In other words, it expresses the truism that, in a stationary state, the inducement to new savings must have ceased . . .'[36]

The truth is that, for all his talk of the 'progressive' state (pp. 252, 260), Walras's analysis is permeated with stationary assumptions. For example, in his treatment of depreciation and insurance allowances, Walras assumes that two technologically determined constants, f_{Z_j1} and f_{Z_j2}, exist, which, when multiplied by p_{Z_j}, yield the charges on these two accounts for each period. Thus, $p_{z_j} - (f_{Z_j1} + f_{Z_j2})\, p_{Z_j}$ becomes the net revenue which, when divided by p_{Z_j}, must be equal for all capital goods. Such treatment implies the constancy of the *numéraire* prices of capital goods through time, unless it be assumed that depreciation and insurance allowances are spent in the period they accrue to renovate the capital stock; but Walras specifically refers to the building up of reserve balances for future replacement (p. 243).

Moreover, Walras is guilty of ambiguous analysis in his presentation of the 'progressive' state. The difficulty begins in a curious passage on pages 244–5 in which he expresses the opinion that in a stationary state no sales or purchases (*vente et achat*) of capital goods would occur since such goods could be exchanged only in the ratios of their net returns over cost, and no benefit would accrue. He thus assumes that replacement of capital goods is automatic and occurs without exchanges (or that they are eternal – but he had just finished discussing depreciation), and unnecessarily imposes the condition of net savings on his model before an interest rate can arise.[37]

Let $E = X_e p_e$, and assume for simplicity that income is earned exclusively by the sale of one factor service z, by the rent of one capital good Z, and that only one consumer good y exists. Then revenue, R, is defined (p. 248) as

$$R = Q_z p_z$$

and E, the excess of revenue over consumption (p. 249), as

$$E = R - (Q_z - \bar{X}_z) p_z - X_y\, p_y$$

or

$$E = \bar{X}_z\, p_z - X_y\, p_y.$$

(Net) savings are defined (p. 250) as

$$E - (f_{Z1} + f_{Z2})p_Z$$

or the excess of E over depreciation. After defining E as a function of prices, he then refers to it as 'positive savings' (p. 252) and later in the same paragraph he imposes the condition that the demand for new capital goods is positive. He continues, 'in making this assumption we restrict ourselves to the study of the production of new capital goods in a progressive society and we neglect that of the consumption of existing capital goods in a retrogressive society.' Since the condition that E be positive is not sufficient to guarantee that a society's capital is not being consumed, the passage is unclear. Are 'new capital goods' to be only those purchased over and above automatic depreciation demands? If this is not what he means, there is no reason why $X_e p_e$ should not include depreciation and insurance allowances, free for investment in any direction, and the reinvestment of such allowances allows the determination of the interest rate in a stationary state. Indeed, by virtue of the peculiar nature of expectations in the model, Walras's theory of investment is rigorous only for a circular flow.[38]

Secondary Conditions

The secondary equilibrium conditions, omitted to this point, may be introduced in this way: (*1*) In the consumers' sphere, the marginal utility functions must decline monotonically.[39] (*2*) In the entrepreneurial sector, two conditions must be met: (*a*) the surplus in every industry must not be negative, a condition already imposed by the condition that price equal average cost and that cost curves be U-shaped, smooth, and continuous; and (*b*) the demand curve for each factor service must fall as quantity increases, all other factors remaining constant in amount; moreover, the marginal product of any group of factors must diminish.[40]

Lastly, since the number of independent and consistent equations in the model is equal to the number of unknowns, the Walrasian model meets the *prima facie* condition for solution.[41]

III CRITICISM OF THE WALRASIAN MODEL

In this section brief consideration is given to those faults of the model which have a direct bearing upon its conversion to or the interpretation of the Leontief model. The treatment is not meant to be exhaustive.

The Static Nature of the Model

As in all static analysis, Walras's model cannot say what the path of approach to a new equilibrium level (given a change in the data) will be, or even whether a new equilibrium can be achieved. Also, although Walras was well aware of the existence of the two separate problems of the conditions for a unique solution for his model and the conditions for its stability,[42] his failure to derive 'laws of change' from stability conditions for his model is a major defect; but it has been corrected by Hicks, who shows that the probability of an unstable system is quite low.[43]

Another failure of Walras is more important for our purposes and cannot be corrected by a mere addendum. This flaw is his failure to allow the plans of individuals to extend over more than one period (except under the implicit stationary assumptions), by which stratagem he excludes uncertainty and thereby all those considerations which Hicks calls speculation in its broadest sense.[44]

Hicks's analysis of an economy in which intertemporal considerations mark the decision-making process is achieved by assuming price expectations for every period within the economic horizon. His analysis fits neatly into that of Walras, in so far as they both adopt a 'market day' during which the system achieves equilibrium.[45] The approach to this temporary equilibrium is assumed to be a smooth one.[46] The important difference is that in Walras demands and supplies are assumed to be functions only of current prices, while, in Hicks, consumers and entrepreneurs enter the market with plans based not only upon current prices, but expected prices for a number of periods within their horizon. Moreover, current supplies will have been shaped largely by decisions made in past periods. At the end of the period, demands and supplies for commodities and services will in most likelihood have changed in so far as expected prices on the part of many buyers and sellers will have failed to materialize; consequently, a new equilibrium will have to be established for the succeeding period, and in this manner the process of change can be reduced (in an oversimplified way) to a system of temporary equilibrium points interrelated by expected and realized prices.

The great dependence of the investment process upon future price and interest rate expectations as well as output expectations casts serious doubts upon any empirical predictions made from a static theory of the Walrasian type. Less serious doubts are cast upon current flows, but recognition must be made of the possibility that current production may reflect future expectations.

Moreover, unless a theoretical framework allows money to become a type of asset held (from the point of view of the present period) for itself, it degenerates into a purely passive factor. Before money can approach the phenomenon it has become under modern economic conditions – the most liquid asset available – it must be placed within the matrix of uncertainty. Only in a speculative economy can there arise the excess demand for money and consequent excess of goods and factor services which are the negation of Say's Law.

The establishment of the successive equilibrium conditions from period to period with given expectations (that is to say, assuming that producers and consumers have firm plans and expectations, and that they maximize such plans through time by equalizing the marginal rates of substitution and the reciprocals of the ratios of discounted expected prices), is merely the first step in the dynamization of Walras. The difficult problem of relating changes in expectations to changes in realized magnitudes still remains, and one need not be a pessimist to doubt that such a theoretical schema could be made operational. In the current period of 'de-price-ation' of our theoretical hard coin, more mechanical approaches to investment determination have been evolved as operational alternatives, the latest of which is found in Leontief's dynamic model.[47]

The Purely Competitive Nature of the Model

With the introduction of Chamberlin's monopolistic competition analysis and its implications, it no longer suffices to have the technological functions, utility functions, and amounts and distribution of resources in order to establish a determinate equilibrium. The dissolution of the atomistic competition of factor and commodity markets and/or the breaking-up of the latter by recognizing the heterogeneity within as well as between them, make the construction of the model theoretically indeterminate without added assumptions – perhaps derived from the institutional study of groups of closely competing producers – which would define patterns of reaction to changes in price, output, products and selling costs. Indeed it seems certain that, even apart from the oligopolistic type of indeterminacy, the mere recognition of products and of selling costs as added variables will cause trouble without added assumptions which would define and limit the nature of their variation.

I believe that determinacy can be salvaged for the 'large group case', i.e., where oligopolistic interdependence is absent, at least under the special assumption of *given* products and *given* total selling outlays for each firm. However, this can be done only at the cost of overlooking certain oligopolistic elements almost bound to be present. For any realistically defined group of competitors, a whole hierarchy of competitive relationships will ordinarily exist, with products A and B more closely competitive (higher cross-elasticities of demand) than products A and C, so that mutual interdependence will in fact characterize the relation between firms producing A and B. The introduction of an explicit spatial variable into the analysis provides a simple example of such a hierarchy, the firms being linked in the familiar chain competition of Chamberlin, and often grouped into somewhat separated clusters of competitors with recognized interdependence. The 'large group' model depends upon the assumption that with the reduction of price by one competitor, increases in sales will be drawn from other competitors in such a way as to be a negligible proportion of each of these competitors' sales; spatial proximity, and more generally, proximity in '*economic* space', serves to upset this assumption.

Triffin, at some points in his analysis, seems not to question the importance of the large-group analysis, for he refers to its development as 'the most important contribution of the new theories to the clarification of the general theory of value', and goes on to recommend its inclusion in the Paretian system in order to 'give validity to the equilibrium conditions there evolved.'[48] Indeed, he seems to give it more importance than Chamberlin, who, after a consideration of chain competition, says that 'the general conclusion must be that the considerations relevant to competition between small numbers are much more generally applicable than might at first be supposed.'[49] My own view is that the inclusion of the 'large-group case' for a majority or the totality of the industries in an economy, while moving the model closer to reality, does not move it far enough. It is more realistic than the purely competitive framework of Walras, but it is still an essentially unrealistic framework because of its neglect of oligopolistic relationships.

The Omission of Important Sectors from the Model

A last defect to be discussed here is the exclusion from Walras's model of two important sectors of the economy: (*1*) the international trade sector and (*2*) the governmental sector. The need for simplicity and the interferences of institutional barriers are sufficient grounds for excluding the international trade sector. The failure of governments to follow traditional economic motivation provides the theoretical grounds for the exclusion of their transactions from the model. If taken into account, both sectors must appear in the data portion of the model.

IV THE LEONTIEF OPEN SYSTEM

The Walrasian system may be summarized in matrix form as presented in Table 1.1. Four sectors are included in the table: industry Y_k, industry Z_j, the household sector H, and the investment sector I (entries in the row of this latter, representing depreciation payments and inventory depletions, have been omitted). The rows of the table enforce the equalities of supplies and demands for goods and services (equations 1.2, 1.3, 1.4.1, 1.4.2). The columns require that prices equal average cost (equations 1.6), that the amount of *numéraire* demanded by households equals income minus purchases of other goods and securities (equation 1.1.6) and that savings equal investment when investment goods are priced at average cost (equation 1.7).

By adding in equations 1.1.1–1.1.5, 1.4.3 and 1.4.4 to define all $\bar{X}$ and X terms except $X_{y'_k}$ and X_{Z_j}; equation 1.8 to determine these latter, p_u and p_e; and equations 1.5 to determine the coefficients of production, the system is reproduced in its entirety.

Given the data postulated and assuming the essential reality of the model, such a schema would be useful if questions such as these were posed: given a change in the tastes of consumers, the number and distribution of resources and money, and/or technology, what would be the new prices and quantities which would have to rule in the new equilibrium?

These uncontrollable phenomena are not the autonomous forces which are of greatest interest to the policymaker.[50] Moreover, the

Table 1.1 The Walrasian system

	Y_k		Z_j		H		I			
Y_k	$a_{y_k y_k} \bar{X}_{y_k} p_{y_k}$	+	$a_{y_k Z_j} \bar{X}_{Z_j} p_{y_k}$	+	$X_{y_k} p_{y_k}$	+	$X_{y'_k} p_{y_k}$	=	$\bar{X}_{y_k} p_{y_k}$	
	+		+				+			
Z_j	....		....		....		$X_{zj} p_{zj}$	=	$\bar{X}_{zj} p_{zj}$	
	$a_{z_j y_k} \bar{X}_{y_k} p_{zj}$	+	$a_{z_j Z_j} \bar{X}_{Zj} p_{zj}$	+	....	+	....	=	$\bar{X}_{zj} p_{zj}$	
	+		+				+			
H	$a_{y'_k y_k} \bar{X}_{y_k} p_{y'_k}$	+	$a_{y'_k Z_j} \bar{X}_{Z_j} p_{y'_k}$	+	....	+	....	=	$\bar{X}_{y'_k} p_{y'_k}$	$\equiv Y$
	+		+				+			
	$a_{u y_k} \bar{X}_{y_k} p_{u'}$	+	$a_{u Zj} \bar{X}_{Zj} p_{u'}$	+	....	+	....	=	$\bar{X}_u p_{u'}$	
	+		+							
I	0		0		....		....	=	0	
	=		=		=		=			
	$\bar{X}_{y_k} p_{y_k}$		$\bar{X}_{Zj} p_{Zj}$		$Y - X_e p_e$		$X_e p_e$			

analysis above indicates that certain of the variables must in fact be converted to data in view of the inability of the static model to approach the realistic processes which in fact determine them. Specifically the demands for capital goods and inventories must be removed from the system and placed among the data of the model. Also, the equation expressing the equality of the demand for and supply of cash balances must be removed from the system of equations and be replaced by an identity expressing the truism that the industrial demand for cash balances equals the industrial *requirements* for cash balances. Lastly, equation 1.8 expressing the necessary equality of the price of each capital good and money to the prices of their services capitalized at the rate of interest are dropped in so far as the structuring of the investment and the price of money are determined outside the system. Thus, in the Leontief open system, the price of money, the interest rate, structured investment, saving, and the need for cash balances are treated as data.

A second reason for removing variables from the Walrasian system intrudes when the question of operationality is posed: certain of the relationships which are conceptually capable of derivation are incapable of empirical derivation. For example, it is manifestly impossible to obtain preference maps for all consumers or even market demand functions. Therefore, consumers' final demands for goods and services and supplies of factor services must be relegated to the data sector, and with these latter must go the prices of such factor services. The equations in the Walrasian system which express the equilibrium condition that demand for each factor service equal its supply are converted in the Leontief matrix to identities expressing the fact that the total requirements for such factor services (netting out household demand) are the sums of the industrial requirements for them. In this way, equations 1.1 are deleted and equations 1.3 are converted to identities.

Although opening the model in this way allows us to escape some of the theoretical criticisms and empirical difficulties mentioned above, it all but emasculates Walrasian equilibrium economics. The traditional economic problems are outside the model: the determination of the amounts of each commodity taken by consumers, the amounts of saving and its structuring among capital goods, the determination of factor service supplies, the demand for cash, and the determination of the interest rate.

But in the absence of thorough-going knowledge about the nature of technological functions, it is necessary to make some simple assumptions. Leontief falls back upon the assumptions of Walras's early model: (*1*) fixity of coefficients, which rules out substitution among the factors; and (*2*) linearity, or the absence of economies of scale.[51]

The substitution of given constants for system 1.5 removes the last trace of general equilibrium except the most simple general interdependence. It marks the abandonment of marginal productivity theory;[52] it allows the system to be split down the middle so that prices and quantities can be considered independently. Economic choice and decision-making have been reduced to zero extent.[53]

Moreover, the economic concepts of supply and demand are replaced by the engineering concept of 'requirements'. The questions posed to the model are of the nature, 'Given the final demand for goods y_k and Z_j, what physical quantities of goods and services would be required to support such levels of demand, given constant coefficients of production?' or 'Given a price of money and the ability of *numéraire* prices to change only with changes in the technological structure or the prices of primary factors, what price changes in money terms would result from a given change in the data?'

What remains of the system may be summarized below:

$$\text{I.}\quad a_{y_k y_k}\bar{X}_{y_k} + a_{y_k Z_j}\bar{X}_{Z_j} = \bar{X}_{y_k} - -X_{y'_k}$$

$$0 = \bar{X}_{Z_j} - X_{Z_j}$$

II. $a_{y_k y_k} p_{y_k} + a_{z_j y_k} p_{z_j} + a_{y'_k y_k} p_{y'_k} + a_{u_{y_k}} p_{u'} = p_{y_k}$

$$a_{y_k z_j} p_{y_k} + a_{z_j z_j} p_{z_j} + a_{y'_k z_j} p_{y'_k} + a_{u z_j} p_{u'} = p_{z_j}$$

Among the data are:

$$a_{y_k y_k}, a_{y_k z_j}, a_{z_j y_k}, a_{z_j z_j}, a_{y'_k y_k}, a_{y'_k z_j}, a_{u_{y_k}}, a_{u z_j}$$

$$X_{y_k}, X_{y'_k}, X_{z_j}$$

$$p_{z_j} = 1, p_{y'_k}, p_{u'}$$

And derivable are:

$$\bar{X}_{y_k} a_{z_j y_k} + \bar{X}_{z_j} a_{z_j z_j} \equiv X_{z_j}$$
$$\bar{X}_{y_k} a_{y'_k y_k} + \bar{X}_{z_j} a_{y'_k z_j} \equiv X_{y'_k}$$
$$\bar{X}_{y_k} a_{u_{y_k}} + \bar{X}_{z_j} a_{u z_j} \equiv X_u$$

where X_{z_j}, $X_{y'_k}$, and X_u are the industrial *requirements* of primary factor services. Given the price of money, p_u, these prices may be converted into money prices, and $p_{u'}$ into the interest rate.

To interpret changes in the parameters of an empirical input–output table as enforcing changes on its variables in a process of equilibration is to stretch the concept of equilibrium beyond an economist's recognition. Nor is this a mere quarrel over words: the very failure of any equilibrating mechanisms, in the Walrasian sense, to operate, leads to the inclusion of 'leakages' as substitutes in order to obtain reasonable results.[54]

Rather than force such an interpretation upon the Leontief model, appeal must be made to the inertia and stability which characterize the adjustments of a modern economy within small limits. For example, if we assume that oligopolistic market structure is representative of the majority of our industries, it is reasonable to assume that small changes in the bill of goods will be accompanied by increased or decreased output in an approximately linear manner and that prices will be reasonably stable. What we are saying, in effect, is that an engineering approximation for small changes yields a reasonably accurate economic result.

Although the Leontief matrix may be an historical derivative of the Walrasian general equilibrium system, its right to a position as an operational logical derivative may be questioned. Disaggregation and interdependence are necessary but not sufficient for such a claim. Unfortunately, each economist must judge on a largely intuitive basis the relative importance of such adjustments present in the Walrasian and absent in the Leontief framework, and therefore the range of applicability of the latter; which is to say that he must judge between those situations in which he may appeal to the stability of inertia and those in which the stability of some rational adjustment pattern involving economic magnitudes is necessary to his analysis.

Notes

1. The edition quoted in this article, unless otherwise noted, is the definitive fifth edition, Paris, 1926.
2. Hicks points out the similarity between Walras's and Marshall's approaches to the theory of exchange before they reach the parting of the ways which produced the analyses associated with their names. See [11]. Schumpeter credits Marshall with the core of a general equilibrium system. See [36], pp. 920n., 952.
3. See J. A. Schumpeter [34], I, 41, and again [36], p. 242: Hicks [13], pp. 60–1; and Marshall [25], p. 850. Jaffé notes that Marshall urged Edgeworth to close the *Economic Journal* to Walras [15], 328.
4. Leontief's work may be found in [20] and [21]. The process of empirical implementation by the Bureau of Labor Statistics is best summarized, and the 50 × 50 models are published, in W. Duane Evans and M. Hoffenberg [7].
5. This last assumption was not made by Walras, who included raw materials in his schema. Its inclusion is recommended by its realism in delineating circular eddies in the production of goods and its greater simplicity in terms of economizing algebraic symbols. Moreover, it moves the model closer to Leontief's empirical models, which of necessity deal with industrial aggregates, with a consequent increase in the probability that 'products' have some consumers' use.
6. Walras's *libre concurrence* is defined loosely as a quantitative characteristic of a market, present in greater or lesser degree, with the stress on 'perfection'. See [41], pp. 44–5, 231.
7. Both of these assumptions are unrealistic and may be avoided:
   (*a*) Greater symmetry between consumption and production theory, as well as a closer approach to reality, can be had by the adoption of a suggestion by Edgeworth, Wicksell, and others. In so far as substitution and complementarity between goods exist, a total utility function, analogous to the production function, must be defined, the marginal utility functions becoming partial derivatives of the total function rather than the first derivatives of the separate utility functions above. See K. Wicksell, [42], I, 48–9, and P. Samuelson, [32], p. 93. This procedure can be extended to include those factor services which have alternative uses as consumption goods.

Only with the introduction of such relationships between goods can the problem of stability in consumers' markets really be attacked, for without them relationships between purchases depend upon the specific characteristics of the marginal utility functions, and generalization is impossible. Note the unsatisfactory nature of Walras's laws of variation on pp. 140–3 and 236–7 of [41] compared with Hicks's analysis of Part I of [13].

The substitution of such a function raises the question of determinateness of equilibrium. Schumpeter quotes an article by Amoroso (in Italian) which convinced Schumpeter that no difficulties arise in this regard under acceptable assumptions. The writer has not been able to find a translation of the article. See [36], p. 1007.

(*b*) The related Hicks-Allen reintroduction of indifference curve techniques may be employed to free the analysis from the assumption of cardinally measurable utility. See [13], pp. 11–52.

8. Walras adopts a rigorous distinction between capital goods, including land, persons, and produced means of production, and the revenue from such objects, that is, the harvest of services from such capital goods. The distinction is not carried as far as durable consumers' goods in his mathematics, a division familiar in American literature because of Irving Fisher's definition of national income. For Walras's treatment, see [41], pp. 176–84, 209.

The assumption (unrealistic except for labor services) of alternative consumption uses for factor services allows Walras to escape the pitfalls of a disutility theory of supply. In addition, other reasons might be mentioned.

Schumpeter says that, under Walras's original assumption of fixed coefficients, some factors might be unable to find employment and so might bid prices down, thus destabilizing the entire system in so far as such price-lowering were insufficient to induce their employment. But such analysis is incorrect in so far as substitution among goods would be possible to stimulate employment of such factors; and, were prices of such factors to decline to zero, they would merely become free goods.

A third reason serves to emphasize the role of marginal utility in the system, in so far as use may be made of such marginal utilities to structure a given *numéraire* value of savings among capital good alternatives. See Walras [41], p. ix.

9. See [41], p. 250. The concept is Walras's method of making specific capital goods homogeneous in the consumer's eye from the point of view of investment in future income, not in the sense referred to in the last paragraph of note 8. Walras felt it necessary to show that the equilibrium conditions in the capital market would assure the maximization of satisfaction for consumers. See [41], chs 26 and 27 for his attempted proof.

The introduction of perpetual income by Walras is, of course, a simplification. The prices of future income for any period of the future may be computed from the equation,

$$p_{e'} = \frac{(1+i)^n - 1}{i(1+i)^n}$$

for one unit of *numéraire*, where n is the term of the annuity.

10. But see section headed 'The Investment Sector' below. The term, q_{ie}, is defined as the sum of all capital goods services possessed multiplied by their respective prices net of depreciation and insurance allowances. Thus, X_{ie}, the demand for units of perpetual net revenue is *numéraire* is made commensurable with the stock of such a commodity in the consumer's possession.
11. [41], pp. 298–9. In Walras, of course, the manner of establishing this equilibrium was by *tâtonnements* from '*prix criés au hasard*'. Although Walras does not introduce this concept of a 'market day' explicitly until his money model, it underlies the analysis of all his models. See, for example, his discussion of the approach to equilibrium in his simplest production model, [41], pp. 214–15.
12. Uncertainty is specifically ruled out on p. 300 of [41].
13. J. R. Hicks, [12], 1–9, esp. 6–7. In *Value and Capital*, Hicks accepts the existence of money balances in a stationary economy; however, the arguments are repeated in the explanation of the reason for a short-term rate of interest. See [13], pp. 164–6, and in particular the statement that 'It is only if this obstacle (the trouble and cost of converting money into short-term securities) were removed, if safe bills could be acquired without any trouble at all, that people would become willing to convert all their money into bills, so long as any interest whatever were offered.'

    For similar criticisms of the existence of money in static equilibria, see P. N. Rosenstein-Rodan. [31], especially 269–72; D. Patinkin, [27], p. 136; and perhaps the first raising of the problem in F. H. Knight, [17], p. xxii.

    It is difficult to understand Schumpeter's cryptic reference to Hicks's analysis in [36], p. 1022n. He seems to imply that Walras, in stating the absence of uncertainty and in defining the difficulty, anticipated Hicks's criticism.

    This consideration is the nearest that Schumpeter comes to dealing with the basic problem of integrating money with value theory. With the introduction of uncertainty into the model, money may be endowed with a direct utility or with an indirect utility such as that used by Walras. The problem is, however, does the rationale of Walras's model allow such a solution? In the midst of his encomium on Walras's achievement, he does not stop to answer this conundrum.

    In his criticism of Cassel's quantity theory approach to the value of money, Schumpeter berates Cassel for his having missed Walras's message, without stopping to question whether in Walras's own schema the only role for money is that of a veil. (See [36], p. 1081n.) Further, in his discussion of the Austrians' attempts to apply a direct marginal utility approach to money, Walras's indirect utility approach is not discussed. (See [36], pp. 1089–90.)
14. A. Marget, [24], pp. 159–61. This quotation was specifically an answer to Hicks. Its date is explained by the fact that Hicks first published his criticisms in [10]. See also a similar solution in G. S. Becker and W. J. Baumol [2], p. 367.
15. [27], pp. 135, 140.

16. [27], p. 140.
17. [41], p. 303.
18. Jaffé pointed this out in his brief abstract [15], pp. 327–28. Other mistakes in Patinkin's presentation escaped attention: his statement that $p_{u'}$ in Walras represents the general price level, instead of the price of money services; and his misquoting of Walras to show that Walras maintained that the general price level had little effect on the demand for money whereas Walras was speaking of the effect of the interest rate. See Patinkin [27], p. 144, and [28], p. 12.
19. See [36], p. 1076n.
20. For a study of the evolution of Walras's monetary theory, see A. Marget [23].
21. Samuelson [32], pp. 177–22, presents a theory of the demand for money in which substitution is possible, and which, because of the interspersed references to Walras, the reader may believe to be the Walrasian theory. In so far as it ignores the 'postponable inventory' approach, it is not Walrasian. Moreover, it is explicitly a theory of money for an economy with some uncertainty.

    In the course of his discussion of the (ordinal) utility of money to a consumer, he says 'Possession of an average amount of it yields convenience in permitting the consumer to take advantage of offers of sale, in facilitating exchanges, in bridging the gap between receipt of income and expenditure, etc. The average balance is both used and at the same time not used; it revolves but is not depleted; its just being there to meet contingencies is valuable even if the contingencies do not materialize *ex post*' (p. 118).

    Only the first of the conveniences listed seems to afford a sufficient basis for endowing money with a continuous utility, and it cannot occur in the Walrasian economy with its lack of uncertainty. Similarly, the inclusion of contingency balances is impossible.
22. See Haberler's criticism of the 'excessive utilization of a marginalistic psychology in a field where a frankly institutionalistic analysis is much more fruitful' in the treatment of the Cambridge equation and in Keynes's liquidity preference [9], p. 164.

    Strictly speaking, this treatment alters the equilibrium conditions to the proportionality of marginal utilities to prices plus marginal expenditures on cash balances; however, to save space, this complication will not be introduced into the body of the paper.
23. See O. Lange [19], pp. 178–81. Schumpeter, as indicated above, is overlaudatory in his estimate of Walras's achievement in monetary analysis. See [36], pp. 1020–6, 1082–3, 1100, 1108.
24. Patinkin [28]; Becker and Baumol [2], pp. 359–60. For a bibliography and summary of this recent controversy, see Becker and Baumol [2].
25. [28], pp. 23–7.
26. [29], pp. 44–8. We shall limit discussion to system 1.1 and its micro-complement in this article.
27. [29], p. 50, n. 1. In view of this, cf. the incorrect analysis of 'Walras's error' in [27], pp. 143–4.
28. Or, less clumsily, $\bar{X}_{y_k} = f_{y_k}(z_j, y_k, y'_k, u)$, etc. The clumsy Walrasian

notation is retained because of the use of the coefficients throughout the input–output analysis to follow in Section IV. A drawback to their retention is that they might imply fixity of proportions for factor services, which is not the case.

Parenthetically, Walras worked with a single industry production function, assuming that under perfectly competitive conditions all firms would be using the same processes. In his fixed coefficient model, he made a 'symmetry assumption' in that he postulated that each of the n firms in an industry produces $\frac{1}{n}$ of the industry's output. His reasoning, then, about entrepreneural behavior was in terms of collective behavior.

Triffin's insistence upon Walras's placing the individual entrepreneur in the foreground, thus making the latter's analysis an ideal starting place for monopolistic competition analysis, seems farfetched. This becomes particularly confusing when Marshall's analysis is compared unfavorably with that of Walras in its preoccupation with the industry, for the substitutional theory of the former contains of necessity a great deal more scope for the analysis of the firm than the latter's neglect of such considerations. See R. Triffin [38], pp. 6–12.

In support of this argument, see Schumpeter's thesis that Marshall recognized the intermediate cases between pure competition and monopoly as the ordinary cases and so shaped his analysis to deal with them. Interestingly enough, Walras is contrasted with Marshall in regard to the former's desire to scrap all aspects of economic life not essential to his schema, and the latter's bent 'on salvaging every bit of real life he could possibly leave in'. [36], pp. 974–5, 997.

29. For Walras's treatment, see [41], p. 298.
30. The foregoing is essentially a reconstruction of Walras's extremely condensed and ambiguous chap. 29, [41], pp. 297–308, except for the handling of cash balances which is analogous to that for households already discussed. He does not define the functions as has been done here, but merely assumes rigid coefficients. Moreover, he does not introduce them as variable coefficients in his brief treatment of marginal productivity.
31. These conditions are not present in Walras, who developed his models with constant coefficients for money and inventory services (as well as for substitutional factors), and omitted them from his discussion of variable coefficients.

    The refinements are suggested by Barone, Palander, and Frisch. See G. Stigler [37], pp. 356–63; O. Lange [18], p. 67, and Schumpeter [36], p. 1030.
32. It must be recalled that these are *industry* production functions.

    Walras's statement of this segment of entrepreneurial equilibrium is the one included in the 1900 edition of his book. In it the conditions were stated (clumsily) in terms of the equality of the value of the marginal product and the price of the factor services; moreover, he explicitly stated that the total product will be exhausted in the distributive process [41], (4th edn, p. 375). However, this condition, combined with the requirement that entrepreneurs exhaust their total revenues in cost payments (system 1.6) and that all their product be sold (equations 1.4.1 and 1.4.2) render the model overdetermined unless the *firms'* production functions are such as to yield constant costs or are U-shaped cost curves with entrepreneurs operating at the minimum cost point. See E. H. Phelps-Brown [30], pp. 194–209.

In the definitive edition, however, after stating the conditions in the manner given above he converts the conditions to the proportionality of marginal products and prices – the minimum cost conditions. Seemingly, the reason for the change (although he makes no explicit mention of this) is a belief that he had escaped the need for a linear homogeneous industry production function.

Schultz has assumed that Walras's production function in his analysis of variable coefficients is linear and homogeneous, in so far as the coefficients are defined as averages, quantity of product is not included as an explicit variable (economies of scale are absent), and all factors are included in the function and are substitutable for all others. However, Walras explicitly rejects this last assumption on [41], p. 375. See H. Schultz [33], pp. 543–4.

Walras's error is to assume that the imposition of the no-profit constraint renders average costs constant. With n maximum profit equations for n factors, plus the production function, the system is overdetermined. If marginal cost equals average cost, then by substituting from the maximum profit equations for the prices of factor services, the price of the product may be eliminated from the no-profit constraint and the proportionality of marginal products is sufficient. However, if average costs depart from marginal costs this cannot be done if the conditions of maximum profits are to be retained. Walras eliminates the wrong equation when he drops the condition that price of the product equals marginal cost (price of the factors divided by their marginal products). Rather, the imposition of no-profits makes one of these redundant. Stigler does not seem to catch this error [37], p. 371. Therefore, Walras's treatment is correct only for linear homogeneous industry functions.

The assumption of linear homogeneity for the firm is not necessary, for Wicksell has demonstrated that as long as U-shaped cost functions for the firm are assumed, the characteristics of linear homogeneity are met at the lowest point of the cost curve. See Wicksell [42], I, 124–33. Moreover, the acceptance of U-shaped cost curves for the firm allows the Walrasian system to escape the criticism that it includes nothing to limit the size of the firm except the assumption that each of the n firms in any industry produces $\frac{1}{n}$ of the total output. See J. Viner [39]; E. Chamberlin [6], Appendix B, pp. 231–59; P. Samuelson [32], pp, 78–9; and G. Stigler, [37], pp. 380–7.

If every firm has the same cost curve, if entry is free, and if the no-profit condition is imposed, each will operate at the minimum point of its cost curve, and the *industry* production function will possess the characteristics of linear homogeneity, its costs per unit constant. It would be preferable, however, to introduce the firms separately and to aggregate for the industry totals, since the industry production function has no operational meaning in pure competition.

Walras relegated his marginal productivity reasoning to a five-page supplement outside the body of his models. Given his pioneering role in the establishment of marginal utility, his neglect of its distributive analogue is understandable and may be used to point up the strategic position that marginal utility holds in his system. See, for example: 'It (the entire system of exchange and production) rests entirely on the theory of exchange, and the theory of exchange resolves into the twin conditions that, in a state of equilibrium, each exchanger obtains a maximum of utility and demand equals supply. . . .' ([41], p. xiv).

This preoccupation with utility allows him to slip into the fallacy of causal direction a little later when he says 'from the moment the final degree of utility determines the price of products, it determines also, by that same fact (*par cela même*) the price of productive services, or rent, wages, and interest. . . .' ([41], p. xvi).

Finally, note the asymmetry of the maximization reasoning in the following: 'Let us note, then, that we find here, in the desire to avoid losses and make profits the determining reason of demand for productive services and the supply of products by entrepreneurs, as we have already in the desire to obtain maximum satisfaction of desires, the determining reason for the supply of factor services and the demand for products by (consumers)' ([41], p. 194).

33. The equation for the *numéraire* may be omitted, for if each equation of system 1.3 be multiplied by the respective price, if each equation of system 1.6 be multiplied by its respective equilibrium demand, and if both systems thus treated be added, the resulting equation reduces to a statement that income minus consumers' expenditure on consumers goods other than the *numéraire* minus the value of investment equals consumer expenditure on the *numéraire*. Since savings and investment are identities in this context, the equation reduces to 1.1.6. Therefore, we may omit 1.6.1 for good $y_k = 1$ as redundant.
34. [37], p. 254. In the original paper this analysis was developed, but is omitted from this article in the interests of conserving space.
35. Walras points out the problem but suggests no method of attack. See [41], p. 293.
36. [42], I, pp. 171, 226.
37. See the quotation from Wicksell above and the references there cited. See also Hicks [11], pp. 345–47; C. Bresciani-Turroni [3]; and Stigler [37], p. 249. Stigler interprets $X_e p_e$ as gross investment.
38. See the comments of Schumpeter in the preface to the Japanese edition of *The Theory of Economic Development* [35], pp. 158–63; and the comments of Stigler [37], p. 255.
39. If a single function is substituted for Walras's individual functions, the secondary stability conditions do not necessarily imply declining marginal utility for each commodity. See Hicks [13], p. 306.
40. Since single product firms have been assumed throughout, the condition that each product must be produced at increasing marginal cost may be

omitted, for it is assured by the second stability condition above. Also, the condition that as the production of a commodity (or group of commodities) increases, its marginal opportunity cost in terms of other commodities must increase, may be omitted. Lastly, for the firm, we may neglect the necessity for the average cost of all products to rise in order to insure that it would not be necessary to stop production of any one product, given a positive or zero surplus for the group of products as a whole. See Hicks [13], pp. 78–88, and J. Mosak [26], pp. 97–8.

41. The equality of unknowns and independent, consistent equations is not sufficient to obtain a unique, non-negative solution of mathematical systems. However, A. Wald [40], has shown that Walras's system is capable of this as it meets these conditions: (*1*) factors are in non-negative supply; (*2*) coefficients of production are non-negative; (*3*) at least one factor service enters into the production of each commodity; (*4*) the demand functions are continuous and positive for any finite price, and reflect marginal utility functions much more influenced by the goods for which they are defined than by quantities of other goods. A remaining condition specifies the unique solution of prices under conditions of fixed coefficients.
42. For a lengthy discussion of the stability considerations in Walras, see Schumpeter [36], pp. 998–1026.
43. Briefly, given mild substitution as the most usual relationship between goods in the consumers' sphere, complementarity in the relations between factor services, and substitution as the most usual relationship between factors and products, the only strong deterrent to the achievement of perfect stability is the possibility of destabilizing (noncompensatory) income effects in the commodity markets; however, the possibility of these effects dominating the economy becomes a slight one in view of the foregoing relationships. See Hicks [13], pp. 55–77 and 99–109, but see Samuelson [32], pp. 257–83, for a demonstration that stability of equilibrium is of necessity a problem in dynamics, in the sense of rates of change in the variables.
44. Hicks [13], pp. 56–7. It matters very little whether we call Hicks's economy 'dynamic', a system of 'intertemporal equilibrium', or, in Baumol's terminology, 'statics involving time' ([1], p. 6); what is important is that the introduction of intertemporal considerations points up a major defect in Walras's model which has important implications for any attempt to make it empirically useful.
45. A significant difference does exist, however. Walras's markets are futures markets, with the contracts made on the market day having relevance to consumption and production for the forthcoming 'week'. Hicks's market is one in which production of the preceding week is sold for consumption and further production in the coming week. In the former, planned production and consumption always equal realized production and consumption; consequently, on the succeeding market day, plans do not have to be altered in view of disappointed expectations, as they do in Hicks's spot markets. Therefore, some of the instability existing in Hicks's model has been eliminated in Walras's, but

only at the expense of reality, for 'futures' contracts can generally exist only when uncertainty is slight and expectations are held with a very high degree of probability. See Hicks [13], chap. X.

46. Walras, [41] pp. 298–99, and Hicks, [13], pp. 119–27.
47. This recent publication (in [21], chs. 3 and 6), essentially a disaggregated accelerator approach to the determination of investment in each industry, points up the conflict which has almost been resolved in favor of the accelerator principle, understandably in view of the need for operational theories.

    Nevertheless, those who have doubts about its ultimate effectiveness, may agree with the writer that Hicks's Additional Note C, [13], p. 337, deserves quoting in part, even though, ironically, it must be turned against the later work of its author: ' . . . By my hypothesis of essentially instantaneous adjustment, I reduced the purely mechanical part of my dynamic theory to the simpler terms – it is now quite evident that I over-simplified it. But in so doing I did leave myself free to make some progress with the less mechanical parts – expectations and so on. I still feel this procedure has its uses, and I should be sorry to abandon it altogether in favor of a pure concentration upon mechanism. It may well be that for econometric work a theory of Professor Samuelson's type is all we need; it gives a superb model for statistical fitting. But for the understanding of the economic system we need something more, something which does refer back, in the last resort, to the behavior of people and the motives of their conduct. It may well be that ways will be found by which we can retain these advantages as well as the advantages of a mechanical theory; but I do not think that they have been found just yet.'
48. 'Triffin [38], p. 78 and pp. 101–2.
49. Chamberlin, [6], pp. 102–3.
50. See Leontief [20], pp. 147–8, and [22], p. 214.
51. Recent work by Samuelson, Georgescu-Roegen, Cameron, and Klein has been shaped toward the definition of conditions under which the first of these assumptions may be dropped in favor of substitutability. If it be assumed that (*1*) each sector produces only one product. (*2*) only one nonproduced factor (usually labor) exists, and (*3*) pure competition rules throughout, then the coefficients in the Leontief matrix can be looked upon as economically determined from linear homogeneous production functions. Klein goes one step further and develops production functions yielding the same results for jointly produced commodities. These attempts to force the Leontief system into an equilibrium framework are opposed to the more pragmatic outlook developed below. See N. Georgescu-Roegen [8], p. 220; B. Cameron [4], pp. 19–27, [5], p. 191; and L. Klein [16], pp. 131–6.
52. Leontief [20], p. 38.
53. Cf. B. Cameron: ' . . . all economising activities are assumed to be absent or pre-determined', [5], p. 191.
54. For an illustration of the problem of convergence and the difficulties which arise in models which include households in the matrix, see W. Isard and R. E. Kuenne [14], pp. 299–300.

References

[1] Baumol, W. J., *Economic Dynamics*, Macmillan, New York, 2nd edn (1959).
[2] Becker, G. S., and W. J. Baumol, 'The Classical Monetary Theory: The Outcome of the Discussion', *Economica* (NS), 19 (1952), pp. 355–76.
[3] Bresciani-Turroni, C., 'Theory of Saving', *Economica* (NS), 3 (1936), pp. 1–23.
[4] Cameron, B., 'The Construction of the Leontief System', *Review of Economic Studies*, 19 (1950–1), pp. 19–27.
[5] Cameron, B., 'The Labor Theory of Value in Leontief Models', *Economic Journal*, 62 (1952), pp. 191–7.
[6] Chamberlin, E. H., *The Theory of Monopolistic Competition*, Harvard University Press, Cambridge, Mass., 6th edn (1949).
[7] Evans, W. D., and M. Hoffenberg, 'The Interindustry Relations Study for 1947', *Review of Economics and Statistics*, 34 (1952), pp. 97–142.
[8] Georgescu-Roegen, N., 'Leontief's System in the Light of Recent Results', *Review of Economics and Statistics*, 32 (1950), pp. 214–22.
[9] Haberler, G., 'The General Theory', ch. 14 in S. Harris (ed.), *The New Economics*, Dennis Dobson, London (1947).
[10] Hicks, J., 'Gleichgewicht und Konjunktur', *Zeitschrift für Nationalökonomie*, 4 (1933), pp. 441–55.
[11] Hicks, J., 'Léon Walras', *Econometrica*, 2 (1934), pp. 341–7.
[12] Hicks, J. 'A Suggestion for Simplifying the Theory of Money', *Economica* (NS), 2 (1935), pp. 1–19.
[13] Hicks, J., *Value and Capital*, Oxford University Press, London, 2nd edn (1946).
[14] Isard, W., and R. Kuenne, 'The Impact of Steel Upon the Greater New York – Philadelphia Industrial Region: A Study in Agglomeration Projection', *Review of Economics and Statistics*, 35 (1953), pp. 289–301 (reprinted as Chapter 4 in this volume).
[15] Jaffé, W., 'Walrasiana: The *Eléments* and Its Critics', *Econometrica*, 19 (1951), pp. 37–51.
[16] Klein, L., 'On the Interpretation of Professor Leontief's System', *Review of Economic Studies*, 20 (1952–3), pp. 131–6.
[17] Knight, F. H., *Risk, Uncertainty and Profit*, University of Chicago Press (1971).
[18] Lange, O., 'On the Economic Theory of Socialism', in B. J. Lippincott (ed.), *On the Economic Theory of Socialism*, University of Minnesota Press (1948).
[19] Lange, O., 'The Rate of Interest and the Optimum Propensity to Consume', reprinted in American Economic Association, *Readings in Business Cycle Theory*, Allen & Unwin, London (1950).
[20] Leontief, W., *The Structure of American Economy, 1919–1939*, Oxford University Press, New York (1953).
[21] Leontief, W., *et al.*, *Studies in the Structure of the American Economy*, Oxford University Press, New York (1953).
[22] Leontief, W., 'Recent Developments in the Study of Interindustrial Relationships', *American Economic Review*, 39 (1949), pp. 211–25.

[23] Marget, A., 'Léon Walras and the Cash-Balance Approach to the Problem of the Value of Money', *Journal of Political Economy*, 39 (1931), pp. 569–600.
[24] Marget, A., 'The Monetary Aspects of the Walrasian Theory', *Journal of Political Economy*, 43 (1935), pp. 145–86.
[25] Marshall, A., *Principles of Economics*, Macmillan, London, 8th edn (1948).
[26] Mosak, J., *General Equilibrium Theory in International Trade*, Principia Press, Bloomington, Ind. (1944).
[27] Patinkin, D., 'Relative Prices, Say's Law, and the Demand for Money', *Econometrica*, 16 (1948), pp. 135–54.
[28] Patinkin, D., 'The Indeterminacy of Absolute Prices in Classical Economic Theory', *Econometrica*, 17 (1949), pp. 1–27.
[29] Patinkin, D., 'A Reconsideration of the General Equilibrium Theory of Money', *Review of Economic Studies*, 18 (1950–1), pp. 42–61.
[30] Phelps-Brown, E. H., *The Framework of the Pricing System*, University of Kansas Press, Lawrence, Kansas, 1949.
[31] Rosenstein-Rodan, P. H., 'The Coordination of the General Theories of Money and Price', *Economica* (NS), 3 (1936), pp. 257–80.
[32] Samuelson, P., *Foundations of Economic Analysis*, Harvard University Press, Cambridge, Mass. (1947).
[33] Schultz, H., 'Marginal Productivity and the General Pricing Process', *Journal of Political Economy*, 37 (1929), pp. 505–51.
[34] Schumpeter, J., *Business Cycles*, McGraw-Hill, New York (1939).
[35] Schumpeter, J., 'Preface to the Japanese Edition of *The Theory of Economic Development*', reprinted in R. V. Clemence (ed.), *Essays of J. A. Schumpeter*, Addison-Wesley, Cambridge, Mass. (1951).
[36] Schumpeter, J., *History of Economic Analysis*, Oxford University Press, New York (1954).
[37] Stigler, G., *Production and Distribution Theories*, Macmillan, New York (1951).
[38] Triffin, R., *Monopolistic Competition and General Equilibrium Theory*, Harvard University Press, Cambridge, Mass. (1949).
[39] Viner, J., 'Cost Curves and Supply Curves', *Zeitschrift für Nationalökonomie*, 3 (1931), pp. 23–46.
[40] Wald, A., 'On Some Systems of Equations of Mathematical Economics', *Econometrica*, 19 (1951), pp. 368–79.
[41] Walras, L. *Eléments d'Economie Politique Pure*, R. Pichon & R. Durant-Auzias, Paris, 5th edn (1926).
[42] Wicksell, Knut, *Lectures on Political Economy*, Routledge, London (1934).

2 The Architectonics of Léon Walras*

In that fascinating area of surmise designated 'the sociology of knowledge' must lie answers to more than one seeming conundrum in the dynamics of economic doctrine. One day, perhaps, we shall know how a perspicacious Ricardo could have framed a theory whose pessimism derived from the neglect of a revolution in technology so thorough-going as to be unique in the history of man. And perhaps it will explain that perverse principle of insularity which so often militates against the spread of economic ideas between the Continent, Great Britain and the United States.

Surely Walras's progress in Great Britain and the United States is a most peculiar case in point. Eighty years have elapsed since the three major pioneering features of his theory – the development of marginal utility theory, a thorough-going use of mathematical analysis, and the concept of general equilibrium – first appeared; the complete structure has stood for half a century. During most of this time the corpus of Anglo-American economics has granted full recognition to the role of the *Elements* in the development of the theory of the consumer; for the last twenty years it has been an area undergoing extensive revision on both the micro- and macroeconomic levels.

Few would question the failure at mid-century of Walras's bitter prophecy: ' . . . mathematical economics will rank with the mathematical sciences of astronomy and mechanics; and on that day justice will be done to our work' (p. 48). But none will deny that the trend from Dupuit and Cournot passes through Walrasian coordinates before eventuating in the mathematical metaphysics of modern economics.

Lastly, the concept of general equilibrium for long has been recognized as one of the most «original» legacies of the nineteenth

* A review article of William Jaffé's translation of Léon Walras, *Elements of Pure Economics*, published for the American Economic Association and the Royal Economic Society by Allen and Unwin, London, 1954. The article was published originally in *Kyklos*, 10 (1956), pp. 241–49, and reproduced with permission.

century, a complete identification has occurred in the minds of American and British economists with the name of Walras, and, with the emergence of Keynesian theory, the popularity of Schumpeter's schema, the refinements of Hicks, and the development of input–output analysis by Leontief, for at least twenty years it has assumed a topical interest for them.

Has there ever been another example in the field of economics of pathfinding originality in so many directions which were to assume such independent and interdependent importance for the better part of the forthcoming century? Who could have doubted, had he been able in 1875 to foresee these developments in Anglo-American economics, that Walras's work would have been the very center of thesis-generation for the academic dialectic? Yet, for the benefit of a notoriously unilingual British-American profession, Professor Jaffé's English translation of Walras's *Elements* has only just appeared.

The reviewer cannot resist speculating about the impact an English translation might have had in the Marshallian Age of the 1920s. His own guess is that it would have been a minor challenge to the dominance of partial analysis in a policy-minded profession, as the limited reception that greeted Bowley's *Mathematical Groundwork of Economics* in 1924 attests. But suppose the Walrasian 'vision' and framework of reference had infiltrated the curriculum to the extent that they were familiar to the young men of the mid-thirties: would the Keynesian innovations have assumed such revolutionary proportions? Might not the profession have maintained a more dignified posture if the generality of its practitioners had been able to compare the data, variables, and equations of both systems? Was it not the novelty to a Marshallian generation of the 'generality' of the Keynesian theoretical structure as well as its operational aggregates that led to an exuberance perhaps a bit overweening? On the other hand, in the strong light of the morning after, might not the present be a propitious time for the appearance of an English *Elements* from the viewpoints of the tranquillity of the Anglo-American theoretical front and the relevance of a 'classical' model to the post-war world?

The present review will treat Professor Jaffé's extraordinary efforts in Section I, scrutinize some of the more important chambers in Walras's stately mansion in Section II, and, in Section III, study the functional relevance of the model from the viewpoint of today's needs in economic analysis.

I ON JAFFÉ'S INTERPRETATION

The reviewer's summary judgment of Professor Jaffé's achievement is that it will stand as a model of the art of translation in the field for years to come. The translation itself is painstakingly accurate and smooth in rendition. Both translator and his labors are unobtrusive in the body of the work. His decision to include the 'Note on Mr. Wicksteed's Refutation of the English Theory of Rent' which Walras published in the third edition but eliminated in the fourth and definitive editions is commendable.

Mr Jaffé's more distinctive contributions as a Walrasian scholar have been relegated to two sections in the back of the book: the first, a sixty-page, chapter by chapter commentary and interpretation, and the second, a collation of the four editions (the so-called 'definitive edition' of 1926 being considered essentially a reprinting of the 1900 fourth edition). Balance and judgment are evident again in the decisions taken on those differences between editions which are important to the student of the evolution of Walras's thought and on the line of praticability between quoting differences or merely citing them.

The reviewer's enthusiasm for the commentary is less marked, although it should be added immediately that most of these differences concern emphasis and interpretation and therefore reflect professional biases. For example, given the context of Walras's model and his 'vision' of economic processes, is it true to say that 'normal profits' are implicit in Walras's cost function? (p. 526).

Also, in defending Walras's initial use of constant coefficients of production and explaining their conversion to variables, Jaffé says:

> . . . When, however, Walras considers the consequences of net aggregate capital formation with the total amount of land remaining constant, then a new type of production function becomes inevitable involving substitutability between the different kinds of productive services. The coefficients must now be variable. It is for this reason that Walras delays introducing the marginal productivity theory until Part VII where he is concerned with economic progress; and there he insists that the object of this theory is to determine the magnitudes of the variable coefficients, not their prices (p. 527).

And later: ' . . . When population and capital increase on a fixed territory, though the state of the industrial arts remains unchanged, the coefficients of production must change' (p. 550).

There is a confusion here: the mere existence of net capital formation and a limitational factor does not invalidate fixed coefficients as long as prices and product mix can change and the input structures of all industries are not equal or of unrealistic similarity. If this contention were true it would mean that serious constraints would have to be put upon the relative amounts of initial factors available at the beginning of the analytical period unless one were willing to allow the model to be cluttered up by free factor services. Jaffé fails to indicate that Walras was thinking loosely in terms of a market basket of goods with a fixed structure increasing in absolute amount: the emphasis in Walras's treatment is on the multiplication of goods of a fixed type, not on a change in the ratio of factor availabilities[1] (pp. 382–3).

On page 528 Jaffé states that since all firms within an industry produce a homogeneous product by processes using fixed coefficients, all must have identical production functions 'which is a sufficient (though not necessary) condition for equal rates of output for all firms at equilibrium under conditions of pure and perfect competition' (p. 528). While it is true that the aggregate production function Walras adopts is unambiguously defined under these conditions, it is not clear to the reviewer that Walras is relieved of the need to define a second assumption distributing output among firms.[2] In the absence of the ability to stipulate that factors be distributed among firms in such a way as to equate marginal products, Walras, as Stigler indicates,[3] assumes the n firms produce $1/n$ of the final output. And what determines n?

On pp. 530–1 Jaffé differentiates between Walras's '*excédent*' of gross income over consumption and gross savings, $D_e p_e$, though they are the same (with the qualification that Walras *imposed* the condition on the latter that it exceed capital consumption).

One of the most difficult parts of economic literature is found in Walras's exposition of his savings-investment theory. Here, more than in any other part of the book, is the guidance of the neophyte by the scholar most welcome and requisite. Yet in this material Jaffé has concentrated upon two chapters (26 and 27) in which Walras felt constrained to 'demonstrate' that a 'maximization' of utility for society in the investment process would occur if net revenues from capital goods divided by selling prices were equal to a common factor of proportionality. Jaffé's preoccupation is not, of course, to attempt to justify Walras's welfare conclusions but rather to get to the root of the demand for capital goods when they are not of direct utility to the purchaser or at least not purchased for consumer use. The reviewer does not understand why this should be a problem if Walras's assumption that perpetual net revenue has a utility to the consumer is allowed: under these conditions capital goods could hardly be 'adventitious' as Jaffé once believed. But in any event it seems to the reviewer that a disproportionate amount of space has been devoted to a marginal problem.

Lastly, his generous speculations about Walras's possible meanings in discussing the welfare implications of pure competition seem unnecessarily labored. For example:

> On another interpretation, Walras was not wrong at all. In fact, there is very little justification for following Wicksell in attributing to Walras any hypothesis of interpersonal comparisons of utility. Though Walras supposes that *rareté* (marginal utility) can be defined as a cardinal magnitude, nowhere does he allude to any actual addition of the utilities enjoyed by different persons; and though he later refers to average *raretés* for the market as a whole and makes use of this concept in his theory of money, the addibility implied in the averaging process is always purely formal, never real (p. 511).

The last phrase is hard to interpret: in one sense the cardinal measurability assumption of Walras could be excused on grounds that he never really totalled up utility magnitudes. Devoted indeed is the man who seeks to defend Walras's welfare propositions: he is due every sympathetic understanding.

But these are peccadilloes pointed up in the best tradition of reviewmanship. The reviewer found the explanation and clarification of marginal productivity in Walras concise and extremely helpful, and more typical of the quality of the commentary than the above.

In summary, the profession has been rendered a fine service by Professor Jaffé who has every reason for pride in the manner in which he carried out his commission.

II THE WALRASIAN MODEL

If the availability of Walras's *Elements* does nothing else it will explode once and for all the notion that the use of mathematics in economics is a sufficient guarantee of expositional clarity. For Walras's analysis is often vague in the non-deliberate (Marshallian) sense, ambiguous, unfinished. The reader must continually remind himself that marginal cost did not exist for Walras (pp. 41–2, 382–4) and that although utility maximization provided the motivational assumption for consumer behavior Walras always backed away from assuming profit maximization as the analogous entrepreneurial motivation (pp. 41, 225, 255). He will wonder over the fact that values in exchange are treated initially as undefined *absolute* magnitudes (pp. 70, 87, 178, but see p. 145). He will shudder at attempts to *deduce* the effects of price changes upon demand before maximization reasoning, stability conditions, or other postulational bases have been introduced (p. 91). He will ponder long over two paragraphs in which the term 'extensive utility' is successively defined as a *kind* of utility, one *attribute* of utility, and the *quantity* taken of a good at price zero (p. 115). Boldly asserted welfare statements of the following type will appal the modern reader: ' . . . maximum satisfaction will be achieved by *each* party when the ratios of the *raretés* of the several commodities demanded to the *rareté* of the one commodity originally held are equal, not to prices as they are first cried, but to the true prices arrived at by arbitrage' (p. 164) [Italics supplied]. Whole sections, such as section 255 in his capital formation theory, are mare's-nests of confusion: at one point Jaffé figuratively throws up his hands and exclaims, 'This sentence and the rest of the mathematical exposition in section 263 is obscure to the point of almost complete incomprehensibility', then proceeds to work out his own proof.

But the perservering reader will watch a master-architect erect a structure magnificent in its imaginative conception, in its ambition, in its symmetry, if not in its detail. For Walras's *chef d'œuvre* like Marx's is one in which it would be difficult to find one structural component which some critic has not challenged with analytical hammer and chisel and found weak. Yet the structure stands without benefit of the buttressing the Marxist structure receives from its functional political utility, the more fascinating for its flaws.

The appearance of Professor Jaffé's work is a propitious occasion for sketching briefly the nature of the attacks upon and the current status of what the reviewer considers the three major components of Walras's most comprehensive model: the utility analysis of consumer choice, the theory of savings and investment, and the theory of money.

Walras's use of a utility function for each good independent of the amounts of other goods consumed was attacked by Edgeworth, Fisher, and Wicksell, each of whom introduced a single function and dealt with partial derivatives of it. More important, of course, has been the attack upon the cardinal measurability hypothesis begun by Pareto and revived by Hicks and Allen in the mid-thirties, although Walras is quite clear in indicating that his assumption of cardinal measurability is on an 'as if' basis (p. 177). Today, even with the work of von Neumann and Morgenstern, Marschak, and Friedman and Savage, it remains true that for the analysis of choice among certain prospects, as opposed to uncertain prospects, the use of a method of measurement relying upon a monotonic transformation of 'satisfaction' does allow the replacement of hypotheses about diminishing marginal utility not capable of test by hypotheses about rates of substitution *now* meaningful. However, a main issue in the cardinalist-ordinalist debate appears to have been a false issue: the question is not whether a given variable can, in the nature of things, be measured by a linear transformation or a monotonic transformation, for since all measurement is contrived the true question is whether or not a method has been devised to measure a magnitude in the first manner. It is true that until some manner of measurement unique to a linear transformation is contrived Walras's hypotheses are meaningless in the Samuelson sense.[4] However, on the pragmatic grounds that conceptually it is quite conceivable that such measurement may be derived, Walras's assumption looks less antique today than it did ten years ago.[5]

This reviewer confesses that even with the removal of the strain of coping with a non-maternal tongue Walras's savings-investment theory is still less than clear to him. As merely one example the following point is made: Walras is strongly critical of theories which do not possess well-identified markets in which price-quantity determination occurs. For example, in the preface he rejects Böhm-Bawerk's theory of interest rate determination on the grounds that since interest is the premium paid for present over future goods the latter's failure to specify the market in which future goods were priced invalidated the theory (pp. 45–6). Of course the mere fact that Böhm-Bawerk spelled out a theory for the existence of interest in no way precluded him from obtaining the interest rate in a market for funds in his theory of its *determination*. But it is one instance of the importance Walras attached to identifying real world markets in which his theoretical magnitudes were determined approximately.

The rate of return on capital goods, i_k, is determined on the market for capital goods – to Walras, the stock market. On the other hand, the money rate of interest, i, is determined on the market for money services. These are two separate markets which yield

$$i_k = i \tag{2.1}$$

only as an equilibrium condition. The consumer's demand for money is the demand for inventory services in physical units, X_g, times the price of these services, p_g, divided by the interest rate, or

$$D_m = \frac{X_g p_g}{i} \tag{2.2}$$

As definitional identity we have the following relationship for the selling price of goods:

$$P_g \equiv \frac{p_g}{i_k} \tag{2.3}$$

Substituting (2.3) into (2.2)

$$D_m = \frac{X_g \, P_g \, i_k}{i} \tag{2.4}$$

The demand for cash will be sufficient to command the desired number of money units only if (2.1) is true. Therefore, the consumer's demand for money is valid only for equilibrium and of limited use for determining cash balances.

Space limitations allow only a passing reference to other criticisms in this least satisfactory portion of the theory: Schumpeter has noted the implicit stationary state assumptions which accord so poorly with the 'progressive' state (net investment); Hicks has indicated the need to drop Walras's assumption of automatic reinvestment of depreciation allowances to meet Wicksell's objection that in the absence of net investment the interest rate would be zero; and others point out the confusing restrictions put upon the excess of income over consumption.[6]

Attacks upon the monetary component of the theory center upon the endowment of money with an indirect utility allegedly inconsistent with the conditions of the model. Del Vecchio began the attack by 1909; Hicks pointed out the necessary conditions to obtain cash balances in a circular flow; and Patinkin, denying that the absolute price level affected any other demand but that for money, making a variation of the price level incapable of correcting excess demands for goods, attempted to show that the absolute price level was indeterminate.[7] On this latter point Walras specifically anticipated the criticism and correctly indicates that as long as the budget equation has an income term from the lending of money, and so long as changes in absolute prices have an income effect, the price level does affect the demand for real goods and is determinate (pp. 326–7). All of the foregoing assumes, of course, that the conditions indicated by Hicks exist to make the holding of cash balances rational.

III ON GENERAL EQUILIBRIUM

The most damning criticisms of general equilibrium theory are aimed, not at its structure, but its function. The 'Why' of the theory transcends the 'What'. This is not merely to rehearse the old plaints that it is a static framework attempting to cope with a dynamic world: economists seem always to have to relearn the basic lesson that the only excuse for static frameworks is to deal with change. The plain fact of the matter is that the Walrasian system is too cumbrous a mechanism to cope adequately with the problems it seeks to solve.

There would be little point in repeating the ABCs of static analysis so ably presented by Hicks and Samuelson[8] except that we are so prone to ignore them. The employment of the concept of equilibrium to establish the existence of a (preferably) unique set of values satisfying a set of human desires and actions is merely the first step on the analytical road. It is more important to use these relationships to establish the quantitative or qualitative nature of the changes enforced upon all variables by changes in the data of the system or variables converted to data. These 'laws of change' are the desiderata of the analysis: they are not implicit in the notion of an equilibrium system but depend upon more detailed knowledge of the nature of the system. In a Marshallian demand–supply analysis stability conditions are sufficient to determine the effects of an increase in demand upon quantity while more detailed knowledge of the relative values of the slopes is necessary before the effect upon price is determinate. The amount of detailed knowledge of these functions required to evaluate the Jacobians for mere qualitative changes speedily becomes so great as to render the whole apparatus useless unless simplifications are made.

Compare Walras's elementary attempts to trace out these laws in the exchange economy:

> Thus an increase in the utility of (B) for some individuals will result in an increase in the price of (B) [a result which is correct given Walras's explicit stability condition that excess demand decrease with a rise in price and increase with a decline in price]. It may also result in a change in the prices of (C), (D), . . . These secondary effects, however, will be less appreciable than the primary effect if there are a great many commodities other than (B) on the market and if, in consequence, the quantity of each commodity exchanged for (B) is very small. Besides there is no way of knowing whether the prices of (C), (D), . . . will rise or fall; nor can we even know that they will change at all . . .
> The effect on the prices of (C), (D), . . . will be that some will rise, others fall, and still others remain unchanged (p. 179).

Other attempts to cope with primary and secondary effects of equilibrium displacements will be found on pages 246 and 260.

It is to pay high tribute to Walras to say that he understood that herein lay the ultimate aim of static analysis. He saw beyond the more immediate need of stability conditions to their use in evaluating the impact of shocks and the overriding need to seek out laws of variation:

> The theorist has the right to assume that the underlying price determinants are invariant over the period he has chosen to use in his formulation of the law of equilibrium prices. But, once this formulation has been completed, it is his duty to remember that the forces are by their nature variable, and consequently he must formulate the law of variation of equilibrium prices . . . (pp. 146–9).

The confession of the futility of his system to cope with this assignment is quite explicit:

> Theoretically, all the unknowns of an economic problem depend on all the equations of economic equilibrium. Nevertheless, even from the viewpoint of static theory, it is permissible to consider

> some of these unknowns as especially dependent on the equations which were introduced at the same time as the unknowns when the problem of their determination was first raised. It is all the more legitimate to do this when we pass from the static to the dynamic point of view, or, better still, when we pass from the realm of pure theory to that of applied theory or to actual practice, for then the variations in the unknown quantities will be effects of either the first or the second order, that is to say, effects which need or need not be taken into consideration, according as they arise from variations in the special or the general data (pp. 307–8).

When it is realized that Walras's *dynamics* implied merely that the data of a system were ever-changing through time so that at any particular time a series of superimposed comparative statics adjustments was being effected, the completeness of the confession is manifest.

It is no accident, then, that admirers of the structural aesthetics of general equilibrium economics have been so hard pressed to find a functional apology for it. In the last analysis, granted its analytical sterility, the general equilibrium system finds its major uses in the more pedestrian but vital field of pedagogy: it serves as a convenient map of the whole economic territory – obstacles to be met, destinations which are desirable, routes which offer analytical promise. And, lastly, it impresses the student as no other system will with the meaning of interdependence. Let not Ambition mock such useful toil!

Notes

1. The assumption of common input structures is adopted by D. H. Robertson in [14], pp. 229–30 as a basis for a similar criticism. This severe restriction is quite correctly criticized by J. R. Hicks in [4] and by implication in the [5], pp. 12–14. Also W. Leontief in [9], pp. 39–40 discusses factor 'substitution' via changes in relative amounts of goods produced requiring different amounts of factors.
2. For the most recent discussion of the difficulties involved in defining an aggregate production function – in particular the fact that it is impossible to get a unique relation between the sum of firm inputs and total output in the absence of a law for distribution of inputs between firms – see C. G. F. Simkin [17], pp. 32–9 and the bibliography therein cited.
3. G. J. Stigler, [18], p. 240n.

4. P. A. Samuelson [15], pp. 4–5.
5. For a partial bibliography of these recent developments see J. von Neumann and O. Morgenstern [19], pp. 15–3; M. Friedman and L. Savage [2]; [3]; A. Alchian [1]; and J. Marschak [12].
6. See Schumpeter's preface to the Japanese edition of *The Theory of Economic Development* [16]; J. R. Hicks [6], pp. 345–7, and references in latter.
7. See A. Marget [10], p. 594, and [11]; J. R. Hicks [7], esp. 6–7; Don Patinkin [13], and earlier studies therein cited.
8. P. A. Samuelson [15], *passim*, especially chs. 1 and 2, and J. R. Hicks [8], parts I and II, especially chs. IV and V.

References

[1] Alchian, A., 'The Meaning of Utility Measurement', *American Economic Review*, 43 (1953), pp. 26–50.
[2] Friedman, M. and L. Savage, 'The Utility Analysis of Choices Involving Risk', *Journal of Political Economy*, 56 (1948), pp. 279–304.
[3] Friedman, M. and L. Savage, 'The Expected Utility Hypothesis and the Measurement of Utility', *Journal of Political Economy*, 60 (1952), pp. 463–74.
[4] Hicks, J., 'Marginal Productivity and the Principle of Variation', *Economica*, 12 (1932), pp. 79–88.
[5] Hicks, J., *Theory of Wages*, St. Martin's Press, New York (1966).
[6] Hicks, J., 'Léon Walras', *Econometrica*, 2 (1934), pp. 345–7.
[7] Hicks, J., 'A Suggestion for Simplifying the Theory of Money', *Economica* (NS), 2 (1935), pp. 1–9.
[8] Hicks, J., *Value and Capital*, Oxford University Press, London, 2nd edn (1946).
[9] Leontief, W., *The Structure of American Economy, 1919–1939*, Oxford University Press, New York (1953).
[10] Marget, A., 'Léon Walras and the "Cash-Balance" Approach to the Problem of the Value of Money', *Journal of Political Economy*, 39 (1931), pp. 569–600.
[11] Marget, A., 'The Monetary Aspects of the Walrasian System', *Journal of Political Economy*, 43 (1935), pp. 145–86.
[12] Marschak, J., 'Rational Behavior, Uncertain Prospects, and Measurable Utility', *Econometrica*, 18 (1950), pp. 111–41.
[13] Patinkin, D., 'A Reconsideration of the General Equilibrium Theory of Money', *Review of Economic Studies*, 18 (1950–1), pp. 42–61.
[14] Robertson, D., 'Wage Grumbles', reprinted in W. Fellner and B. F. Haley (eds), *Readings in the Theory of Income Distribution*, Blakiston, Philadelphia (1946).
[15] Samuelson, P., *Foundations of Economic Analysis*, Harvard University Press, Cambridge, Mass. (1947).
[16] Schumpeter, J., 'Preface to the Japanese edition of *The Theory of Economic Development*', reprinted in R. Clemence (ed.), *Essays of*

J. A. Schumpeter, Addison-Wesley, Cambridge, Mass. (1951).

[17] Simkin, C.G.F., 'Aggregate Production Functions', *Economic Record*, 31 (1955), pp. 32–9.

[18] Stigler, G., *Production and Distribution Theories*, Macmillan, New York (1941).

[19] von Neumann, J. and O. Morgenstern, *Theory of Games and Economic Behavior*, Princeton University Press, 2nd edn (1947).

[20] Walras, Léon, *Elements of Pure Economics*, Allen & Unwin, London (1954).

3 On Hicks's Concept of Perfect Stability in Multiple Exchange*

I HICKS'S DEFINITION

The concept of static stability and its conditions, as defined by Professor Hicks in *Value and Capital*,[1] have received a great deal of attention in the literature.[2] Most of the discussion has been conducted in terms of the validity of a purely static stability analysis and its relationship to dynamic stability properties. This paper poses the question whether there is not a fundamental ambiguity in the definition of Hicksian perfect stability for a given market equilibrium which makes the concept treacherous analytically.

Since our criticism will apply to a multiple exchange economy with four or more goods, we shall phrase Hicks's definitions in such terms. Assume that such an economy exists with n goods ($n > 3$), including money as good n. Assume, further, that the general market equilibrium of the economy is described by $n - 1$ independent and consistent excess demand equations yielding money prices as equilibrium solutions. Specifically, the vector $(P_1, P_2, P_3, \ldots P_{n-1})$, or (P), is an equilibrium price set for a given set of data.

Now, let a perturbation in the market for good 1 send price up from P_1 to P_1'. Were the economy a two-good system, Walrasian stability conditions would require that excess demand for good 1 fall (which would mean that excess demand for good 2 would rise identically as P_2 fell identically). In an economy with more than two goods, however, where the obverse of excess demand in any one market is found, not always in one other market, but perhaps in two or more, an extension of this Walrasian stability condition must cope with the question: if we desire to study the effect of price P_1' upon the excess demand for good 1, what conditions should we allow to rule in other markets? Should all other prices be readjusted to bring about

* Published originally in the *Quarterly Journal of Economics*, 73 (1959), pp. 309–15, and reproduced with permission. The author is indebted to W. J. Baumol and R. E. Quandt for reading the manuscript and making several suggestions for improving it.

equilibrium in all markets but those for goods 1 and n (money) before the excess demand for good 1 is studied? Should no other price be so adjusted? Or should some subset of markets be re-equilibrated before examining the excess demand for good 1?

Since the function of a general analysis is to take explicit account of the interdependencies of all markets, the logic implied by the selection of the analytical framework would tend toward an affirmative for the first response. Hicks affirms that 'it is what happens when all other prices are adjusted that is really most important'.[3] But he also asserts an interest in the cases of less-than-complete price adjustment, and, consequently, his stability analysis takes the form of successive market equilibrium adjustments.

Let us return to the case above to illustrate his method. First, the amount of excess demand is noted at P_1' before any markets are restored to equilibrium; then the market for good 2 is restored to equilibrium and good 1's excess demand at P_1' noted; next, the markets for goods 2 and 3 are brought into equilibrium, and good 1's excess demand noted; the prices of goods 2, 3 and 4 have their equilibrating changes, and excess demand for good 1 is once more studied; and so forth, until the markets for goods 2 through $n - 1$ are back in equilibrium, and the final strength or weakness of the forces tending to pull P_1' back to P_1 are gauged.

If, at every step of the succession outlined above, excess demand is negative,[4] Hicks terms the equilibrium point one with perfect stability; if any one of the successive steps yields a positive excess demand, but the last step (when all markets but those for goods 1 and n are in equilibrium) yields a negative excess demand, the equilibrium point is said to be imperfectly stable.

To be a useful concept, perfect stability should be invariant to the order in which goods markets are adjusted. After all, the numbering of goods in the economy we have assumed is essentially random, and no significance attaches to it. Therefore, we pose the question: is Hicks's concept of perfect stability independent of the order in which a set of markets is adjusted? Unfortunately, as it is defined and used in chapter V, the answer is 'no', and the value of the concept becomes questionable.

II SOME DIFFICULTIES

Before we demonstrate rigorously the truth of this charge, we may indicate it intuitively in this fashion. Suppose we had an exchange economy with four goods: 1, 2, 3, and money. We desire to investigate the stability of good 1's equilibrium, starting from a position of zero excess demand in all markets. When good 1's price is raised arbitrarily, assume that excess demand for good 1 becomes −6 units, goods markets 2 and 3 unadjusted. Skipping over the intermediate step to the final stage, we adjust markets 2 and 3 simultaneously to be in equilibrium. We shall assume that excess demand rises to −3 with these adjustments, and, therefore, that the initial equilibrium for good 1 was at least imperfectly stable and possibly perfectly so.

But which? We can approach the last adjustment in two ways: first, by adjusting market 2, leaving 3 unadjusted; second, vice versa. Nothing in our analysis specifies the need for any particular path. Thus, if we follow the first, we might decrease excess demand for good 1 from −6 to −10, in which case stability would be perfect. But if we follow the second path, a rise in excess demand from −6 to +1 might occur, and the equilibrium would be imperfectly stable. Table 3.1 summarizes these hypothetical results.

Therefore, if we discover a particular sequence of market adjustments which yields perfect stability, no guarantee exists that another sequence will not yield imperfect stability for a given good's market equilibrium.

This can be shown more rigorously in the following fashion. Given the same economy discussed above, let us define demand and supply in Wicksteedian fashion, so that supply will equal the amounts of each good existing in the economy, and changes in excess demand will be the same as changes in demand. Then, the independent excess demand equations can be written:

$$E_i = X_i - Q_i = F_i\,(P_1, P_2, P_3) - Q_i,\ i = 1, 2, 3 \tag{3.1}$$

Table 3.1 Gross effects upon market 1's excess demand of alternative adjustments

Excess demand for good 1	*Market(s) adjusted*	*Market(s) unadjusted*
−6	. . . .	2, 3
−10	2	3
+1	3	2
−3	2, 3	. . . .

where E_i is excess demand, X_i is the quantity demanded, and Q_i is quantity supplied. With a small change in P_1, changes in excess demand (equals changes in demand) will be defined as below:

$$\frac{dX_1}{dP_1} = \frac{\partial X_1}{\partial P_1} + \frac{\partial X_1}{\partial P_2}\frac{dP_2}{dP_1} + \frac{\partial X_1}{\partial P_3}\frac{dP_3}{dP_1}$$

$$\frac{dX_2}{dP_1} = \frac{\partial X_2}{\partial P_1} + \frac{\partial X_2}{\partial P_2}\frac{dP_2}{dP_1} + \frac{\partial X_2}{\partial P_3}\frac{dP_3}{dP_1} \tag{3.2}$$

$$\frac{dX_3}{dP_1} = \frac{\partial X_3}{\partial P_1} + \frac{\partial X_3}{\partial P_2}\frac{dP_2}{dP_1} + \frac{\partial X_3}{\partial P_3}\frac{dP_3}{dP_1}$$

The first step in the succession considered above is to evaluate dX_1/dP_1 when all other prices are held constant ($dP_2/dP_1 = dP_3/dP_1 = 0$), i.e., when $dX_1/dP_1 = \partial X_1/\partial P_1$. For perfect stability this term must be less than zero.

Next, if we adjust P_2 so that dX_2/dP_1 becomes zero, i.e., so that market 2, originally in equilibrium, remains so despite P_1's change the solution for $(dX_1/dP_1)2$ – the change in (excess) demand for good 1 when the market for good 2 is equilibrated – is given by the ratio of two Jacobians:

$$\left(\frac{dX_1}{dP_1}\right)2 = \frac{\begin{vmatrix} \frac{\partial X_1}{\partial P_1} & \frac{\partial X_1}{\partial P_2} \\ \frac{\partial X_2}{\partial P_1} & \frac{\partial X_2}{\partial P_2} \end{vmatrix}}{\begin{vmatrix} \frac{\partial X_2}{\partial P_2} \end{vmatrix}} = \frac{J_{1,2}}{J_2}. \tag{3.3}$$

Similarly, if P_2 and P_3 are adjusted to yield zero (excess) demand changes in the good 2 and 3 markets, and if $(dX_1/dP_1)_{2,3}$ denotes the change in (excess) demand in market 1 under these conditions, the value of the latter will be found to be:

$$\left(\frac{dX_1}{dP_1}\right)2,3 = \frac{\begin{vmatrix} \frac{\partial X_1}{\partial P_1} & \frac{\partial X_1}{\partial P_2} & \frac{\partial X_1}{\partial P_3} \\ \frac{\partial X_2}{\partial P_1} & \frac{\partial X_2}{\partial P_2} & \frac{\partial X_2}{\partial P_3} \\ \frac{\partial X_3}{\partial P_1} & \frac{\partial X_3}{\partial P_2} & \frac{\partial X_3}{\partial P_3} \end{vmatrix}}{\begin{vmatrix} \frac{\partial X_2}{\partial P_2} & \frac{\partial X_2}{\partial P_3} \\ \frac{\partial X_3}{\partial P_2} & \frac{\partial X_3}{\partial P_3} \end{vmatrix}} = \frac{J_{1,2,3}}{J_{2,3}}. \tag{3.4}$$

which, for perfect or for imperfect stability, must be less than zero.

To summarize: in Hicks's definition, for perfect stability to rule at a given equilibrium point in the market for good 1, it is necessary that the following conditions hold:

1. $(dX_1/dP_1) - \partial X_1/\partial P_1 < 0$.
2. $(dX_1/dP_1)_2 < 0$.
3. $(dX_1/dP_1)_{2,3} < 0$.

Now, in order for perfect stability to be invariant to the ordering of market adjustments, it must be shown that the three conditions above imply

3′ $(dX_1/dP_1)_3 < 0$.

That is, it must be shown that the change in (excess) demand brought about by a rise in P_1, after the market for good 3 is brought back to equilibrium, is always negative. Solving equations 3.3 for dX_1/dP_1 when dP_2/dP_1 and dX_3/dP_1 equal zero, we obtain as a solution, employing our shorthand:

$$\left(\frac{dX_1}{dP_1}\right)3 = \frac{J_{1,3}}{J_3}. \tag{3.5}$$

If we now suppose conditions 1, 2, and 3 to hold, it can be shown that condition 3′ is not implied. For example, suppose the partial derivatives in all the Jacobians had the following values:

$$J_{1,2,3} = \begin{vmatrix} -1 & -9 & 4 \\ 1 & -3 & 1 \\ 2 & 1 & -7 \end{vmatrix} = -73. \tag{3.6}$$

Condition 1 is then met by the − 1 value in the first row, first column (see equation 3.4). Condition 2 also holds as demonstrated below:

$$\frac{J_{1,2}}{J_2} = \frac{\begin{vmatrix} -1 & -9 \\ 1 & -3 \end{vmatrix}}{-3} = \frac{+12}{-3} = -4 \tag{3.7}$$

obtained by substituting into equation (3.3)

And the third condition is met by substituting into equation (3.4)

$$\frac{J_{1,2,3}}{J_{2,3}} = \frac{-73}{\begin{vmatrix} -3 & 1 \\ 1 & -7 \end{vmatrix}} = \frac{-73}{+20}. \tag{3.8}$$

But if we substitute into expression 3.5 to see if condition 3′ is implied, we obtain:

$$\frac{J_{1,3}}{J_3} = \frac{\begin{vmatrix} -1 & 4 \\ 2 & -7 \end{vmatrix}}{-7} = \frac{-1}{-7}. \tag{3.9}$$

Therefore, it is impossible to use Hicks's partial equilibrium definition of perfect stability unless one postulates the further and sufficient condition that excess demand diminish in the stipulated way for every possible ordering of market adjustments. An exception would occur, of course, if some dynamic consideration specified an ordering of the market adjustments.

An abandonment of the partial approach of Hicks for the general equilibrium framework of the economy, so that the equilibrium set (P) can be considered as a single equilibrium point in n-dimensional space, can lead us to condition 3′. For let us now consider the stability of equilibrium in goods markets 2 and 3. Specifically, for market 2 let us obtain (dX_2/dP_2), $(dX_2/dP_2)_1$, and $(dX_2/dP_2)_{1,3}$, and let us state the Hicksian conditions for perfect stability:

4. $\left(\frac{dX_2}{dP_2}\right) = \frac{\partial X_2}{\partial P_2} < 0.$

5. $\left(\frac{dX_2}{dP_2}\right) 1 = \frac{J_{1,2}}{J_1} < 0.$

6. $\left(\frac{dX_2}{dP_2}\right) 1, 3 = \frac{J_{1,2,3}}{J_{1,3}} < 0.$

Lastly, we obtain analogous conditions for perfect stability in market 3:

7. $\left(\frac{dX_3}{dP_3}\right) = \frac{\partial X_3}{\partial P_3} < 0.$

8. $\left(\frac{dX_3}{dP_3}\right) 1 = \frac{J_{1,3}}{J_1} < 0.$

9. $\left(\frac{dX_3}{dP_3}\right) 1, 2 = \frac{J_{1,2,3}}{J_{1,2}} < 0.$

For perfect stability to characterize the general equilibrium of this economy, conditions 1–9 must hold. For all of these conditions to hold, as Hicks points out in the Mathematical Appendix,[5] it is necessary and sufficient that the following Jacobians alternate in sign negatively and positively:

$$J_i, J_{ij}, J_{ijk} \quad i \neq j \neq k, i = 1, 2, 3. \tag{3.10}$$

Suppose that this condition holds. Then it can be said that for any market, perfect stability will be independent of the ordering of the adjustments. That is, if in every market (except the *n*th) a random ordering of such adjustments produces perfect stability, then it can be said that no other ordering would have produced imperfect stability. For example, suppose we had chosen to adjust market 2 instead of market 1 to obtain condition 8 above. We should then have obtained:

$$\left(\frac{dX_3}{dP_3}\right) 2 = \frac{J_{2,3}}{J_2} \; . \tag{3.11}$$

The denominator of this expression is negative by virtue of condition 4, and the numerator is positive by virtue of conditions 1, 5, 9, and 3. Therefore, expression 3.11 itself is negative. Similarly, any other 'neglected' adjustment could be shown to be negative.

In conclusion, therefore, Hicksian stability in the multiple exchange economy is a general equilibrium concept which cannot be stated in the simple, partial manner that Hicks employs in his analysis of Chapter V without serious objection to its meaning and, therefore, its utility.

Notes

1. See [1], esp. pp. 62–77, 99–109, 315–19, 325, and 333–7.
2. See P. A. Samuelson [6], chs. IX, X, and XI; and [5]; J. L. Mosak [4]; O. Lange [2], Appendix; and L. A. Metzler [3].
3. [1], p. 66. See also Samuelson [6], pp. 272–3.
4. Note that the criterion is the summation of the successive impacts on excess demand, not the sign of the marginal impact. That is, suppose the excess demand in the market for good 1 is −10 initially, and the impact of the adjustment of the market for good 2 is + 8. Since excess demand for good 1 will remain negative, at −2, the perfect stability of good 1's equilibrium is not endangered.
5. Pp. 315–16. It should be stressed that the present paper is not criticizing Hicks's mathematical statement of stability conditions presented on these pages.

References

[1] Hicks, J., *Value and Capital*, Oxford University Press, London, 2nd edn (1946).
[2] Lange, O., *Price Flexibility and Employment*, Principia Press, Bloomington, Ind. (1944).
[3] Metzler, L., 'Stability of Multiple Markets: The Hicks Conditions', *Econometrica*, 13 (1945), pp. 277–92.
[4] Mosak, J., *General Equilibrium Theory in International Trade*, Principia Press, Bloomington, Ind. (1944).
[5] Samuelson, P., 'The Relation Between Hicksian Stability and True Dynamic Stability', *Econometrica*, 12 (1944), pp. 256–7.
[6] Samuelson, P., *Foundations of Economic Analysis*, Harvard University Press, Cambridge, Mass. (1947).

4 The Impact of Steel upon the Greater New York–Philadelphia Industrial Region*

For a long time the inadequacy of the traditional Weberian agglomeration analysis[1] has been fully recognized. Palander, in particular, has pointed up its shortcomings.[2] More recently, a fresh and more general approach has been made by P. Sargant Florence and others.[3] This approach, primarily an attempt to identify from processed census data meaningful geographical associations in industry, of necessity does not concern itself with the development of a theoretical framework.[4]

It is the purpose of this paper to attempt some development in the theory of agglomeration and of the broad spatial clustering of all economic activities by grafting on to the sounder elements of location theory a modified regional input–output schema. Viewed from a different standpoint, we attempt an impact study, wherein the direct and indirect repercussions of the location of a basic industry in a region are evaluated.[5] In this sense, crucial elements of regional development theory are attacked. It is not our intent here to present a comprehensive analysis of the forces affecting the location of a basic industry. In other studies this has been attempted for the iron and steel industry.[6] Rather, we accept the conclusions of these studies as a starting point and attempt to appraise the direction and strength of the force generated by the basic industry upon all types of economic

* This work was co-authored with Walter Isard and was published originally in the *Review of Economics and Statistics*, 35 (1953), pp. 289–301. It is reproduced with permission of Elsevier Science Publishers.

The study was done partly under the auspices of the Social Science Research Center, the University of Puerto Rico, and of the Center of Urban and Regional Studies, the Massachusetts Institute of Technology. The basic projections were completed in February 1952 for investment use by the firm of Minot, De Blois and Maddison, who generously financed the work in the initial stage.

activity within the region. This may be defined as the 'broad agglomeration effect'.

This effect during the coming decade will be studied with reference to the Greater New York–Philadelphia urban-industrial region.[7] Some attention will be given to interregional repercussions, particularly upon the closely linked region of Puerto Rico. Throughout the study it is assumed that aggregate demand remains at a high level during the next decade, that no technological revolution in the production of steel takes place, and that no violent shocks such as a major destructive war impinge upon the economy.

THE SCALE OF BASIC STEEL OPERATIONS

Past studies of the costs of assembling raw materials, production, and delivery of steel to market demonstrate that if there exists a market capable of absorbing the total output of a modern integrated steelworks, with its associated semifinishing mills, steel-producing capacity should be located in the vicinity of the market as closely as site conditions, transportation facilities, and socio-cultural factors permit. (There are of course exceptions when special conditions obtain, as when coal and ore sources coincide or are adjacent.) On the basis of these studies, it had been concluded that capacity on the eastern seaboard should be expanded. This conclusion has been borne out by the recent construction of the United States Steel Corporation's Fairless Works at Morrisville, Pennsylvania, the announcement of National Steel's intention to build at Paulsboro, New Jersey, and reports of the persistent interest of other steel companies in the Delaware Valley.

The cost rationale for the above conclusions and their subsequent fulfilment is presented succinctly in Table 4.1. Since the steel industry is a transport-oriented industry, transport costs alone need be presented.[8] The advantage of a Trenton location is clear-cut and marked.

From the standpoint of sound regional planning, the mere specification that an industry should be situated in an area on the basis of a cost advantage is insufficient; the scale of its operation must also be given. For such a market-oriented industry as steel, this entails the estimation of future demand for steel in the area. *Iron Age* estimated the consumption of steel by metalworking industry alone in the New York–Trenton–Philadelphia industrial area in 1948 as 5.1 million

Table 4.1 Transportation costs on ore, coal and finished products for selected producing locations serving New York City*
(in dollars)

Location	Ore	Coal	Finished products	Total
	Transportation costs on:			
New London (hypothetical)	3.68	5.42	8.80	17.90
Pittsburgh	5.55	1.56	12.40	19.51
Cleveland	3.16	3.85	14.00	21.01
Sparrows Point	3.68	4.26	8.40	16.34
Buffalo	3.16	4.27	11.60	19.03
Bethlehem	5.56	5.06	5.80	16.42
Trenton	3.68	4.65	4.80	13.13

* The costs on iron ore and coal are calculated per net ton of steel.
Source: Isard and Cumberland [14], p. 257.

tons.[9] (In 1948, for the nation as a whole, the metalworking industry received 46 million of the 66 million tons of steel shipped by steel mills.) The *Iron Age* estimation procedure, however, is subject to serious criticism.[10] The much more reliable Census data record consumption by metal-fabricating industries of 4.0 million tons for 1947 after appropriate adjustment of state data to correspond with our area.[11]

If the adjusted Census figure is blown up by 43 per cent to account crudely for consumption by non-metalworking industry; if, in addition, allowance is made for a share of the coastwise trade and export markets of the United States;[12] and if the expected growth of effective demand by steel consumers in the area is considered, an optimist might predict steel consumption of at least 9 to 11 million tons annually by the early 1960's. On this basis he might contend that an almost equivalent amount of new capacity could be justified. Less sanguine observers, cognizant of the intense competition from Sparrows Point, Bethlehem, and Buffalo, might with good reason restrict their estimates of new capacity by 1962 to 5 or 6 million tons.[13]

For purposes of this article, a range of estimates would unduly complicate computations, as is easily seen later. The use of an average also is undesirable, since in terms of planning capital outlay and development of the greater metropolitan regions within the area it is more desirable to have firm estimates of minimum expansion effects than infirm estimates of most likely effects.[14] Hence, here, as throughout this paper, we attribute expansion to an industry in our area only when a sound basis exists for such a prospect. Our minimum estimate for new basic steel capacity for this area, exclusive of steel capacity induced by the new steel demand incident to the agglomeration effect in its broadest sense, is 3.0 million tons. This new steel capacity, in effect a displacement of extra-areal capacity, estimated as of autumn 1951, was based on the announced intention of the United States Steel Corporation to construct 1.8 million tons of capacity, and on our considered judgment – particularly in the light of National Steel's proposal to erect 1 million tons of capacity – that at least 1.2 million tons of additional new capacity was justified.[15]

From here on, the problem decomposes into two major parts: (*1*) to estimate the resulting direct expansion of existing and influx of new steel-fabricating activities within the area, and (*2*) to quantify the direct and indirect repercussions of the new steel production and associated new steel-fabricating operations upon all activities in the area.

DIRECTLY INDUCED STEEL FABRICATION

Several attempts have been made to identify the forces dictating the juxtaposition of industries representing consecutive stages in a production process or related in some other vertical or horizontal manner.[16] The approach of these attempts becomes more valid and useful when location theory and historical trend analysis supplement the location quotient and coefficient of geographical association upon which these studies heavily depend. In associating steel-fabricating activity with basic steel capacity, we have treated individually the steel-fabricating industries listed below in Table 4.4 by three-digit Census classification. For each such activity: (*1*) we compare, for each of 69 industrial areas as defined by *Iron Age*, the number of production workers engaged in it and the capacity to produce the major basic steel items consumed by it;[17] (*2*) we observe the rates of growth of the given activity during the period 1939–47 in those areas where capacity for producing the major steel items consumed by the activity increased considerably more rapidly than for the nation as a whole, and we compare such rates with those of other areas;[18] and (*3*) we consider particular features and characteristics of the production process in the given activity, with special reference to transportation costs on raw materials and the finished product, economies of scale, industrial organization, and other location factors whose effect might easily and objectively be identified.

To illustrate the procedure, industry number 343, Heating Apparatus (Except Electric) and Plumbers' Supplies (SIC code) is analyzed below. This industry produces the following major products: cast iron heating boilers; cast iron radiators; residential steel heating boilers; warm air furnaces; gas, coal, wood and kerosene domestic heating stoves and cooking stoves, ranges and appliances (except electric); water heaters (except electric); steam and hot water unit heaters; floor furnaces; tanks for water heaters; oil burners, domestic and industrial; plumbing fixture fittings and trim; cast iron and steel kitchen sinks; cast iron and steel bathtubs. In Table 4.2 are recorded, for 69 industrial areas and for the balance of each state not included in the industrial areas, capacity for hot-rolled sheet and strip production in 1948 (column 1) and *Iron Age* estimates of production workers in 1948 (column 2). It should be noted that of the steel consumed by industry 343 in 1947, 86 per cent consisted of sheet and strip.[19]

Careful scrutiny of the table reveals two types of relationships. The fairly large concentrations of employment in the New York–Newark–Jersey City area (5876), Bridgeport–New Haven (2853), Hartford–New Britain (1988), and other southern New England districts, indicate a strong market pull. This, too, is borne out when the disproportionately small capacity in the combined San Francisco–Oakland–Los Angeles area is contrasted with the large amount of employment in industry 343 (12 123).

Less clear-cut cases of strong market pulls are Philadelphia–Camden (3909 employed), and the entire state of Wisconsin (13 002), including Milwaukee (3768) and Sheboygan (5500). However, each one of the areas already mentioned has some sheet and strip capacity or, aside from New England points, is no more than about 200 miles from a major sheet- and strip-producing point. This testifies to the pull of sheet- and strip-producing points which is even more forcibly illustrated by Chicago–Gary (12 959 employed); Cleveland (7510); Louisville, somewhat over 100 miles from Cincinnati–Covington, (6820); Cincinnati–Covington (3960); Detroit (3875); and Pittsburgh (3437). In short, both market and sheet- and strip-producing points exert significant pulls upon the location of the industry. This suggests that with the addition of new sheet and strip capacity at the Fairless Works and elsewhere in the Delaware Valley, there should be a net positive attraction for industry 343, especially since, relative to the

Table 4.2 Production employment in industry 343 and hot-rolled sheet and strip capacity, 1948, by selected areas*

Industrial area	*Capacity (in tons)*	*Employment*	*Industrial area*	*Capacity (in tons)*	*Employment*
Chicago–Gary	3 879 700	12 959	Minneapolis–St Paul	0	1 073
Youngstown	2 638 350	650	Seattle	0	943
Pittsburgh	2 614 520	3 437	Fall River–New Bedford	0	935
Detroit	2 451 000	3 875	Boston	0	885
Baltimore	1 195 000	2 031	Moline	0	717
Buffalo	1 155 380	2 540	Worcester	0	640
Cincinnati–Covington	950 000	3 960	Dallas–Ft Worth	0	611
Balance of Ohio	793 000	791	Rochester	0	610
Birmingham	725 200	2 210	Utica	0	610
Cleveland	651 000	7 510	Ft Wayne	0	565
Sharon–New Castle	600 000	1 110	Syracuse	0	565
Balance of Kentucky	374 000	300	San Diego	0	500
St Louis	328 500	5 361	Akron	0	500
Canton	314 000	625	Benton Harbor	0	435
San Francisco–Oakland	238 926	2 918	Balance of Alabama	0	410
Mansfield–Marion	205 000	2 040	Balance of Texas	0	381
Los Angeles	113 284	9 205	Springfield–Holyoke	0	361
Bridgeport–New Haven	150 000	2 853	Trenton, N. J.	0	330
Philadelphia–Camden	105 100	3 909	Allentown–Bethlehem	0	275
Anderson–Muncie	95 000	855	Balance of Missouri	0	220
Columbus	61 500	2 715	Balance of Michigan	0	210
Indianapolis	55 000	1 477	Balance of Georgia	0	200
Atlanta	25 500	620	Toledo	0	186

New Castle, Ind.	25 000	0	Flint	0	175
Kansas City	23 500	426	Balance of Indiana	0	175
Houston	9 000	68	Balance of Washington	0	150
Balance of Minnesota	7 120	785	Evansville, Ind.	0	109
Louisville	0	6 820	Balance of Connecticut	0	80
New York–Newark–Jersey City	0	5 876	Dayton	0	65
Sheboygan	0	5 500	Albany	0	50
Milwaukee	0	3 768	Binghamton–Elmira	0	50
Balance of Wisconsin	0	3 734	Balance of California	0	40
Lansing	0	2 830	Balance of Maryland	0	25
Hartford–New Britain	0	1 988	Balance of New York	0	0
Peoria	0	1 875	Balance of Pennsylvania	0	0
Erie	0	1 760	Balance of New Jersey	0	0
South Bend	0	1 599	Hagerstown	0	0
Rockford	0	1 540	Danville, Ill.	0	0
Lancaster–York	0	1 450	Madison, Wis.	0	0
Springfield, Ill.	0	1 376	Balance of Kansas	0	0
Wichita	0	1 255	Mobile	0	0
Reading	0	1 206	New Orleans	0	0
Balance of Illinois	0	1 167	Balance of Louisiana	0	0
Grand Rapids	0	1 097	Beaumont	0	0
Scranton–Wilkes–Barre	0	1 095			

* Heating apparatus (except electric) and plumbers' supplies.

other major population focal points, such as Chicago–Gary and San Francisco–Oakland–Los Angeles, our area is deficient in its per capita representation of this industry.

A study of relevant historical material is also illuminating. In Table 4.3 are presented Census data on the rates of growth of industry 343 for the United States and for Maryland and California, 1939–47. (Because of data shortcomings, a comparison over a longer historical period is not justified.) Maryland and California were during this period among the fastest growing states in terms of sheet and strip capacity,[20] and they embrace or are peripheral to major market areas. The growth rates for California and Maryland well exceed that for the nation. This suggests that, in any given large market area, growth of industry 343 is at least partially related to the existence and growth of sheet and strip capacity in that area. This adds further support to the thesis that installation of sheet and strip capacity in the Delaware Valley will induce additional employment in this industry, particularly since the existing Maryland capacity can be considered peripheral to our area.

Theoretical considerations reinforce the empirical suggestions. The products of industry 343 tend to be heavy and bulky and to bear high transport costs relative to their value if not produced at the market; similarly, transport costs on the raw material, steel, are high, although to a lesser degree, if production does not take place at a steel source. A situation which involves the coincidence of both market and raw material, as in our area, strongly argues for significant expansion of employment in this industry. Thus, since this industry is a relatively fast-growing one, as indicated in Table 4.3, it seems logical that a decreasing proportion of the area's demand for the products of this industry will be met from plants in the Western Pennsylvania–Ohio area, and an increasing proportion from plants within the area.

Having established on three counts that the coexistence of market and capacity in our area warrants expansion in industry 343 during the coming decade, we confront the problem of estimating the specific magnitude of this expansion. No neat estimating formula exists. Judgment based on knowledge of the area and conversations with informed persons must be resorted to. In line with such knowledge and conversations and in view of the fact that in 1948 our area had 12 968 production workers whereas the Chicago–Gary area had 12 959 (exclusive of the very substantial employment in the 200-mile area about Chicago–Gary), we judge that 20 per cent represents a

Table 4.3 Production employment in industry

	1939	*1947*	*Per cent rate of growth 1939–47*
United States	78 330	126 725	61.8
Maryland	1 128	2 499	121.5
California	4 652	9 373	101.5

firm minimum rate of increase in production workers in industry 343 incident to our area's increased ability to compete as a result of new steel capacity. This 20 per cent increase represents an increment of 2594 workers.

Similarly, for every other three-digit steel-fabricating activity, an estimate of numbers of new production workers was derived.[21] These estimates are listed in Table 4.4.[22] The total number of new pro-

Table 4.4 Estimates of new production employees in the area by steel-fabricating activity

Industry number	*Title*	*Per cent rate of growth*	*New production employees*
341	Tin cans and other tinware	10.0	923
342	Cutlery, hand tools, and general hardware	0.0	0
343	Heating apparatus, (except electric) and plumbers' supplies	20.0	2 593
344	Fabricated structural metal products	4.6	1 132
346	Metal stamping, coating and engraving	10.0	2 408
347	Lighting fixtures	2.5	1 911
348	Fabricated wire products	10.0	670
349	Miscellaneous fabricated metal products	17.5	2 695
351	Engines and turbines	8.0	723
352	Agricultural machinery and tractors	10.0	294
353	Construction and mining machinery and equipment	10.0	252
354	Metal-working machinery	13.5	2 137
355	Special-industry machinery (except metalworking machinery)	16.0	5 204

(continued on page 86)

Table 4.4 *continued*

Industry number	*Title*	*Per cent rate of growth*	*New production employees*
356	General industrial machinery and equipment	25.0	5 566
357	Office and store machines and devices	10.0	1 225
358	Service-industry and household machines	20.0	4 512
359	Miscellaneous machinery parts	25.0	5 666
36	Electrical machinery, equipment and supplies	5.0	8 025
371	Motor vehicles and motor-vehicle equipment	20.0	7 331
372	Aircraft and parts	3.0	871
373	Ship and boat building and repairing	5.0	1 989
374	Railroad equipment	3.0	364
375	Motorcycles, bicycles and parts	0.0	0
379	Transportation equipment (n.e.c.)	7.0	32
38	Professional, scientific, and controlling instruments, photographic, etc.	0.0	0
39	Miscellaneous manufacturing industries	10.5	4 803
	Total		61 326

duction workers is 61 326 which, when divided by the 9166 production workers to be associated with the 3.0 million tons of new steel capacity, yields a ratio of 6.7. Despite its serious shortcomings, a comparison of this ratio with the 8.9 national ratio of production workers in steel-fabricating activities to production workers in steel manufacture suggests the conservative nature of our estimates, especially since of all the areas in the United States our area represents the most concentrated market.[23]

In setting forth these estimates, we do not claim that each will be realized. It is well recognized that unpredictable events and inadequate knowledge of the interaction of basic underlying forces as they affect an individual industry may have led us astray in individual instances. However, as a whole, we feel that these estimates represent a set of firm minimum figures upon which subsequent analysis may rest.[24]

REPERCUSSIONS OF INITIAL ACTIVITIES

To quantify the direct and indirect repercussions of the operation of the above new steel-fabricating industries as well as of the new steel capacity, a modified regional input–output schema is employed. The entire set of inputs absorbed directly by the steel and steel-fabricating activities constitutes, in a sense, the 'bill of goods' that must be furnished to our area. This bill of goods is constructed: (*1*) by multiplying the cents' worth of every input required per dollar output of a given steel-fabricating activity by the dollar volume of that activity corresponding to the estimated new employment;[25] (*2*) by multiplying the cents' worth of every input per dollar output of steel by the dollar volume of new steel output corresponding to our estimate of new steel capacity; and (*3*) by summing horizontally for each of the 45 different industrial inputs the requirements by steel and each fabricating activity. The totals for each of the 45 inputs are recorded in column 1 of Table 4.5.[26]

In designing our study we made an important modification of the traditional input–output matrix. Since in regional analysis it is important to catch the local multiplier effect resulting from the generation of new income, households have been introduced into the structural matrix as an industry. Hence, the labor and other household services required by the new activities are recorded.[27]

After the bill of goods had been constructed, it was necessary to determine the fraction of the total amount of each input minimally expected to originate in the area. (It is clear that of the bill of goods requirements some will be produced within the area, while others, such as coal, will come from outside the area.) This step was necessary since in this study the expansionary effect upon the economy of the area is maintained by only the outputs which are produced within the area. It is not to be denied that the area may expand as a result of new demands for its products by population elsewhere, for example as a result of new demands by the population in the Appalachian coalfields consequent to additional demand for coal within the area. However, the inability to apply an interregional model[28] which would quantify these feedbacks compels us to neglect this indirect

Table 4.5 Direct and indirect repercussions of new basic steel capacity

	Input requirements of initial steel and steel-fabricating activities (in $ thousand) (1)	*Minimum percentage of input requirements to be produced in area (2)*	*First round expansions in area (in $ thousand) (3)*	*Second round expansions in area (in $ thousand) (4)*	*Third round expansions in area (in $ thousand) (5)*	*Sum of round expansions in area (in $ thousand) (6)*	*Total new employees corresponding to round expansions (7)*	*Total new employees in initial steel and steel fabricating activities (8)*	*Overall total of new employees (9)*
1. Agriculture & fisheries	50.0	0	0.0	0.	0.	0.	0		0
2. Food & kindred products	294.6	60	176.8	17 660.	8 249.	42 492.	1 833		1 833
3. Tobacco manufactures	0.0	0	0.0	0.	0.	0.	0		0
4. Textile mill products	3 864.7	10	386.5	406.	39.	1 280.	142		142
5. Apparel	1 285.6	75	964.2	10 124.	3 461.	21 155.	2 302		2 302
6. Lumber & wood products	5 610.7	5	280.5	93.	36.	450.	64		64
7. Furniture & fixtures	1 753.4	33	578.6	802.	198.	2 000.	234		234
8. Paper & allied products	4 818.7	40	1 927.5	1 674.	1 297.	6 574.	426		426
9. Printing & publishing	425.5	90	383.0	5 929.	3 014.	14 617.	1 667		1 667
10. Chemicals	10 626.4	45	4 781.9	3 599.	1 630.	12 077.	601		601
11. Products of petroleum & coal	10 936.6	25	2 734.2	2 547.	1 118.	7 634.	228		228
12. Rubber products	8 381.5	15	1 257.2	355.	102.	1 879.	169		169
13. Leather & leather products	647.7	20	129.5	679.	194.	1 371.	150		150
14. Stone, clay, & glass products	9 031.7	15	1 354.8	441.	139.	2 083.	268		268
15. Iron & steel	121 170.5	50	60 585.3	13 566.	2 965.	78 335.	6 093	11 666	17 759
16. Nonferrous metals	33 997.4	20	6 799.5	1 667.	381.	9 063.	505		505
17. Plumbing & heating supplies	3 192.4	25	798.1	248.	50.	1 189.	118	3 640	3 758
18. Fabricated structural metal products	3 480.7	40	1 392.3	312.	33.	1 809.	151	1 420	1 571
19. Other fabricated metal products	31 770.9	40	12 708.4	2 146.	561.	16 121.	1 537	10 060	11 597
20. Agric'l, mining, & const. machinery	3 651.3	5	182.6	46.	11.	251.	22	707	729
21. Metalworking machinery	7 389.1	25	1 847.3	270.	43.	2 210.	289	2 705	2 994

22. Other machinery (except electric)	28 463.6	40	11 385.4	2 675.	551.	15 384.	1 486	28 607	30 093
23. Motors & generators	11 265.9	20	2 253.2	226.	42.	2 560.	301		
24. Radios	4 562.2	30	1 368.7	428.	101.	2 026.	192		
25. Other electrical machinery	21 773.9	50	10 887.0	2 011.	432.	13 903.	1 427	10 392	12 312
26. Motor vehicles	50 530.8	10	5 053.1	742.	260.	6 421.	389	8 770	9 159
27. Other transportation equipment	2 605.5	20	521.1	276.	69.	958.	117	4 605	4 722
28. Professional & scientific equipment	3 221.4	50	1 610.7	801.	287.	3 123.	416		416
29. Miscellaneous manufacturing	5 116.8	60	3 070.1	2 888.	982.	8 418.	845	6 108	6 953
30. Coal, gas, & electric power	7 767.0	50	3 883.5	1 843.	2 693.	11 079.	1 100		1 100
31. Railroad transportation	13 575.8	75	10 181.9	6 010.	2 390.	21 532.	3 308		3 308
32. Ocean transportation	457.3	75	343.0	331.	170.	1 021.	110		110
33. Other transportation	4 179.4	95	3 970.4	8 422.	2 836.	19 694.	2 394		2 394
34. Trade	13 969.8	95	13 271.3	36 585.	11 855.	83 642.	13 874		13 874
35. Communications	1 790.7	90	1 611.6	2 409.	1 283.	7 305.	1 191		1 191
36. Finance & insurance	3 086.2	90	2 777.6	9 472.	5 062.	25 252.	2 329		2 329
37. Rental	3 018.8	95	2 867.9	26 222.	9 603.	55 680.	909		909
38. Business services	5 338.5	95	5 071.6	2 385.	2 406.	13 384.	1 305		1 305
39. Personal & repair services	396.9	95	377.1	14 399.	5 088.	24 212.	4 443		4 443
40. Medical, educ., & nonprofit organization	000.0	90	0.0	9 811.	2 160.	17 271.	4 370		4 370
41. Amusements	000.0	90	0.0	3 677.	1 066.	6 591.	1 100		1 100
42. Scrap & miscellaneous industries	8 388.2	50	4 194.1	2 054.	727.	7 411.	771		771
43. Undistributed	103 638.6	50	51 819.3	5 875.	6 019.	69 236.	7 208		7 208
44. Eating and drinking places	000.0	95	0.0	16 916.	3 903.	29 551.	3 705		3 705
45. Households	348 281.0	82	285 590.4	63 002.	80 894.	509 578.			
Totals	903 807.7		521 377.2	282 024.	164 400.	1 177 822.	70 089	88 680	158 769

sionary effect. This is not inconsistent with our accumulated bias toward underestimation, and hence is fully in line with the objectives of this article.

The determination of the above fraction for each industrial activity was based on the small amount of location theory directly relevant, on the accumulated production and derived consumption data for our area, on data of flows of commodities into and out of the area, and, finally, upon judgments of informed persons within the area. These fractions are recorded in percentage terms in column 2 of Table 4.5. Again many subjective decisions had to be made, and in each case we attributed to the area the production of only that fraction of the inputs required of an activity for which we felt there was a firm basis. For example, because the new industry will draw upon existing agricultural labor in the area, because some of the best agricultural land will be diverted into industrial, commercial and residential uses, and in view of the past relative decline of agriculture in the area, it was felt that there was no firm basis for projecting an expansion of agriculture incident to steel development. Hence a percentage of zero is recorded for agriculture, even though the likelihood exists that there will be some expansion. Similarly, in the case of chemicals, a 45 per cent figure represents an understatement in view of the rapid expansion of, and the natural advantages for, chemical production in the area. Observe, too, that the highest percentage figure associated with any activity is 95, even though it is likely that for certain service industries the percentage figure will be higher.[29]

Applying the percentages of column 2 to the corresponding items of column 1 yields column 3, which records (as firm minimums) the first round expansion of each industrial activity in the area which we anticipate will be required to maintain the new steel capacity and its associated fabricators in reasonably full operation. It should be noted that in column 3 of Table 4.5 the requirements of steel and of the diverse outputs of steel-fabricators are considered to represent demand which is over and above the demands for steel and fabricated steel products which initially justified the erection of new basic steel and steel-fabricating capacities; these industries must produce beyond their initial capacity. In effect, we have shaped the analysis so as to leave these activities within the structural matrix.

To produce the first round of expansions requires in turn a whole series of inputs. One can derive this series of inputs: (*1*) by multiplying the cents' worth of each input per dollar output of a given activity by the dollar volume of output of that activity recorded as the first round expansion; and (2) by summing horizontally over all activities the requirements so derived for inputs from each industrial category. However, not all of these second round input requirements will originate in the area. Multiplying these second round input requirements by the corresponding percentages in column 2 of Table 4.5 yields firm minimum estimates of the second round expansions for the area, which are recorded in column 4. In turn, third round input requirements and third round expansions (column 5) are derived, and so forth.[30] Since the rounds converge rather rapidly, it was necessary for our problem to compute only the first six rounds of expansions with rough extrapolations to cover the infinite number of succeeding rounds. By industrial category, the sum of the first six rounds plus extrapolations are listed in column 6 of Table 4.5.[31]

For each industrial category (except the household sector), the value of output per worker in 1947 was used to obtain a crude estimate of new employment[32] corresponding to the sum of the round expansions of column 6. These estimates include nonproduction employment. They are listed in column 7 of Table 4.5, and total 70 089 workers. To them must be added the employees in the initial new capacity of steel[33] and steel-fabricating activities (column 8, Table 4.5). This yields column 9 of Table 4.5, and a total of 158 769 workers.[34] Finally, new employees expected to be engaged in construction (8574)[35] and government activity (12 884)[36] must be included. The overall new employment thus amounts to 180 228 workers.

It should be emphasized that the above total is a gross figure and relates to change in employment only in activities which expand as a result of new steel capacity. Theoretically, one should subtract from this figure the decline in employment in activities which are forced to contract as a result of labor shortages, high wages, and the like. However, contractions are likely to be more severe in the short run than in the long run when the new employment openings will have had an opportunity to exert a positive effect upon labor migration into the area. Since this study is concerned with conditions a decade hence, at which time we do not feel that such contractions will be of major proportions, and since we have already allowed for some contractions when we weighed expanding sectors against declining sectors for an industrial category in deriving the percentages of column 2 of Table 4.5, we retain the gross figure of 180 228 as an overall firm minimum estimate of new employment. The reader may prefer to make a small downward adjustment.

In terms of the population engaged in the labor force (roughly 43 per cent) which characterized the area in 1950,[37] the total new population resulting, on the basis of the above gross employment figure, is 419 000. To the extent that the new wage jobs replace existing wage jobs and cause the decline of the industries associated with the latter, and to the extent that the above per cent increases in the future, the net increase in population supported by the new industrial stratum will be smaller.

BASIC QUALIFICATIONS

In appraising the above results, we need not discuss the qualifications springing from the use of constant production coefficients. They have been fully discussed elsewhere.[38] However, the inclusion of constant consumption coefficients in the structural matrix requires some discussion. To the extent that the new income generated does accrue as income to new job-holders (either from in-migration or natural population growth)[39] and does not increase wages of existing job-holders, and to the extent that the age and sex composition, ethnic and racial backgrounds, savings habits and other socio-cultural characteristics of the new job-holders and their families are similar to those of the population as a whole in 1947, the use of constant coefficients is not so tenuous a procedure as might appear at first sight. Also, it might be argued that in the light of the inadequacies of existing consumption studies, no better alternative exists.

In addition, we have abstracted from regional production coefficients and have assumed that production practices (as well as consumption habits) for the area are identical with those for the nation as a whole. This assumption is necessary since only national coefficients exist. However, from our knowledge of the resource characteristics of our area and its industrial composition, it is considered unlikely that the use of national coefficients entails serious distortions.[40] Nor, because of obvious difficulties, have these coefficients been adjusted industry by industry for increases in labor productivity or for changes in technology on the horizon; hence, our results reflect the 1947 productivity-technological structure.[41]

Since our projection is made for a decade hence, by which time the construction and equipping of our new plants and facilities will have been completed, we exclude input requirements to build up this capital stock. Our analysis is aimed at the delineation of current inter-sectoral flows to be expected in the early 1960s.[42]

One major problem which is not readily apparent needs to be bared. To be of practical use all types of input–output analysis require some leakage factor or device which reflects social frictions and the retarding effects of a whole array of noneconomic factors, as well as resource limitations and other economic resistance to expansion. In the typical input–output study, this is implicitly obtained by excluding sufficiently important sectors from the structural matrix (households, government, foreign trade, inventory and capital formation) and placing them in the bill of goods. In our analysis, the restoration of households to the structural matrix without compensating changes would yield a leakage factor insufficient to guarantee a realistic convergence by rounds. In other words, the sum of the rounds would yield an expansion effect which far exceeds the upper limits of a range of reasonable levels. The system would be insufficiently anchored.[43]

The counterbalancing factor in our analysis is the leakage introduced through the omission of interregional feedback effects. In other words, no expansionary impulse comes from the increased income in other areas incident to their supply of inputs to our area. Though neglect of the interregional feedback effects is a shortcoming which calls for improved models,[44] nonetheless in our crude framework it provides anchorage and serves the desirable purpose of compensating, if not overcompensating, for the elimination of the customary household leakage. This point is strongly supported by the fact that if the bill of goods of the 1947 BLS national interindustry study were increased by amounts corresponding to the outputs of the new steel and steel-fabricating capacities (which form the bill of goods in the problem of this paper) the resulting expansions in all activities of the nation which would be required would yield a dollar total somewhat greater than that obtained in this paper for the subsequent expansions of activities in our area. Put otherwise, for every new employee in the initial steel and steel-fabricating plants, we estimate that on the average another 0.79 persons would be employed in the subsequent expansions of activities in our area. In comparison, where the same bill of goods increase is injected into the BLS national model a somewhat greater number of persons would be employed in the required expansions of all activities in the nation for every employee corresponding to this bill of goods increase.

SOME INTERREGIONAL ASPECTS

With all the required qualifications, it is firmly believed that the empirical results achieved have significance as firm minimum projections. The potential usefulness of the results and the associated technique, however, has not been fully developed. It has already been observed that the area will not furnish all of the inputs required to permit each round of expansion. This circumstance permits the isolation of the magnitude of those commodities which must be imported into the region. For regions well placed competitively for supplying these inputs, this leads to at least a partial analysis of the effects upon these regions and of their potentialities for development. An excellent example of such a region in terms of our area is Puerto Rico. In order to sketch the procedure in adding this partial interregional dimension to the analysis (the full development of this procedure must be left to a later paper) we now pose the question: to what extent can Puerto Rico benefit from the expansionary movement in the area?

To answer this question requires the identification of those activities for which Puerto Rico has a competitive advantage relative to the area. It is beyond the scope of this paper to present finely detailed cost studies. Suffice it to indicate that Puerto Rico with its abundant cheap labor supply and niggardly endowment of mineral resources tends to excel in the cheap labor-oriented type of industry, such as certain types of textiles, assembly work, and the like. The gross industrial categories of the 50-industry BLS classification precludes isolation of the specific textile and other activities which must be examined. Not so, however, with the 192-industry classification, which forms the basis of the BLS 192 × 192 matrix which is currently available for use.

A recent report containing suggestions for a planned industrial promotion program for Puerto Rico[45] has clearly indicated the feasibility of expansion and/or establishment of activities which fall in, among others, the following categories of the 192-industry classification: industry number 30, spinning, weaving and dyeing; industry 31, special textile products; industry 69, footwear (excluding rubber); industry 139, radios and related products; and industry 140, tubes (electronic). Accordingly, we have asked what new demands for the products of these industries will ensue from the steel-induced expansion of the area. These new demands are easily approximated, since per dollar output of each of the 45 industrial categories employed in

Table 4.6 Requirements of the products of selected cheap labor-oriented industries

Industry number	*Name*	*Induced demand (in $ thousand)*
30	Spinning, weaving, and dyeing	9,731
31	Special textile products	1,471
69	Footwear (excluding rubber)	4,560
139	Radio and related products	2,448
140	Tubes	163

the previous sections we can obtain the cents' worth of inputs required from each of the five cited industries.[46] These new demands are recorded in Table 4.6. However, it should be remembered that portions of these new demands will not only be provided from the area and from Puerto Rico, but also from other regions which may be in a position to compete in these lines. Therefore, the determination of the quantities of these new demands which will be supplied by Puerto Rico must be left to market area analyses and related research.

In conclusion, we have attempted to develop more adequate techniques to anticipate the impact of the location of a basic industry upon the various economic activities of a region. Our effort to achieve an improved agglomeration and regional development analysis has not involved pure theory alone. Rather it has involved a weaving together of diverse theoretical threads and existing empirical material in order to obtain a realistic fabric of an induced development. As they now exist, analytical tools for area study are far from adequate, and it is for this reason that we have been compelled to aim not at the most probable effects of new basic industry but rather at the firm minimum. In this guise, and only in this guise, are these tools currently useful. It is hoped that with continued research, still more valid techniques can be developed.

Notes

1. C. J. Friedrich [9], chs. 5–7.
2. T. Palander [25], ch. 8; see also E. M. Hoover, Jr. [10], pp. 89–99; and

O. Engländer [5]. For a defense of Weber's agglomeration theory, see E. Niederhauser [24].

3. National Resources Planning Board [23], ch. 5; and P. S. Florence [8], esp. ch. IV.
4. Florence's interpretations may be construed by some as implicit theorizing. In particular, objection may be made to a conversion of his coefficient of geographical association into a general indicator of economic spatial dependence between industries. It should be noted that such a conversion can be questioned in view of the statistical shortcomings of the coefficient. See Robert E. Kuenne, [21], ch. 2.
5. In many respects our impact study for a region resembles the pioneering, but neglected, study of Barfod's for a city. Børge Barfod [2]. Because of the larger quantities of data currently available, we have been able to adopt a more comprehensive approach.

   Among other studies, somewhat related to ours, are: M. C. Daly [4] Federal Reserve Bank of Kansas City [7]; E. Ullman [27].
6. W. Isard [11]; W. Isard and W. Capron [13]; W. Isard and J. Cumberland [14]; John H. Cumberland [3]. For the analysis of locational factors in another basic industry, see John Krutilla [20].
7. This region includes the following counties: Fairfield and New Haven in Connecticut; Kings, Queens, Richmond, New York, Bronx, Westchester, Orange, Putnam, Rockland, Suffolk, and Nassau in New York; Bergen, Passaic, Morris, Essex, Hudson, Union, Middlesex, Monmouth, Ocean, Burlington, Atlantic, Salem, Gloucester, Camden, Mercer, Hunterdon, Somerset, Warren, and Sussex in New Jersey; and Chester, Delaware, Philadelphia, Montgomery, and Bucks in Pennsylvania.

   Ideally, the area should be defined in such a way as to include all of those counties and parts of counties which together comprise a continuous spatial entity and which will undergo noticeable change as a result of the steel development. Practically, because of data inadequacies, only whole counties can be treated. From a consideration of the nature and direction of transportation facilities, industrial and residential land use, population distribution, and similar characteristics, the above listed counties seem to form the most appropriate spatial entity for our problem, although undoubtedly some contiguous areas will be directly affected, too.
8. See Isard [11], pp. 203–4; and Isard and Capron, [13], pp. 120–3.
9. O. Johnson [19]. The New York–Trenton–Philadelphia industrial area, considerably smaller than our Greater New York–Philadelphia urban-industrial region, includes the following counties: Fairfield and New Haven in Connecticut; Bronx, Kings, Nassau, New York, Queens, Richmond, and Westchester in New York; Bergen, Burlington, Camden, Essex, Gloucester, Hudson, Mercer, Middlesex, Passaic, and Union in New Jersey; and Bucks, Chester, Delaware, Montgomery, and Philadelphia in Pennsylvania.
10. For Connecticut the *Census of Manufactures*, 1947, vol. I, pp. 244–5, reports consumption by metalworkers of 509 210 tons as against 1 194 694 tons in *Iron Age*; for New York, 2 109 010 tons as against

3 232 356 tons; for New Jersey, 1 130 845 tons as against 1 684 699 tons; and for Pennsylvania, 5 600 224 tons as against 5 498 654 tons.

11. The adjustment procedure was as follows: for those sectors of our area in Connecticut, New York and Pennsylvania, the consumption for each (as given in *Iron Age*) was taken as a percentage of the appropriate *Iron Age* state figure. These percentages were multiplied by the appropriate Census data for the state, and the products were added to the total consumption listed in the *Census* for New Jersey.
12. Exports from the port districts of New York and Philadelphia of iron and steel mill products and semi-manufactures were 2.6 million tons and 451 800 tons respectively in 1947, and 1.6 million tons and 283 000 tons respectively in 1948. US Army, Board of Engineers for Rivers and Harbors, [28]; and [29] pt. 2, 1947, pp. 231, 285–366, and pt. 2, 1948, pp. 255, 315–97.
13. In view of such competition we postulate that only 50 per cent of the new steel demand arising as a result of this regional development will be supplied by producers in the area. For discussion of this point see Section 3 and column 2, Table 4.5 below.
14. It should also be noted that a secondary objective of this study was to isolate and pinpoint those localities in our area which would experience highest rates of population growth and physical development and hence would be most attractive for nonspeculative real estate investment, *ceteris paribus*. For this objective, too, firm minimum estimates are more desirable insofar as they reduce the element of risk.
15. It is interesting to note that the United States Steel Company in discussing the operation of the Fairless Works stated: 'Initial operation of the plant starting in 1952 will require delivery of 3 300 000 gross tons of coal and foreign ore annually by water. . . . The rolling mills which will be installed at the Fairless Works in 1951–52 will have sufficient capacity to consume the output of probable future open hearths and blast furnaces requiring an additional annual tonnage of 3 700 000 gross tons of iron ore and coal' [36], pp. 2–3. These statements, in addition to others, suggest a capacity well beyond the 1.8 million tons currently under construction.
16. Florence [8]; and National Resources Planning Board [23], ch. 5.
17. Though the quality of the *Iron Age* data is poor, they are used since there are no other data on three-digit steel-fabricating activities for such a fine breakdown of the United States.

    Because of the aforementioned reasons and those cited in the source quoted in note 4, we have not computed the coefficient of geographical association. Rather, we have carefully inspected these two sets of data.
18. Moreover, because costs of relocation are significant, consideration was given to the rate at which each activity has been expanding nationally. It was felt that, *ceteris paribus*, greater employment could be anticipated for our area in the faster growing than in the slower growing industries.
19. See [30], vol. I, p. 238.
20. Hot-rolled sheet and strip capacity grew from 498 000 tons in 1938 to 1 195 000 in 1948 for Maryland; and for the same years from 85 100 tons to 352 210 tons for California. These represent respectively a more than

doubling and tripling effect for these states. Data from American Iron and Steel Institute [1].

21. The lack of any clear-cut patterns in the three-digit industries of the two-digit categories, industry groups 36, 38 and 39, strongly argued for the use of the two-digit aggregates of these three-digit industries.
22. It should be noted that expansion in metal furniture trades has been omitted. Difficulties of separating metal furniture from non-metal furniture manufacture, both in the above steps and the steps to follow, argued for excluding this activity, although about 2500 new production workers in metal furniture trades might be anticipated. This contributes to the minimal nature of our projections. It still further balances any overestimates that may have crept into the analysis, especially in the industry groups, 354–9, where a re-examination of procedure after computations were completed indicated that we had made less conservative estimates for these three-digit industries than for other three-digit industries.
23. For curiosity's sake alone, a multiple linear correlation between production workers and four explanatory variables was made for the 69 industrial areas of Table 4.2 (balances of states were excluded). The four variables were: (*1*) market proximity, as given by an index measuring distances of any given area from the eleven largest market centers in the United States; (*2*) distance of that area from a region of major steel capacity (arbitrarily defined as one capable of producing more than 1 250 000 tons annually); (*3*) population (1950) of that area; and (*4*) steel capacity of the given area.

    A multiple correlation coefficient of 0.82 was obtained. However, only 11 of the estimates derived from the estimating equation were within plus or minus 10 per cent of the actual employment data. Injection of new values for the variables to account for new steel capacity in the New York–Trenton–Philadelphia industrial area led to an estimate of 45 000 new production workers in metal-fabricating activities for this area. However, no significance should be attributed to this finding, since extreme items have seriously distorted the influence of steel capacity upon fabricating employment, and since, in any case, multiple correlation analysis has little meaning in this particular context. For a more detailed presentation of the correlation analysis, see Kuenne [21], ch. 2.
24. To the extent that the input structures of these individual activities tend to resemble one another, errors of over- and under-estimation tend to cancel themselves when we proceed to quantify subsequent repercussions. In the limit, when input structures are identical, the industries may be aggregated into one industry without loss of accuracy.
25. These cents' worth of inputs are simply the technical coefficients in the Leontief input–output model. The coefficients for a 50 × 50 matrix were developed for 1947 by the Bureau of Labor Statistics. See Table 5 in W. D. Evans and M. Hoffenberg [6].

    The dollar volume corresponding to new employment in steel and each fabricating activity was crudely obtained by multiplying the dollar volume per production worker for steel and for each activity, as derived by the 1947 *Census of Manufactures* for the United States as a whole, by the anticipated new employment. This derivation of input requirements assumes linear homogeneous production functions. In addition it should be borne in mind that the production worker data of the *Census* and *Iron Age* diverge. Hence, unavoidable error is introduced, since new employment estimates of three-digit categories are stated in terms of *Iron Age* production workers while the dollar volume expansion of each activity is based on *Census* production worker data.

26. An error of computation crept into the construction of this bill of goods. Production worker employment in industry 343 was overstated by 477. In view of the crudeness of the analysis and the smallness of the error, recomputation of the round effects did not seem to be warranted.
27. New construction was removed from the Bureau of Labor Statistics structural matrix because its inclusion is questionable. Given the data currently available, any separation of construction into that on current and that on capital account involves subjective judgments. As a consequence, results obtained on the requirements of construction when construction is included in the structural matrix tend not to be superior to those obtained by the crude rule-of-thumb method which will be employed later.
28. See W. Isard [15]. Experimentation with an interregional model is currently being undertaken.
29. In the case of households a figure of 82 per cent was used in order to allow for some outflow to persons residing in other areas and to recognize leakages through savings and other channels.

    A useful check on the percentages was furnished by data derived for 1939 in connection with regional input–output work of the Harvard Economic Research Project (see W. Isard [17]. In this work, for each of a group of industries somewhat similar to ours, estimates were made of the consumption of outputs by states. Then, subtracting for any given activity the consumption estimate from the actual production for a given state yields a surplus or deficit which can be expressed as a percentage of total consumption of that activity in that state. This percentage gives some indication of the degree to which a state supplies its own needs for the outputs of any given industry. The percentages derived for 1939 for the states of Connecticut, New York, New Jersey and Pennsylvania tend to have either smaller deficits than our percentages would yield or actual surpluses.
30. Under certain circumstances, different sets of percentage reduction factors can and should be used from round to round to estimate expansions. See W. Isard and M. J. Peck [16].
31. These, of course, represent the dollar value of each industry's product in 1947 prices.
32. Data were obtained from the *Census of Manufactures*, 1947; *Census of Business*, 1948; *Minerals Yearbook*, 1947; *Survey of Current Business*, July 1949; various issues of the *Monthly Labor Review*; and from the Harvard Economic Research Project.
33. The 11 666 estimate of number of total workers for the 3.0 million tons of initial new capacity is obtained from data available on the Fairless Steel Works. See Pennsylvania State Planning Commission [26], p. 1.
34. Column 9 of Table 4.5, when converted into dollar figures, approximates

the column vector X, where $X = (I-RA)^{-1}Y$. In the term on the right-hand side of the equation, Y represents the bill of goods column vector whose elements are the outputs of the initial steel and steel-fabricating activities. It is multiplied by the inversion of the matrix $(I-RA)$ where A is the coefficient (structural) matrix constructed to include households, and where R is the diagonal matrix whose elements are the set of percentages of column 2, Table 4.5.

35. We estimated construction employees: (*1*) by dividing the total national labor force in 1947 less government and construction workers into the total of construction workers for the nation; and (*2*) by multiplying this ratio by the total of the estimates in column 9.
36. We estimated government employees: (*1*) by dividing the total national labor force in 1947 less all government employees (state, local, and federal) into the total of state and local government employees in 1947 for the nation; and (*2*) by multiplying this ratio by the total of the estimates in column 9 plus new construction employees. This calculation, which excludes an allowance for federal employment, is in keeping with the probable less-than-proportional expansion of government employees, and also in line with our 'firm minimum' objective.
37. According to the *Census of Population*, 1950 (general characteristics reports by states), the percentages for New Jersey, for the New York–Northeastern New Jersey metropolitan area, and for the Philadelphia metropolitan area are 43.5, 44.0, and 42.2 respectively.
38. Among others, see [12]; and W. Leontief [22] pp. 38–41.
39. This is not at all unlikely since our problem concerns a long-term adjustment. In the long run labor from other regions can be attracted to these industries in the area, and growth of the area's population can take place. As a consequence the new income is likely to be absorbed by new job-holders.
40. These distortions are certainly not likely to be any greater than those arising in general input–output analysis from insufficient disaggregation of industrial categories into component activities, from insufficient attention to varying product-mix of firms encompassed by any one component industry, and from differential efficiencies among such firms. The latter, too, pervade our analysis.
41. The reader may care to adjust our employment estimates by a net labor saving coefficient.
42. Included, however, is an allowance for 'average' new construction activity. This activity plus replacement of depreciated facilities engages the labor of the new employees in construction.
43. This point is neatly borne out by operations with the Leontief balanced regional model. In the Leontief model, households are removed from the bill of goods and put into the structural matrix without any compensating changes. Hence the use of the resulting inverted matrix gives the following unrealistic results: an increase of one dollar in the bill of goods requirements for the output of agriculture and fishing necessitates an expansion of $7.59 in outputs for all industries in the nation; a similar increase in the bill of goods requirements for food-processing necessitates expansions amounting to $6.64 for all industries; and the same

increase for iron and steel foundry products leads to a total expansion of $7.03 for all industries; and so forth. Clearly, these expansionary effects are extreme and point up the inadequate leakage of the Leontief model. On the other hand, when households are retained in the bill of goods, similar one dollar increases in the bill of goods requirements for the output of each of these three activities necessitate respective expansions of $1.73, $2.14 and $1.89 in the outputs for all industries.

These statements, however, do not constitute a criticism of the Leontief balanced regional model. The Leontief model was designed to obtain the percentage change of the output of each activity from its 1939 level when the bill of goods demand for any given activity is changed in order to make comparisons among these percentage changes. It was not designed to obtain absolute changes. The data resulting from various operations with this model are presented in percentage terms, and not in dollars.

44. Ideally, as improved models are developed to encompass this interregional feedback effect, those elements of each industrial category which do not respond readily in approximately proportionate manner to general impulses and consequent changes in their outputs and the outputs of related industries should be removed from the structural matrix. For example, that set of inputs which households require but which are relatively unrelated to income and must be considered as functionally related to other variables should be taken out of the matrix and transferred to the bill of goods. Thus, the construction of better models necessitates the inclusion of more realistic response patterns in the matrix.
45. W. K. Joelson [18].
46. To determine the cents' worth of any one of these five inputs per dollar output of a given industrial category of the 50-industry breakdown requires the summation of the amounts of the specified input which went into each of the components of the given industrial category (as recorded in the 192 × 192 BLS Interindustry Flow Table) and the division of this sum by the gross output of this industrial category.

References

[1] American Iron and Steel Institute, *Annual Reports*, various years.

[2] Barfod, Børge, *Local Economic Effects of a Large-Scale Industrial Undertaking*, Copenhagen (1958).

[3] Cumberland, J., *The Locational Structure of the East Coast Steel Industry with Emphasis on an Integrated New England Steel Mill*, PhD Dissertation, Harvard University (1951).

[4] Daly, M. C., 'An Approximation to a Geographical Multiplier', *Economic Journal*, 50 (1940), pp. 248–58.

[5] Engländer, O., 'Kritisches und Positives zu einer allgemeinen reinen Lehre von Standort', *Zeitschrift für Volkswirtschaft und Sozialpolitik*, Neue Folge, 5 (1926).

[6] Evans, W. D., and M. Hoffenberg, 'The Interindustry Relations Study for 1947', *Review of Economics and Statistics*, 34 (1952), pp. 97–142.
[7] Federal Reserve Bank of Kansas City, 'The Employment Multiplier in Wichita', *Monthly Review*, 37 (1952).
[8] Florence, P. S. *Investment, Location and Size of Plant*, Cambridge University Press, Cambridge, UK (1948).
[9] Friedrich, C. J., trans., *Alfred Weber's Theory of Location of Industries*, University of Chicago (1929).
[10] Hoover, E. M., Jr., *Location Theory and the Shoe and Leather Industries*, Harvard University Press, Cambridge, Mass. (1937).
[11] Isard, W., 'Some Locational Factors in the Iron and Steel Industry Since the Early Nineteenth Century', *Journal of Political Economy*, 56 (1948), pp. 203–17.
[12] Isard, W., *et al.*, 'Input–Output Analysis: Discussion', *Papers and Proceedings of the American Economic Association*, 39 (1949), pp. 226–40.
[13] Isard, W., and W. Capron, 'The Future Locational Pattern of Iron and Steel Production in the United States', *Journal of Political Economy*, 57 (1949), pp. 118–33.
[14] Isard, W., and J. Cumberland, 'New England as a Possible Location for an Integrated Iron and Steel Works', *Economic Geography*, 26 (1950), pp. 245–59.
[15] Isard, W., 'Interregional and Regional Input–Output Analysis: A Model of a Space-Economy', *Review of Economics and Statistics*, 33 (1951), pp. 318–28.
[16] Isard, W., and M. J. Peck, 'Location Theory and International and Interregional Trade Theory', *Quarterly Journal of Economics*, 67 (1953) pp. 97–114.
[17] Isard, W., 'Some Empirical Results and Problems of Regional Input–Output Analysis', ch. 5 in W. Leontief, *Studies in the Structure of the American Economy*, New York (1953).
[18] Joelson, W. K., 'Suggestions for a Planned Industrial Promotion Program', Economic Development Administration, San Juan, PR, 1 October 1952.
[19] Johnson, O., *Steel Consumption in 1948: A Report by* Iron Age *to the Metalworking Industry* (1949).
[20] Krutilla, J., *The Structure of Costs and Regional Advantage in Primary Aluminum Production*, PhD dissertation, Harvard University (1952).
[21] Kuenne, Robert. E., *The Use of Input–Output Techniques for the Estimation of Employment in the Delaware Valley*, PhD dissertation, Harvard University (1953).
[22] Leontief, W., *The Structure of American Economy, 1919–1929*, Harvard University Press, Cambridge, Mass. (1941).
[23] National Resources Planning Board, *Industrial Location and National Resources*, US Government Printing Office, Washington, DC (1943).
[24] Niederhauser, E., 'Die Standortstheorie Alfred Webers', *Staatswirtschaftliche Studien*, 14 (1944), pp. 181–7.
[25] Palander, T., *Beiträge zur Standortstheorie*, Almqvist, Uppsala (1935).

[26] Pennsylvania State Planning Commission, *Boom in Bucks County*, mimeographed, undated.
[27] Ullman, E., 'The Basic-Service Ratio and the Areal Support of Cities', mimeographed, undated.
[28] US Army, Board of Engineers for Rivers and Harbors, *Commercial Statistics: Waterborne Commerce*, US Government Printing Office, Washington, DC, various years.
[29] US Army, Board of Engineers for Rivers and Harbors, *Annual Reports* (1947, 1948) US Government Printing Office, Washington, DC.
[30] US Bureau of the Census, *Census of Manufactures, 1947*, US Government Printing Office, Washington, DC (1956).
[31] US Bureau of the Census, *Census of Business, 1948*, US Government Printing Office, Washington, DC (1951).
[32] US Bureau of the Census, *Census of Population, 1950*, US Government Printing Office, Washington, DC (1952–7).
[33] US Bureau of Labor Statistics, *Survey of Current Business*, July 1949.
[34] US Department of Commerce, *Minerals Yearbook, 1947*, US Government Printing Office, Washington, DC (1949).
[35] US Department of Labor, *Monthly Labor Review*, various issues.
[36] US Steel Company, *Statement at Public Hearing by US Army Engineers, March 14, 1951, Trenton, NJ*, mimeographed.

5 Employment Impacts of Achieving Automobile Efficiency Standards in the United States*

I THE PROBLEM DEFINED AND METHODOLOGY EMPLOYED

The United States Congress in Title III of the Energy Policy and Conservation Act (EPCA) of 1975 imposed upon the US automobile industry a set of time-phased miles per gallon (mpg) standards. Failure to attain these standards as an average over all automobile sales in a given year subjects the offending company to graduated civil penalties per car manufactured in that year. The fleet-average mpg standards for each year until 1985 are listed in Table 5.1.[1]

One interesting question surrounding the program is whether in fact automotive manufacturers will be led via the profit motive to attain the standards.[2] This study will assume that an effective set of sanctions will be enforced in order to induce all manufacturers to produce a mix of automobiles which meets or exceeds the standards for each year they are mandated.

This paper addresses the question: What will be the employment impacts upon those industries most importantly affected either directly or indirectly by the industry's steps to achieve the standards? As will be shown, the answer to this question depends upon how much of the mpg efficiency gains are achieved through technological change and how much through change in average automobile weight,

* Written by three authors: Douglas C. Dacy is a Research Staff member, Institute for Defense Analyses and Professor of Economics, University of Texas; Robert E. Kuenne is Professor of Economics, Princeton University and Consultant, Institute for Defense Analyses; and Paul McCoy is Senior Operations Research Analyst, ANSER Corporation, Arlington, Virginia. This paper is a shortened version of a paper by Dacy *et al.* prepared for the US Department of Energy, Conservation and Solar Applications, under contract CO–04–50174–00.

Table 5.1 EPCA mandated MPG efficiency standards, 1978–85

Year	*Standard (mpg)*
1978	18
1979	19
1980	20
1981	22
1982	24
1983	26
1984	27
1985	27.5

technology constant. Our research, therefore, attempts to project the automobile industry's choices between these methods as modelling inputs and to derive from them the employment impact outputs we seek.

To estimate these employment impacts we used *INFORUM* (Interindustry Forecasting Model of the University of Maryland). *INFORUM* is a large-scale econometric model of the US economy at whose core is a 185 × 185 input–output system. A large number of supporting equations serves to endogenize variables that in straightforward input–output analysis are exogenous (consumption, construction, investment, etc)[3] Among the important parameters we have specified for the *INFORUM* runs and which underlie all of our estimates unless otherwise indicated are the following for the forecast period 1978–85:

1. *Per capita* disposable income growth is 2.15 per cent annually. This assumption was derived from a recent forecast from the Wharton Econometric Forecasting Model.
2. The real long-term interest rate is 8 per cent.
3. The real short-term interest rate is 5 per cent.
4. Relative prices remain constant (i.e. if the general price level rises, all prices rise proportionately).
5. Materials inputs into the 185 *INFORUM* industries, as specified by that model's input–output tables, are unchanged except where explicitly noted.

The industry of basic interest is *INFORUM* sector 145, motor vehicles. This sector includes Standard Industrial Classification industries 3711 (motor vehicles) and 3714 (motor vehicle parts). It is

Table 5.2 Potentially strategic input sectors for INFORUM sector 145, motor vehicles

Sector name	*INFORUM number*	*1972 coefficients (1972 dollars of input per 1972 dollars of output)*
Plastic materials and resins	68	0.000473
Tyres and inner tubes	80	0.010945
Rubber products	81	0.004843
Miscellaneous plastic products	82	0.005122
Steel	91	0.067897
Copper	92	0.002780
Lead	93	0.000166
Aluminium	95	0.006948
Nonferrous castings and forgings	99	0.001237
Metal stampings	106	0.079888

significant to note that *INFORUM* sector 145 excludes all activities related to the production of trucks and buses. For any given year in the forecast period 1978–85, therefore, the input coefficients in the input–output column of Table 5.2, for sector 145, depict (in 1972 dollars) the cents' worth of every sector's input per dollar's worth of passenger cars and car parts.

Of the remaining 184 sectors in the model, 10 have been singled out because (*1*) they are important direct suppliers of sector 145 as indicated by the size of their coefficients in the sector 145 column, or (*2*) they are related, at least potentially, to the technological materials substitution likely to occur in the 1978–85 period. These strategic sectors are listed in Table 5.2.

For each of the years of the forecast period *INFORUM* incorporates a different 185 × 185 input–output matrix reflecting foreseeable technological advances in the usage of intermediate inputs. If such advances were discernible, changes in the coefficients were projected by the use of logistic functions in the construction of *INFORUM*. It is important to point out that the *base case* we have adopted – the 'built-in' *INFORUM* base case modified by the parameter changes explicitly noted above – already contains those technological changes in sector 145 that were thought to be discernible in 1975 when the model was updated. Hence it is not a static base case.

To alter the parameters of the base case *INFORUM* model in

manners necessary to perform the policy runs we used a methodology best described stepwise.

Step 1

It was necessary to determine the weight of the notional or 'average' automobile that underlies the 1978 input–output table's sector 145 column of inputs. To do so we estimated the weight of each major material that entered its construction in the 1973 model year (for which the initial *INFORUM* work was performed) and then updated the weight for the 1978 model year by study of the input coefficient changes between 1972 and 1977. The end product of this analysis is to yield an estimate of the weight of the 1978 model year *INFORUM* notional automobile. This calculated weight was checked against the known average car weight in model year 1978 to give credibility to the results derived for *INFORUM*. The execution of Step 1 will be found in Section II of this paper.

Step 2

For the notional 1978 automobile it was necessary to determine its mpg performance. To do so we employed standard automotive engineering formulae fitted to our data. This analysis is in Section III.

Step 3

Next, it was required to estimate the gains in mpg due to technological advances for a car with the weights of the notional automobile for every year of the forecast period 1978–85. The stimulus to adopt these innovations, of course, was the need to meet the efficiency standards mandated. The additions of these mpg increments to the mpg achievement of the 1978 standard automobile is a sum that was needed to calculate the deficit in mpg performance that must be achieved by nontechnological means. The shortfall can be made up by (*1*) shifting the market mixes among small, mid-size, and large models to lower the average weight from the base case value, and/or (*2*) substituting among materials to permit weight reductions. The work of Step 3 is found in Section IV.

Step 4

The next step was to determine the average weight of the three automobile class sizes cited in Step 3, the market mixes that can be expected to emerge in the years 1978–85, and the consequent average weight changes implied. The total weight loss for all automobiles could then be used to determine input reductions from the major industries contributing to the automobile manufacturing sector. This research is summarized in Section V.

Step 5

Residual shortfalls in mpg performance from standards must be eliminated by materials substitution. The resulting calculations are used to change further the input coefficients in the column of *INFORUM* sector 145. This is done in Section VI.

Step 6

Lighter automobiles employing superior technologies will lead to considerable fuel saving. Reduced fuel usage with no relative gasoline price change increases households' potential discretionary spending on other goods and services. It was necessary to estimate the dollar value of gasoline savings to reduce personal consumption expenditures on gasoline in the policy runs of *INFORUM*. The dollar value of gasoline savings was distributed to the final demands for outputs of other industries. These changes, in addition to the input coefficient changes made to reflect the technological, weight and materials substitution changes discussed above, were applied to obtain a *policy case* for an *INFORUM* run. These analyses will be found in Section VII.

Step 7

The results of the policy case runs were compared with those for the base case to compute the employment impacts to be expected from the mandated standards. These comparisons are derived in Section along with a brief conclusion.

II DETERMINING THE WEIGHT OF THE 1978 NOTIONAL AUTOMOBILE

It was necessary as a first step to determine whether the average weight of the automobiles produced with the average inputs specified in the column of *INFORUM* sector 145 for 1978 corresponds with the average weight of automobiles actually produced in that year. The 1978 table's column of inputs from 184 supplier industries to sector 145 is itself a projection on the basis of 1972 data for the 1973 model year. That is, *INFORUM* is the result of intensive work in the years culminating in 1972, with input–output matrices changing over the forecasting period on the basis of information available, updated in succeeding years through 1975.

INFORUM coefficients are stated in money values rather than physical quantities of inputs: they denote dollars of input per dollar of output in 1972 dollars. We needed therefore to convert the input ratios for sector 145's strategic supplying industries into their corresponding weight in pounds in order to estimate the weight of the 1973 model notional car and update it to the 1978 notional car. After deriving these dollars-to-weight conversion factors we took the difference between the 1972 and 1978 coefficients (both defined in 1972 dollars) and obtained the changes in weight for each important input. The algebraic sum of these weight changes was the difference in total weight between *INFORUM*'s notional 1972 and 1978 model year automobiles. Although this indirect derivation is tedious, it was necessary because all of *INFORUM*'s benchmarks are for 1972 rather than 1978.

First, it was necessary to obtain the weight of the average automobile in 1972. Cheslow's ([3], p. 39) detailed distribution of 1973 model domestic passenger cars yielded a weighted mean curb weight of 3600 pounds. This estimate corresponds broadly with Roig *et al.* ([9], p. 5) who quote the sales weight average of *all* cars (including imports) sold in the US in 1975 as 3400 pounds. Yasnowsky and Colby ([11], table 3) cite the weight of a standard 1973 model car as 3810 pounds and of an economy model as 2242 pounds. On the basis of these data we shall assume that the notional 1972 car depicted in the *INFORUM* 1972 input–output table weighed 3600 pounds.

To determine the total value of any of the 10 strategic inputs listed in Table 5.1 per automobile in 1972, we multiplied the relevant input coefficient by the 1972 price of the notional automobile on a producers' value basis (i.e. exclusive of transport charges to the consumer). The *Statistical Abstract, 1974* quotes the average factory price of automobiles in 1972 at $2622.[4]

Yasnowsky and Colby ([11], table 3) list the materials content of the 1972 automobile as percentages of total weight. These data are reproduced in Table 5.3. The percentages are translated into weights in the last column of the table. It was then necessary to allocate these materials weights among the appropriate *INFORUM* supplying

Table 5.3 Materials content, 1972 automobiles, by per cent of total weight and levels for notional automobile

Material	*Percentage*	*Pounds per notional automobile*
Aluminium	1.9	70
Copper	1.1	40
Iron and steel	74.8	2738
Lead	0.7	26
Nickel	0.1	4
Zinc	2.7	73
Glass	2.7	99
Rubber and plastics	12.7	465
Miscellaneous	4.0	145
Total	100.0	3660

Table 5.4 Estimates of weights of materials inputs into the notional 1972 automobile contributed by 10 strategic INFORUM sectors

INFORUM *sector*	*Pounds per notional automobile*	*Pounds per dollar*
68 Plastic materials and resins	40	0.0153
80 Tyres and inner tubes	125	0.0477
81 Rubber products	220	0.0839
82 Miscellaneous plastic products	80	0.0305
91 Steel	1488	0.5675
92 Copper	40	0.0153
93 Lead	26	0.0099
94 Zinc	73	0.0278
95 Aluminium	13	0.0050
99 Nonferrous castings and forgings	57	0.0217
106 Metal stampings	1250	0.4767
Total	3412*	

* Total differs from that in Table 5.3 because of weight contributed by non-strategic industries.

industries. On the basis of information contained in Cheslow ([3], pp. 33–5, 38) and Roig *et al.* ([9], pp. 5, 19) we distributed them as shown in column 1 of Table 5.4.

The figures in column 2 were derived by dividing the data in column 1 by \$2622 – the price of the notional 1972 car – to get pounds of input per dollar of output in 1972 dollars.

To determine the change in weight between the 1972 and 1978 automobile from the data, we had to know the change in pounds of material input per dollar of output in 1972 dollars. This difference can be multiplied by the price of the 1972 car to obtain the total weight difference attributed to each material. Note that

$$\left(\frac{\text{lbs of material input}}{\text{\$ of output}}\right)1972 - \left(\frac{\text{lbs of material input}}{\text{\$ of output}}\right) 1978$$

$$= \Delta\left(\frac{\text{lbs of material input}}{\text{\$ of output}}\right)$$

The second term on the left-hand side is unknown, but an equivalent of the equation can be obtained with the existing data. For we may write

$$\left(\frac{\text{lbs of material input}}{\text{\$ of material input}}\right) 1972 \left[\left(\frac{\text{\$ of material input}}{\text{\$ of output}}\right)1972 - \left(\frac{\text{\$ of material input}}{\text{\$ of output}}\right) 1978\right] = \Delta\left(\frac{\text{lbs of material input}}{\text{\$ of output}}\right)$$

where, as throughout our analysis, all dollar values are measured in 1972 dollars.

The expression before the brackets is simply the inverse of the price of the material per pound in 1972. The brackets contain the difference between the input coefficients for the years 1972 and 1978. Hence, this difference when divided by the price of the material per pound in 1972 yields the gain (+) or loss (−) in the weight of the material input. These changes summed and multiplied by the 1972 price of the automobile give the total weight change. The relevant data for calculating the weight change in automobiles are given in Table 5.5.

INFORUM's 1973 automobile weighs 257 pounds less than the notional car in 1972 at 3407 pounds. To check this figure for accuracy we calculated the average weight of 1978 domestically produced cars for the first quarter of 1978 from data in the *Wall Street Journal* and

Table 5.5 Calculation of weight change in the INFORUM notional automobile between 1972 and 1978

INFORUM *sector*	*Input–output coefficient difference*[a]	*Pounds per dollars of input material*[b]	*Weight difference per 1972 dollar of 1978 automobile due to material input*[c]
68 Plastic materials and resins	0.00	30.60	0.00
80 Tyres and inner tubes	+0.0002	4.38	+0.0009
81 Rubber products	0.00	17.48	0.00
82 Miscellaneous plastic products	0.00	5.98	0.00
91 Steel	−0.0111	8.36	0.00
92 Copper	0.00	5.46	−0.0928
93 Lead	0.00	49.50	0.00
94 Zinc	d	d	0.00
95 Aluminium	0.00	0.72	0.00
99 Nonferrous castings and forgings	0.00	18.08	0.00
106 Metal stampings	−0.0010	5.97	−0.0060
Total			−0.0979
Total weight change in automobile = −0.0979 × $2622 = −257 pounds			

a Column 1 is the difference between the input–output coefficients in 1972 and 1978.

b Column 2 figures are obtained by dividing the number of pounds of the material in the notional car (see Table 5.4) by the input–output coefficients in Table 5.2 times $2622, the price of the notional car in 1972. Note that the input–output coefficient of Table 5.2 times $2622 is the total value of the input material.

c Column 3 = Column 1 × Column 2.

d *INFORUM* does not use zinc as a direct input to its automobile sector. The difference in Column 3 is assumed to be zero.

the curb weight listed for each car in *Consumer Guide: 1978 Cars* [4]. When the curb weight was not available from this source we obtained the figure from dealers. The average curb weight of all domestic cars sold in this period was 3420 pounds. Thus, it appeared that the weight of the notional car implied by the *INFORUM* 1978 coefficients is very accurate, and we accepted the coefficients as they stand.

III ESTIMATING THE 1978 NOTIONAL CAR'S EFFICIENCY

To estimate the Environmental Protection Agency (EPA) mpg efficiency of the notional 1978 automobile we fit a log-linear function of the type mpg = AW^{α}, where W is weight in thousands of pounds, to data for 1976 models of domestic automobiles which were collected and analyzed by Silver [10]. The equation obtained was

$$\ln mpg = \underset{(57.92)}{3.99195} - \underset{(-17.77)}{0.924} \ln W, \; R^2 = 0.861 \tag{5.1}$$

(t-values in parentheses)

or

$$mpg = 54.2\ W^{-0.924} \tag{5.2}$$

Given Equation 5.2 and the notional car's weight (W = 3403 thousand pounds) mpg is calculated to be 17.5 We used this estimate as a lower bound, for it assumes that no progress beyond this efficiency was attained in the automobiles produced in 1978.

For an upper bound we employ a curve used by the US Department of Transportation (DOT) and EPA [6] in 1974 depicting *potential* 1980 EPA standard mpg results as a function of automobile weight. We estimated the function from the graph by fitting an exponential equation to two widely separated points:

$$mpg = 186W^{-1.755} \tag{5.3}$$

This equation estimates that an automobile of notional weight could potentially achieve an mpg efficiency of 21.7 in 1980. We assumed that in 1978 one-third of the gap between the lower bound of 17.5 and this ideal target had been closed, or that the notional car attained 18.9 mpg in 1978.

To summarize, upper and lower bounds of 18.9 and 17.5 mpg respectively were established for the 1978 *INFORUM* notional car. We resorted to such bracketed estimates rather than a point estimate to provide a reasonable range within which the true value lies with high probability.

IV PROJECTING TECHNOLOGICAL ADVANCES IN AUTOMOTIVE ENGINEERING FOR AN AUTOMOBILE OF NOTIONAL WEIGHT

Without weight changes, mpg efficiency is expected to rise year by year through technological advances in engine design, drive line and other mechanical and design alterations. A critical question for the study was how these anticipated changes would improve mpg efficiency each year of the period 1978–85.

To answer this question we relied heavily upon the DOT-EPA study [6]. Once more we chose to establish a range of estimates rather than a point estimate. The size of the bracket is controlled by two assumptions: (*1*) the speed with which such potential mpg efficiency was achievable, and (*2*) the level of current mpg efficiency.

Equation 5.3 is the depiction of the DOT-EPA study's 'potential' 1980 trade-off between weight and efficiency. On that function, a car weighing 3403 pounds (the notional car weight) can be expected to attain a potential of 21.7 mpg in 1980. Suppose that the 1977 mpg for the 1978 model year is the upper bound estimated in Section III of 18.9. An improvement of the mpg by 2.8 by 1980 by technological change did not seem realistic at the time.[5] To err on the conservative side we assumed that the potential 1980 m.p.g.-weight relation of Equation 5.3 applied to 1982 rather than 1980, and therefore that the 2.8 mpg gain would not occur until the end of 1982. We termed this estimate the *optimistic technological case* and used it as an upper bound for 1982 performance. However, if the notional car only achieved 17.5 mpg in 1977, the lower bound on its estimated efficiency, this incremental improvement yields only 20.3 mpg in 1982 (17.5 + 2.8 = 20.3).

It seemed reasonable to assume a *pessimistic technological case* also. For it we assumed that only 70 per cent of the potential gains in mpg efficiency would be achieved in 1982, so that an increment of only 2 mpg would be forthcoming from technological change alone. The pessimistic case receives upper and lower bounds depending on the use of upper or lower bounds for the estimate of 1977 mpg efficiency.

Next, it was necessary to distribute the gains in mpg efficiency over the years 1978–82. There was no guidance available for this task so that we had to make our own assumptions about the timetable. These are given in Table 5.6.

From the data on mpg gains in Table 5.6 we estimated the range of

Table 5.6 Per cent and absolute gains in mpg due to technological advance assumed achieved by the end of each year

Assumed technological advance	*1978*	*1979*	*Year* *1980*	*1981*	*1982*
Optimistic technological case					
Per cent of 1982 goal attained	50	60	75	90	100
gains in mpg*	1.4	1.7	2.1	2.5	2.8
Pessimistic technological case					
Per cent of 1982 goal attained	20	30	45	60	70
gains in mpg*	.6	.8	1.3	1.7	2.0

* These values are obtained by multiplying 2.8 by the percentage immediately above.

mpg efficiencies attainable through technological change for an automobile of notional weight. These values are given in Table 5.7 on the bases of both upper and lower bounds for the estimates of 1977 mpg efficiency derived in Section II.

For the years 1983–5 we applied a similar technique to the DOE-EPA projection of the 1985 limiting theoretical mpg efficiency curve. Given the slippage in timing, we interpreted this theoretical 1985 curve to be relevant to 1987. Again, we fitted an exponential function to two points on the graph and obtained

$$mpg = 91\ W^{-0.969} \tag{5.4}$$

From the equation, we derived a 1987 estimate for the notional car of 3403 lbs of the theoretical limit through technological improvement of 27.9 mpg. In the most optimistic case, technological improvements could add 6.2 mpg (27.9 – 21.7) between 1982 and 1987 and 9.0 mpg (27.9 – 18.9) between 1977 and 1987.

Tables 5.8 and 5.9 are repetitions of the methodologies underlying Tables 5.6 and 5.7 using different assumptions about the speed of technological implementation and beginning m.p.g. efficiencies for the remaining years of each forecast period.

By using the data of Tables 5.7 and 5.9 and the EPCA standards we estimated the annual shortfalls in mpg efficiency that would occur after accounting for technological improvement. These shortfalls are given in Table 5.10.

Table 5.7 Technological advance projections of fuel economy, in miles per gallon, for 1978–82

Assumed technological advance	*1977 (estimate)*	*Year* 1978	1979	1980	1981	1982
Optimistic technological case						
Upper bound (mpg)	18.9	20.3	20.6	21.0	21.4	21.7
Lower bound (mpg)	17.5	18.9	19.2	19.6	20.0	20.3
Pessimistic technological case						
Upper bound (mpg)	18.9	19.5	19.7	20.2	20.6	20.9
Lower bound (mpg)	17.5	18.1	18.3	18.8	19.2	19.5

Table 5.8 Per cent and absolute gains in mpg due to technological advance assumed achieved by the end of each year from 1983 through 1987

Assumed technological advance	*1983*	*1984*	*Year* 1985	*1986*	*1987*
Optimistic technological case					
Per cent of 1987 goal attained	40	50	60	–	100
gains in mpg*	2.5	3.1	3.7	–	–
Pessimistic technological case					
Per cent of 1987 goal attained	10	20	30	–	–
gains in mpg*	0.62	0.24	1.86	–	–

* These values are obtained by multiplying 6.2 by the percentages immediately above.

Table 5.9 Technological advance projections of fuel economy, in miles per gallon, for 1983–5

Assumed technological advance	*1982*	*1983*	*Year* 1984	*1985*
Optimistic technological case				
Upper bound	21.7	24.2	24.8	25.4
Lower bound	20.3	22.8	23.4	24.0
Pessimistic technological case				
Upper bound	20.9	21.5	22.1	22.8
Lower bound	19.5	20.1	20.7	21.4

Table 5.10 Shortfalls in mpg efficiency from implemented EPCA standards after technological advances, 1978–85

	Year							
Assumed technological advance	*1978*	*1979*	*1980*	*1981*	*1982*	*1983*	*1984*	*1985*
Optimistic technological case								
Upper bound	0.0	0.0	0.0	0.6	2.3	1.8	2.2	2.1
Lower bound	0.0	0.0	0.4	2.0	3.7	3.2	3.6	3.5
Pessimistic technological case								
Upper bound	0.0	0.0	0.0	1.4	3.1	4.5	4.9	4.7
Lower bound	0.0	0.7	1.2	2.8	4.5	5.9	6.3	6.1

These mpg shortfalls must be eliminated by reducing the weight of automobiles. To find the required weight reductions necessary to eliminate them, theoretically, one would have to read the indicated amount from a function relating mpg efficiency to weight. The problem was that we have used three such functions in Equations 5.2, 5.3 and 5.4 relating to 1977, 1980, and 1987 respectively. Consider a weight reduction of 1000 pounds from the weight of the notional car. From Equation 5.2 the mpg gain would be 6.6; from Equation 5.3 it would be 4.9; and from Equation 5.4 the gain would be 11.2. For each year between the benchmark years it was possible to estimate the gains in mpg due to a weight reduction of 1000 pounds by interpolation of the relevant equations. These figures are displayed in Table 5.11.[6]

From the estimates in Tables 5.10 and 5.11 we computed the weight reduction *below 3403 pounds* that must be attained to meet the relevant EPCA standards. These are tabulated in Table 5.12. These projections are relative in every instance to the notional weight of 3403 pounds. For example, in the optimistic case the standard car in 1985 will have to weigh 188 pounds less than the notional 1977 car, but to achieve the standard in 1983 it will have to weigh 307 pounds less.[7] In Table 5.13 we present the necessary *incremental* weight losses implied each year after cumulating prior weight reductions.

After accounting for technological progress, we find that no weight reduction will have to occur after 1983 in any case. We estimate that the average car will have to be reduced from about 3400 pounds to between 2700 and 3100 pounds, or by 300 to 700 pounds. These time-phased changes could be achieved by (*1*) shifting from larger toward smaller models, and (*2*) materials substitution in all cars. Our

Table 5.11 Estimated mpg increases for a weight reduction from 3403 pounds to 2403 pounds, 1978–85

	Year							
	1978	*1979*	*1980*	*1981*	*1982*	*1983*	*1984*	*1985*
mpg gain	6.6	6.0	5.0	6.2	7.5	8.7	10.0	11.2

Table 5.12 Weight reductions below notional car weight of 3403 pounds necessary to attain fuel efficiency standards, 1978–85

	Year							
Assumed technological advance	*1978*	*1979*	*1980*	*1981*	*1982*	*1983*	*1984*	*1985*
Optimistic technological case								
Upper bound	0	0	0	97	307	207	220	188
Lower bound	0	0	80	323	493	368	360	313
Pessimistic technological case								
Upper bound	0	0	0	226	413	517	490	420
Lower bound	0	117	240	452	600	678	630	544

Table 5.13 Incremental weight reductions necessary to meet EPCA fuel efficiency standards, in pounds, 1978–85

	Year									
	1977 weight	*1978*	*1979*	*1980*	*1981*	*1982*	*1983*	*1984*	*1985*	*1985 weight*
Optimistic technological case										
Upper bound	3403	0	0	0	80	210	0	0	0	3096
Lower bound	3403	0	0	80	243	170	0	0	0	2910
Pessimistic technological case										
Upper bound	3403	0	0	0	226	187	184	0	0	2886
Lower bound	3403	0	117	123	212	148	78	0	0	2725

next task, therefore, was to estimate how much of the required reductions were likely to come from model shifting.

V PROJECTING WEIGHT CHANGES THROUGH SHIFTS AMONG MODELS

To perform this task we drew upon recent projections of the Automobile Demand Simulation Model of DOE which uses the size-weight categories shown in Table 5.14. This model is programmed to achieve EPCA mandated standards for each year under the assumption that a tax on large cars and a doubling of the current civil penalties of EPCA have been enacted (see note 3). For this parameter set the model projects the model mixes shown in Table 5.15.[8]

The projections indicate only a small change in the composition of size classes. However, automobile size class is not determined by weight alone, so that the data of Table 5.15 cannot be interpreted as implying a nonreduction in automobile weight. Yet, it is clear that the average weight of automobiles must decline if EPCA standards are to be met. The projection of relatively little change in the mix of size classes is consistent with the assumption that the average weight of each size class will decline proportionately. This is a convenient assumption and one which we have adopted for this study. It permitted us to assume that the required weight loss for the average automobile will be achieved by substituting lighter for heavier materials in all automobiles.

VI PROJECTING WEIGHT LOSS THROUGH MATERIALS SUBSTITUTION

The major question is how will materials be substituted to achieve the necessary weight reductions? In reading the available literature and from discussion with knowledgeable persons in the area of automobile efficiency we concluded that only one type of substitution will occur within our time frame to effect meaningful changes. The forecast period is too short and the lead times that characterize the industry too long to justify other substitutions.

The substitution we envisage is that of plastic for metal parts and, most notably, the use of plastic sheet with steel laminate as a substitute for cold-rolled sheet in automobile bodies. Difficulties with

Table 5.14 Automobile demand simulation model size categories

Size		*Estimated weight (1977)*
Small	less than 3050 lbs	2600
Mid-size	3050–3500 lbs	3180
Large	over 3500 lbs	3850

Table 5.15 Projected mixes in automobile demand simulation model, gas-guzzler tax plus double civil penalties parameter set, 1978–85 (*percentages*)

	Year							
Model	*1978*	*1979*	*1980*	*1981*	*1982*	*1983*	*1984*	*1985*
Small	18.2	17.2	16.4	16.6	16.9	18.0	18.7	18.8
Mid-size	32.8	33.5	33.7	33.1	32.3	30.3	28.6	27.6
Large	49.0	49.3	49.9	50.2	50.8	51.7	52.7	53.6

welding aluminium radiators in cars continue to cast doubt on their imminence as replacements for copper radiators, and similar delays seem to be occurring in the much-talked-about use of cast aluminium for cast iron in such parts as cylinder blocks, cylinder heads, intake manifolds, valve bodies and assemblies, and crank-cases.[9]

There are three apparent approaches that may be taken in input–output analyses to capture the effects of weight reduction in automobiles:

1. Reduce steel input to metal stampings and increase the plastics input to automobiles while maintaining the value of metal stampings input to automobiles.
2. Reduce metal stampings input to automobiles and increase the input of plastics to automobiles. This affects metal stampings directly and steel indirectly through metal stampings.
3. Reduce steel input to automobiles and increase plastic inputs while maintaining the value of metal stampings.

Method 1 encounters a major problem. The automotive industry is huge compared to metal stampings – about eight times as large in 1977 measured by value of shipments. If *all* the steel in metal stampings were eliminated, the resulting weight reduction in automobiles would not be sufficient to achieve EPCA standards.

Method 2 raises some serious problems of interpretation. Does a reduction of stampings into automobiles mean that they are lighter weight? If so, does this imply they are proportionately cheaper? And there remains the objection cited in our discussion of Method 1.

Method 3 is less neat in terms of the materials flow concept but there is much to recommend it from the practical point of view. The most important advantage is that all the weight reduction in automobiles cannot be due to reducing the steel content of metal stampings. Much of it must come from a lowering of the weight of other steel components in automobiles. The principal substitutions we have discussed are taken from Cheslow [3] but many less dramatic substitutions of plastics for steel may be undertaken.

Although we prefer Method 3 we felt that Method 1 had merit if for no other reason than to serve as a check on the results of Method 3. Therefore, we employed Methods 1 and 3.

For the proportionate changes in input coefficients we relied upon Cheslow [3]. For each pound reduction in the weight of the automobile, steel input must be reduced by 1.53 pounds and plastics increased by 0.53 pounds. Thus the iron and steel content of the notional 3400-pound automobile would have to fall by 765 pounds and the weight of plastics increase by 265 pounds to obtain an automobile with average weight of 2900 pounds. From Table 5.4, it is calculated that steel content must decline by 51 per cent and plastic content must increase by 221 per cent.

Plastic inputs enter automobile production from two *INFORUM* industries and the total increase must be allocated between them. Of the 0.53 pound increase in plastics per pound decrease in automobile weight, 0.18 pounds is assigned to *INFORUM* sector 68 (plastic materials and resins) and 0.35 pounds is attributed to sector 82 (miscellaneous plastics products).

The required input coefficient changes for iron and steel, plastic materials and resins, and miscellaneous plastics products are achieved by multiplying the automobile weight changes of Table 5.13 by −1.53, 0.18, and 0.35, respectively. In Method 2, which reduces metal stampings in automobiles, the multiplier −1.53 is used to adjust the weight of metal stampings by 100 per cent and by 30 per cent, but for reasons mentioned earlier even the drastic (and totally unreasonable) change is not sufficient to reduce the weight of the automobile by the required amount. For all methods, the materials weight changes were divided by their actual weights in the 1977 national automobile as given in Table 5.4 to obtain the proportionate changes.

In the hypothetical example above for Method 3, when the average automobile weight must fall by 500 pounds, the steel input coefficient is reduced by 51 per cent (from 0.068 to 0.033), the coefficient for plastic materials and resins is raised by 225 per cent (from 0.00047 to 0.0015), and miscellaneous plastic products is increased by 219 per cent (from 0.0051 to 0.0162). The process was repeated for each material for each year that the automobile was required to lose weight. *INFORUM*'s input–output matrices were then ready for a run.

VII ESTIMATING GASOLINE SAVINGS WITH IMPROVED MPG

EPCA mpg standards will lead to large savings in gasoline consumption. Difiglio and Kulash [7] conclude that it could fall from 5.280 million barrels per day ($mmbd^{-1}$) in 1978 to 4.476 $mmbd^{-1}$ in 1985 if a gas-guzzler tax and double civil penalties were enforced. This 11 per cent reduction contrasts with a 2 per cent increase if such policies were not enacted. The gasoline saving attributed to EPCA standards is estimated to be 0.624 $mmbd^{-1}$ in 1985 or 228 million barrels for the whole year. In 1972 dollars this represents a saving of $4.5 billion for final consumers of gasoline.

In fact, it is likely that these estimates overstate the savings, because the EPA test standards met by the automobile companies may diverge from actual road mpg. We have accepted the Difiglio-Kulash estimates but caution the reader that they are quite optimistic. Under this assumption we reduce the final sales of *INFORUM* sector 76 (petroleum refining) by 11 per cent and this saving is spread over all the other sectors in accordance with *INFORUM*'s final demand equation. The saving translates into one of $4–5 billion a year in 1972 dollars. With this adjustment *INFORUM* was ready to run for the optimistic and pessimistic technological cases and the high and low estimates of 1977 mpg efficiency under each case.

VIII THE RESULTS

In Table 5.16 we tabulate the employment impacts for 1980 and 1985 and for Method 3 input coefficient adjustments in the two extreme

Table 5.16 Net employment effects by selected industries in 1980 and 1985 due to meeting EPCA mpg standards under Method 3 assumptions of reduced iron and steel and increased plastics inputs (deviations from base case projections in thousands)*

	1980		*1985*	
	Most optimistic case	*Most pessimistic case*	*Most optimistic case*	*Most pessimistic case*
All economy	60	30	140	140
Mining	0	–1	0	–2
Petroleum and gas	–6	–6	–22	–22
Motor vehicles	1	0	0	0
Iron and steel	7	–15	–6	–36
Metal stampings	6	5	11	11
Plastic materials and resins	0	2	1	6
Miscellaneous plastics products	–3	8	0	21
Wholesale and retail trade	–16	–22	–174	–183
Financing and services	31	22	109	97
Other 178 sectors	40	37	221	248

* The base case assumes that changes in energy efficiency embodied in *INFORUM* will be effective.

policy cases, as defined in Section IV. (*1*) optimistic technological case-upper bound mpg estimate and (*2*) pessimistic technological case-lower bound mpg estimate. These cases respectively provide the most optimistic and the most pessimistic forecasts.[10]

Given the assumptions underlying our most optimistic case, 80 per cent of the gain in mpg will come from technological advance (and 20 per cent from weight reduction), whereas only 40 per cent would be due to that source (and 60 per cent from weight reduction) under the most pessimistic case. The net effect on employment in 1985 is not sensitive to the assumptions about technological change; however, the impacts on some sectors are sensitive to the assumptions.

The major interindustry employment impacts are in iron and steel and petroleum and gas (production). In the former, employment is calculated to fall by nearly 4 per cent in 1985 under the most pessimistic assumptions but less than 1 per cent under the most optimistic. In petroleum and gas (production), the change in 1985 is about 10 per cent of total employment. The decline in wholesale and retail trade amounts to 0.7 per cent (some of which is probably due to employment losses in retail gasoline outlets) but it is more than offset by small gains in other sectors. The two plastics sectors combined gain about 4 per cent in employment under the most pessimistic assumptions.

To test the sensitivity of the results, Method 1 (reducing steel in metal stampings) runs were made but the net aggregate employment effects were not large. In one run, metal stampings lost much employment and total employment was about the same as in the base case.[11] Thus, the results are somewhat sensitive to the manner of introducing materials substitutions into the input–output matrices.

The general conclusion of our study is that if macroeconomic policy manages to keep the US economy growing at the average of past rates, automobile mpg standards policy will have a slight positive impact on employment after taking into account minor shifts among sectors. Even in sectors directly affected by the changes imposed by conforming to such mandated standards, employment impacts are relatively unimportant. Energy policy in this area, therefore, can afford to concentrate upon environmental and other economic concerns (e.g. balance of payments problems) of more impressive dimensions.

Notes

1. The Secretary of Transportation has limited ability to change the statutory 1985 standard in Table 5.1. We assume throughout that all standards listed there will remain fixed at these levels.
2. The EPCA sets the civil penalties for failure to conform to the standards at $5 per 0.1 mpg shortfall from the standard for every vehicle manufactured in the given year. Carmen Difiglio and others in the Department of Energy, using an automobile demand simulation model, find that only if drastic increases in these civil penalties are imposed will all of the standards be met on profit motivation alone. See C. Difiglio and D. Kulash [7] and C. Difiglio *et al.* [8].
3. *INFORUM* is documented in Clopper Almon *et al.* [1]. An explanation and illustration of the model is given in Norman Asher *et al.* [2].
4. Table 928, p. 556.
5. The DOT-EPA study ([6], p. 9) indicates that a lead time of four years is necessary for structural, transmission, and component modifications, and a lead time of six years is necessary for new engine configurations.
6. Note that for the purpose of finding mpg–weight trade-off we accepted

the DOT-EPA curves as being relevant to the years 1980–5, but for the purpose of estimating the speed of technological change we assumed a two-year slippage. We saw no need to postpone application of these curves for mpg–weight trade-off because we assumed the technological changes which have a rapid mpg pay-off for weight loss – elimination of accessories, reduction in aerodynamic drag, and automatic transmission improvements – will be introduced more rapidly than some major changes in engine design.

7. Because technological change is assumed to be rapid in the 1983–5 period, cars could begin to add weight after 1983 and still meet the average standard. We assumed that weight will not increase after 1983 as the figures of Table 5.13 will indicate.
8. The data of Table 5.15 were provided by Dr Carmen Difiglio, Chief, Office of Transportation Conservation Policy, Office of the Assistant Secretary for Policy and Evaluation, Department of Energy.
9. The same difficulties have prevented other analysts who use input–output models to estimate employment impacts from energy programs from adjusting input coefficients for aluminum and copper. The usual applied approach in energy impact studies is simply to reduce the amount of steel in automobiles. We are grateful to Mr Tom Mooney of the US Department of Labor for verifying some of the technological difficulties and the restrictions they impose on the use of input–output analysis.
10. Actual levels of employment and the results of runs for Method 1 can be found along with base case projections in Dacy *et al.* [5], pp. 91–2.
11. See Dacy *et al.* [5], table E.2, p. 92.

References

[1] Almon, C., Jr, *et al.*, *1985: Interindustry Forecasts of the American Economy*, Lexington (1974).

[2] Asher, N., *et al.*, *The Microeconomic Aspects of Energy Conservation*, vol. V, IDA Report R-223, Institute for Defense Analyses, Arlington, Virginia (1976).

[3] Cheslow, M., *Industrial and Economic Impacts of Improving Automobile Fuel Efficiency: An Input–Output Analysis*, working paper 1216-3-2, the Urban Institute, Washington, DC (1975).

[4] *Consumer Guide: 1978 Cars*, Signet Books, New York (1978).

[5] Dacy, D. C., R. E. Kuenne, and P. McCoy, *Employment Impacts of Achieving Federal Energy Conservation Goals Pertaining to Automobile MPG Efficiency, Home Retrofitting and Industrial Energy Usage for the Period 1978–1985*, IDA Paper P-1347, Institute for Defense Analyses, Arlington, Virginia, 1978.

[6] Department of Transportation–Environmental Protection Agency, *Potential for Motor Vehicle Fuel Economy Improvement*, Washington, DC (1974).

[7] Difiglio, C., and D. Kulash, *Methodology and Analyses of Ways of Measuring the Effectiveness of the Use of Fuel Energy Resources:*

Increasing Automobile Fuel Economy via Government Policy, October 1977.

[8] Difiglio, C., *et al.*, *Marketing and Mobility*, Office of the Secretary of Transportation, Washington, DC (1976).

[9] Roig, R. W. *et al.*, *Impacts of Material Substitution in Automobile Manufacture on Resource Recovery*, vol. I, *Results and Summary*, International Research and Technology Corporation, IRT-403-R, Arlington, Virginia, undated.

[10] Silver, S., *Relationship of Price and Fuel Consumption to Weight for U.S. Automobiles*, Institute for Defense Analyses Internal Paper, July 1975.

[11] Yasnowsky, P. N., and D. Colby, *Impact of Gasoline Prices on the Demand for Minerals for the Manufacture of Automobiles*, Bureau of Mines, Department of the Interior, Washington, DC, undated.

6 General Equilibrium Analysis: Its Meaning, Purposes and Limitations*

I THE PHENOMENON OF VARIATION

The need for scientific analysis springs from a fundamental bias of our natural and social environment against sameness and constancy. Physical objects that are similar in specified attributes will vary among themselves in the exact degree to which they possess such attributes. Static variation of this type is compounded with dynamic variation – or *change* – in the same object's attributes over time. The ubiquity of variation, manifest in the very triviality of its assertion, forces upon man, individually and collectively, a concern for its patterns. Upon his ability to accept and shape them to his ends in a nonintuitive fashion is based his sanity and his survival.

For the limited purposes of this book, we may define a 'vision of variation', or, more simply, a 'vision',[1] as a mechanism or model whose structural components are patterns of association among variates, and whose purpose is the prediction of variation. That is to say, by postulating significant covariation relationships among sky conditions and rain; temperature, volume, and pressure of gases; Wall Street and the occurrence of war; time elapsed and the velocity of falling bodies; the amount of money in existence and the price level; drawing to inside straights and the loss of money, and so forth, models can be constructed for the prediction (or back-casting) of the value or attribute of a variate in the model.

In our wholly general definition of 'vision', the witch-doctor's

* Published originally as chapter 1 in Robert E. Kuenne, *The Theory of General Economic Equilibrium* Princeton University Press, Princeton, 1963, and reproduced with permission. Cross-references to parts and chapters refer to the original publication.

magic mechanism to explain illness, the supernatural constructions of religion, the intuitive explanatory mechanisms of a child's fantasy, as well as the sophisticated hypotheses of the nuclear physicist, are equally well covered. The touchstone of the definition is the existence of some predictive mechanism to cope with a field of variation, whether that mechanism proves to be a useful device or not, whether its construction is purposive or intuitive, sophisticated or naïve.

'Science' or 'the scientific method' may be defined as a body of techniques designed to yield criteria of choice for certain classes of these causal visions. It provides objective procedures and standards with which to judge the usefulness of the models against their explanatory pretensions. Science brings analytically uncontrolled ideas under control.[2] It provides the means of filtering out the distortions introduced by ideological commitment, personality structure and the imperfect perception of the holder of the vision. For that subset of visions (*a*) whose concepts may be defined by a set of mental or physical operations, (*b*) whose subject variation is capable of being reproduced or is recurrent under acceptably stable generating conditions, and (*c*) whose variation is observable in a manner relevant to the predictions of the vision, this body of methods can be used to gauge the usefulness of visions by testing against reality their responses to questions for which operations may be defined to obtain answers.[3]

This task of deriving operationally meaningful theorems from visions or models implies the need to deduce in a valid manner a set of predictions for which a set of operations may be specified which, actually or ideally, can test the correspondence of the predictions to empirical fact. If such operations are actually capable of performance, current scientific methods are applicable; if it is believed they are only potentially capable of use, current scientific methods are inapplicable, but future methods may be useful; if they are currently thought to be neither actually nor potentially capable of performance, the vision falls outside of that subset amenable to scientific method. According to this view, 'The function of a scientific hypothesis is to enable us to "predict" phenomena not yet observed, that is, to make statements about phenomena not yet observed that are (*a*) capable of being contradicted and (*b*) will not in fact be contradicted. . . .'[4]

It is possible to agree with the assertion of the operationalist that only those visions which yield such hypotheses are in an ultimate sense useful. Yet visions which are not themselves capable of rendering such services may perform an important, if prescientific, role in shaping other models which are able to do so. This book will be concerned largely with the efforts of economists to evolve models of economic variation as prescientific steps toward other visions which can yield meaningful theorems about reality. The most important factor militating against the analytical usefulness of these models is their 'generality' or ambition: indeed, it will be shown that the ability of a model to yield operational theorems declines rapidly to zero as the number of variables to be determined is increased, even when huge expenditures of ingenuity are invested in the manipulation of the model.

Some of these general models have been framed, it is true, to yield operational theorems, and these, too, will be analyzed. But for the most part the models that we shall study will contain the intuitive, largely untested hypotheses of the economist's domain, many of which are actually or potentially untestable for one reason or another. These models are cumbersome, nonmanipulable, nonoperational, analytically valueless in the direct senses defined above. Why, then, should economists dwell upon the gargantuan visions at all? Why not allow the glacial age of operationalism to kill off these impressive but ill-adapted dinosaurs to the advantage of simpler and analytically more viable organisms? Of what value are general economic models which have never left the stage of prescientific vision and never will, at least in unaltered form?

II THE GENERAL MODEL AND INTERDEPENDENCE

Before attempting answers to these questions, we must discuss at greater length the more distinctive features of the 'general' model. Any formal model requires a fundamental distinction between those phenomena whose variation is assumed to be frozen by analytical controls and those phenomena whose values or attributes are predicted by the model. The former – termed the *data* of the model – lie beyond the analytical ambition of the model, in the sense that they are determining rather than determined. The latter – the *variables* of the model – are determined in value by the constraining interaction of the data and the interrelationships that exist among the data and variables by virtue of the natural or behavioral assumptions of the model, or both. A set of data, relationships and variables, therefore, constitutes a vision or model.

A solution to the model may be *specific* or *nonspecific*. A specific solution defines a *state* of the model: it is the set of values for the variables which results from a specific set of values for the data. A nonspecific solution to the model defines the set of values for the variables for any and every allowable set of values for the data.

Both the interrelationships among data and variables which are part of the structure of the model and the nonspecific solutions of the model may take the form of mathematical functions and should be distinguished from one another. For example, let us anticipate some of the analysis of chapter 2 by stating that the conventional demand function is a nonspecific solution to a model of consumer choice, and relates the amount purchased of a commodity to the data of the model: prices of goods, consumer wherewithal and consumer preferences. The structural interrelationships of the model relating variables and data are derived, in the usual interpretation, from the assumptions of maximizing behavior and of spending neither less nor more than one's income.

The 'generalness' or ambition of such models must be defined in terms of the extent of conventionally defined 'economic' variation that is included in the variable category of the model as opposed to the amount placed in the data category. The greater the proportion of the whole field of variation 'internal' to the model, i.e. whose values are to be determined by the model, the more 'general' the model. The distinctive feature of the models developed and treated in this book, therefore, is that they are among the more ambitious visions of economic variation.

Another common structural feature of such models – apart from the fact that the number of variables is relatively large – is contained in the interrelationships of the model. More specifically, it is the complicated mutual interdependence among the *variables* of the model inherent in these relationships. This is a frequent mark of the general model, the source of much of the advantage it may possess over its less ambitious rivals, but the wellspring of its analytical impotence as well.

In strict logic, general economic systems need not display this interdependence of the variables that we have termed characteristic. Imagine a model concerned with three variables X_i, $i = 1, 2, 3$, and a three-element set of data, Q_k, $k = 1, 2, 3$. If the model related the variables directly to the data in a nonspecific solution without benefit of interrelationships among the variables, we should have:

$$X_i = F_i(Q_1, Q_2, Q_3), \qquad i = 1, 2, 3 \tag{6.1}$$

While this model might have great explanatory power, its several parts do not cohere. Each of the equations in 6.1 may be solved independently of the other two: for each equation there exists an unimpeded flow of causation from the data to a single variable. The system 6.1 is decomposable into its individual equation components, indicating a complete absence of interdependence among the variables. It is, however, a general system within our meaning of the term.

Suppose, however, that the structural interrelationships of the model were not as written in 6.1, but were instead:

$$X_1 = F_1(Q_k), \qquad k = 1, 2, 3 \tag{6.2.1}$$

$$X_2 = F_2(X_1; Q_k) \tag{6.2.2}$$

$$X_3 = F_3(X_1, X_2; Q_k) \tag{6.2.3}$$

We have used a semicolon to separate variables from data in the functional interrelationships, a convention we shall follow throughout this book. Equation 6.2.1 reveals the same quality of separability from the system as that possessed by every equation of 6.1. Given the data set, it may be solved directly for X_1. However, from 6.2.2 it is clear that in addition to the set of data given it is necessary to know the value of X_1 to obtain the solution value of X_2. For the first time we have encountered an eddy in the flow of 'causation' from Q_k to a variable in addition to the direct impact via their determination of X_1. We note, however, that while 6.2.1 and 6.2.2 must be solved together to obtain the value of X_2, 6.2.3 need not enter the computations. To obtain the solution value of X_3, on the other hand, it is necessary to know X_2 and X_1, in addition to the data set. The flow of 'causation' from Q_k to X_3 is even more complicated than that to X_2: in addition to the 'direct' impact of Q_k upon X_3, there is a first indirect effect from X_1, a second indirect effect from the indirect effect of X_1 upon X_2, and a third indirect effect upon X_3 from the presence in 6.2.3 of X_2.

System 6.2 is characterized by a type of interdependence among the variables termed 'triangular' from the configuration of the X_i's on the right-hand side of the equations. The peculiarities of this form of interdependence are well illustrated if the system is 'shocked' at a solution by a slight change in the value of one of the data (say Q_1). If we assume that the equations in system 6.2 are continuous and differentiable, at least in the neighborhood of a specific solution, the changes induced in the solution values of the variables may be obtained from the system in 6.3

$$\begin{bmatrix} 1 & 0 & 0 \\ -F_2^1 & 1 & 0 \\ -F_3^1 & -F_3^2 & 1 \end{bmatrix} \begin{bmatrix} \dfrac{dX_1}{dQ_1} \\ \dfrac{dX_2}{dQ_1} \\ \dfrac{dX_3}{dQ_1} \end{bmatrix} = \begin{bmatrix} F_1^{Q_1} \\ F_2^{Q_1} \\ F_3^{Q_1} \end{bmatrix} \tag{6.3}$$

where $F_i^j = \dfrac{\partial F_i}{\partial X_j}$, i and $j = 1, \ldots, 3$, and $F_i^{Q_1} = \dfrac{\partial F_i}{\partial Q_1}$, $i = 1, \ldots, 3$. These latter notations will be used throughout the book for partial derivatives. We may abbreviate 6.3 to

$$[D_2][dV] = [dK] \tag{6.4}$$

where the subscript 2 identifies [D] as derived from system (6.2), and the symbols V and K are used to denote vectors of variables and constants respectively.

The 'triangularity' of the interdependence of the variables is revealed by the existence of zeros above the main diagonal of $[D_2]$. This implies, for example, that when Q_1 changes slightly, the impact of the induced changes upon X_2 and X_3 need not be taken into account when computing the induced change in X_1. This degree of interdependence contrasts with that revealed by $[D_1]$, the matrix obtained when 6.1 is differentiated with respect to Q_1, which is the identity matrix, all of whose off-diagonal elements are zero and whose diagonal terms are unity. This constellation reflects the need to consider only direct impacts of the change in Q_1 on every variable in the system.

Consider a third model, obtained from 6.2 by substituting the following equation for 6.2.1:

$$X_1 = F_1(X_2, X_3; Q_k), \qquad k = 1, 2, 3 \tag{6.5}$$

If this system were shocked in the same manner as the previous two, 6.4 would become

$$[D_5][dV] = [dK] \tag{6.6}$$

where the subscript identifies the matrix as derived from the model containing 6.5. Further,

$$[D_5] = \begin{bmatrix} 1 & -F_1^2 & -F_1^3 \\ -F_2^1 & 1 & 0 \\ -F_3^1 & -F_3^2 & 1 \end{bmatrix} \tag{6.7}$$

We have not reached what we might term 'perfect interdependence' in the sense that every variable is directly dependent upon every other variable. Nevertheless, we have achieved a degree of interdependence which is qualitatively different from that of system 6.2, for it is now impossible to solve any subset of the three equations with fewer than three equations. This is true because even though X_2 does not depend directly upon X_3, it does depend upon X_1, which in turn is a function of X_3. Therefore, by virtue of indirect relationships, X_2 cannot be obtained independently of X_3: every variable in the system is linked, directly or indirectly or both, to every other variable.

We may express this property of such systems as 6.6 as that of 'indecomposability'. For linear systems of that type, it is an easy matter to establish whether the model does or does not possess it. If it does $[D_5]$ will be indecomposable. Let s symbolize a row in such a matrix, t a column, and a_{st} the element in the sth row and t^{th} column. For a square matrix to be indecomposable it is necessary and sufficient that there exist at least one chain of nonzero, nondiagonal elements of the form $a_{sk} \rightarrow a_{k1} \rightarrow a_{1m} \rightarrow a_{mn} \rightarrow \ldots a_{qs}$ with n or more elements, n being the number of rows (or columns) of the matrix, and with each subscript occuring at least once in the chain.[5] For example, from $[D_5]$ it is possible to form the chain $a_{21} \rightarrow a_{13} \rightarrow a_{31} \rightarrow a_{12}$. It is not possible to find similar chains for $[D_2]$ or, of course, $[D_1]$.

Further, let us form yet another model by substituting the following equation for 6.2.1.

$$X_1 = F_1(X_2; Q_k), \qquad k = 1, 2, 3 \tag{6.8}$$

Then, if the system were disturbed in the standard manner, we should obtain:

$$[D_8][dV] = [dK] \tag{6.9}$$

where

$$[D_8] = \begin{bmatrix} 1 & -F_1^2 & 0 \\ -F_2^1 & 1 & 0 \\ -F_3^1 & -F_3^2 & 1 \end{bmatrix} \tag{6.10}$$

In this model, at least one variable acts directly upon every variable, but $[D_8]$ fails to meet the conditions for indecomposable matrices. This is reflected in the ability to solve for $\frac{dX_1}{dQ_1}$ and $\frac{dX_2}{dQ_1}$ independently of $\frac{dX_3}{dQ_1}$.[6] Thus, merely including at least one variable in each structural relationship of the model is necessary but not sufficient to guarantee indecomposable interdependence.

Lastly, of course, we treat the case of perfect interdependence, where every variable is affected directly by every other variable in the system:

$$X_i = F_i(X_j; Q_k), \qquad i = 1, 2, 3; j \neq i; k = 1, 2, 3 \tag{6.11}$$

For a displacement of equilibrium similar to the previous cases, a system is obtained whose $[D]$ is the following:

$$[D_{11}] = \begin{bmatrix} 1 & -F_1^2 & -F_1^3 \\ -F_2^1 & 1 & -F_2^3 \\ -F_3^1 & -F_3^2 & 1 \end{bmatrix} \tag{6.12}$$

It, of course, meets the conditions of indecomposability.

Systems 6.1, 6.2, 6.5, 6.8, and 6.11, as well as the systems yielding $[D_1]$, $[D_2]$, $[D_5]$, $[D_8]$, and $[D_{11}]$, are general systems in our sense. Yet that feature of general systems which we have signalized as being of greatest interest is not possessed by system 6.1 or by that model from which $[D_1]$ was derived; is found in incomplete forms in systems 6.2 and the $[D_2]$-model, and in system 6.8 and the $[D_8]$-system; is contained only imperfectly in system 6.5 and the $[D_5]$-model; and is found in full measure only in system 6.11 and the $[D_{11}]$-system.

Modern general economic systems tend to stress the indecomposable variety of interdependence, of either the imperfect or perfect form. Models with this characteristic are believed to be more reflective of the circularity in the paths of association among realistic economic variables in modern complex economies. As an index of the strength of this characteristic, in the productive sector of the United States economy Chenery and Watanabe found that on the average, in 1947, about 43 cents of every dollar of gross output were fed back to intermediate uses by industries rather than devoted to final uses.[7]

III THE USES OF GENERAL MODELS

Let us now return to the questions posed at the end of Section I. If, as we shall demonstrate below, most of the potentiality for yielding operational theorems is sacrificed in general models on the altar of interdependence, of what use are they to the economist?

A first benefit to a field of study derives from the 'limiting' general model in that field: the closed model. Suppose the economist were to pose this question to himself: 'If I were to essay the construction of a model whose variables encompassed every phenomenon I believe to be primarily and essentially economic in nature, what variables would I include in the model? Further, what explicit data would be required in my model, and what physical-behavioral assumptions might I employ, to determine this "complete" set of variables?' In the case of the individual economist, of course, a good deal of arbitrariness would exist in his definition of the relevant field of variation, and, for economists collectively, the element of inertial convention in a consensus, if it could be achieved, would be a large one. Moreover, these conventional limits to the 'field of economics' would be expected to change with the consensus, generation by generation. The English classical economists included the level of population as an economic variable in their systems, employing as they did a Malthusian population response to income change to determine changes in it. Today, few economists would regard population levels as predominantly economic in nature.

To the extent that a field of study can achieve a consensus on the 'proper bounds' for its investigators, its workers gain a sense of coherence and unity not otherwise attainable. We shall term any model which seeks to cope with this *whole* field of variation a 'closed' model, and we shall regard it as the limiting general model. In economics, for all practical purposes, the general models of Walras, Wicksell and, to a lesser extent, Pareto, are closed models. They pushed the boundaries of economics outward to include within their variable-sets those phenomena economists are concerned with even today, and to exclude what are still regarded as predominantly extra-economic phenomena. In so doing, the authors of these models helped to determine important components in the visions of economists who followed them; their systems provide maps of the whole economic terrain, within which models of lesser ambition obtain position and scale.

A second advantage may spring from the existence of the closed model. In Schumpeter's terms,[8] the ambitious closed model may be the 'Magna Charta' of a field of study, establishing it as a separate and self-contained body of variation, fully determinate from the data and interrelationships of the model. If, through the use of the techniques discussed in chapter 9, it were possible to prove that with acceptable restrictions upon data and assumptions a solution to the closed model existed, then workers on less ambitious models would be assured that they *need* not go beyond the range of variation in the larger model for their data and assumptions.

Thirdly, the treatment of general models stresses the importance of prescientific acts in the search to understand variation. The vision provides the theorems which, if subject to testing by the scientific method, are grist for the mill of learning. But, in the social sciences, we are often doomed to shape our visions of reality with little hope of testing derived theorems objectively. In our discussion of the operational approach to model construction we did not touch upon the kinds of evidence acceptable in the validation of meaningful hypotheses. In the social sciences much of our evidence is of the 'indirect' type – uncontrolled observation or introspective judgment.[9] If one asserts that he prefers coffee to tea by more than he prefers tea to cocoa, whether or not this is an operational hypothesis hinges upon the hearer's judgment of its intuitive plausibility and his willingness to accept such 'evidence'.

We should be romanticizing the fund of economic 'knowledge' were we to imply that effort in the field was relentlessly subjecting our visions to objective, direct tests against reality, continuously winnowing out those models not found useful, accepting provisionally those whose theorems were not contradicted by such tests, and retesting these latter mercilessly. Rather, the greater part of our analytical frameworks are untested visions – untested, that is, by nonintrospective, controlled methods.

Yet we continue to use these frameworks in analysis and the framing of policy. For example, the vision of the market economy as a vast complex of closely interconnected markets whose prices and quantities are *greatly* interdependent is still just that: a set of hypotheses that have not been subjected to any thorough going objective testing, but so plausible in the light of indirect observation as to make it inconceivable that they are not useful.

We should be foolish, therefore, to ignore general models, including closed models, on the grounds that they yield few operational theorems. Pedagogically, if not analytically, their role is quite important. To teach that everything depends upon everything else may be a simple lesson, but it is an impressive one. Models designed to trace out the vastly complex interdependencies of economic sector with economic sector may yield few testable theorems, even if indirect evidence be accepted in validation, but they may throw brilliant light into dark corners. Consider, for example, the Slutsky-Hicks substitution-income effects concerning leisure and income and their implications for tax policy.

Finally, there exists the possibility that in the future these general systems will be able to yield operational theorems by virtue of some Newtonian breakthrough in the area of multivariable analysis, or, less fancifully, by simplifications of the interrelationships among the variables. While we are pessimistic about the future analytic uses of general systems with much ambition, the possibility of developments in these directions must not be dismissed out of hand. But merely to mention such potentialities leads one to ask what those analytic deficiencies are? Why must such models be approached with the expectation that few testable theorems will be derivable from them? To answer this question, as we shall do in Section V, we must devote some time to an extensive examination of the more important properties of general models.

IV THE PROPERTIES OF GENERAL MODELS

It is clear from the discussion that the potential value of any vision lies in the insights it yields of realistic economic variation. Ideally, but as we have argued in Section III not necessarily, these insights should be operationally meaningful theorems, placing qualitative or quantitative restrictions upon empirical variation, and capable of validation by direct evidence. While we may employ a system to instill broader visions of economic process, and while such employments are of great importance, it is admitted that economic systems usually must yield theorems with empirical content before they can be of continuing analytical utility to the policymaker. Pareto and Osorio spoke of three separate levels of such knowledge concerning economic interdependence: (*1*) knowledge of its existence, (*2*) qualitative knowledge of the movements of variables with changes in the data, and (*3*) quantitative knowledge of such movements.[10] While each level of knowledge contributes useful insights, each of the three levels is on a different plane of importance. Failing the ability to rise from the first level, a system can assume pedagogical or 'background' significance, but can seldom attain an active analytical significance.

The reason for this inability does not spring solely from the inapplicability of the scientific method, as we have defined it, to such models. It is simply that sooner or later we must arrive at the ultimate aim of any program developed to cope with variation – prediction. The ability to yield predictive theorems – to rise above Pareto's first level – springs from the properties of the model. We shall discuss, in turn, a group of the more important properties of economic systems as they relate to the ability to obtain operational theorems from such systems.

Static and Dynamic Models

Because this distinction will be treated at length in chapter 8, we will limit the present introduction to a discussion of the types of theorems potentially derivable from models with these properties. The value of viewing the distinction from this angle is seen immediately in discussing the implications of a static system.

In the literature of economics, statics often seems to be treated as a framework for the study of equilibrium states in which motion has not only ceased to exist but in which change does not exist, even in a virtual sense.[11] In such treatments, dynamic properties are identified with fruitful analysis of change, characteristic of models that are not locked in time. The outlook developed in this chapter reveals that any economic model designed to place limits on variation in the real world must be able to cope with change. Of course, this does not mean that it must yield meaningful theorems about the *time process* of change; but it does imply that it must be able to relate a new state of the model to a change in the data of the model, and thereby to present the old and new states for comparison. Statics, therefore, to be fruitful, cannot imply merely the study of conditions holding at a specific solution to the system.

From this viewpoint, even though statics and dynamics are concerned fundamentally *with the structure of the interrelationships among variables and data of models*, a description of statics and dynamics which compares the nature of the theorems derivable from systems structured in both ways is most useful at this point. In a purist sense, both types of model preclude the derivation of theorems relating to virtual states which are not specific solutions of them. Let us assume such solutions exist in the following discussion.

The differences begin to emerge when such specific solutions are interpreted as 'positions of rest', i.e. as situations in which motion in the variables has ceased. Movement implies the existence of a medium through which it occurs, and therefore time is explicitly or implicitly present when states of models are interpreted in this sense. The fundamental difference between models characterized by these structures is that a specific solution to a static system yields a single solution vector (if the solution is unique), whereas a specific solution to a dynamic system is a set of such vectors linked in a path through time.[12] If we adopt the 'position-of-rest' interpretation of a static specific solution, a time dimension is built into the space which contains the solution, and we move that solution in a repetitive fashion 'parallel' to the time axis; on the other hand, the time dimension for the dynamic solution is integral to its representation, and the path of the variables through it need not be an unchanging one. In the dynamic case, no position of rest need exist in the solution, in that the variables need not approach limiting values as time goes on indefinitely. At this point in our consideration, however, let us assume that the dynamic systems with which we are concerned do yield specific solutions whose vectors do approach limiting values, at least 'in the long-run'.

A dynamic model contains the potential for the derivation of theorems concerning the values of the variables, or changes in those values, before the position of rest has been attained. It has this potential by virtue of the fact that these values are part of the specific soluton: 'solution' and 'position of rest' are not synonomous. A static system can yield theorems about the values of the variables only in a state of rest, or theorems about changes in the values of the variables only between two states of rest: 'solution' and 'position of rest' are synonomous.

It is not, then, the existence of time in a dynamic system and its nonexistence in a static system which is the distinguishing feature of the two types of models. Any realistic use of a static model must involve the interpretation of its solutions against time as a backdrop.[13]

Rather, the difference is reflected in the above distinction between potentially derivable theorems. This distinction keeps clear terms which distinguish the properties of a *model*, and, specifically, the functional interrelationships of a model (static, dynamic), from those used to specify the behavior of its specific solutions through time (stationary, nonstationary).[14]

A stationary solution to a model is a specific solution which does not change through time, while a nonstationary solution is a specific solution which does not approach a position of rest. We have seen that an unchanged static model always yields a stationary specific solution (if a solution exists), while a dynamic model may yield stationary or nonstationary solutions.[15]

Comparative statics is a *method* employing static models analytically by imposing changes upon the data of the model and tracing the changes imposed upon the specific solutions – the displacement of such solutions – by such data manipulation. It is useful to distinguish between changes in the data which are 'infinitesimal' in size and which lead to a new solution in the neighborhood of the old, and changes in the data which are finite in size. Similarly, comparative dynamics is a *method* of using dynamic models with the same implications as those in the case of statics.

Extremum and Nonextremum Properties

Given the underlying data of the model, within whose constraints decision making units are forced to operate, there remains the important task of specifying the motivation of these units. What are the criteria by which decisions among alternative strategies are made? Practically speaking, these assumptions of motivation for firms and consumers must be simple if they are to possess a workable generality. One of the most frequent and convenient of these assumptions in economic models is that decisions are taken as if the unit involved were attempting to reach an *extremum* point, i.e. a minimum or maximum. Such assumptions, together with the data of the model, may be combined to obtain the interrelationships among variables and data discussed in Section II.

In the case of microeconomic models these extremum strivings are frequently interpreted as attempts to maximize consumers' 'utility' and firms' profits. In general models, these assumptions are treated as occupying fully symmetrical roles: the consumer is assumed to act to maximize his subjective welfare, guided in his choices by considerations of more or less, subject to an expenditure constraint. At the same time, the firm is viewed as seeking to maximize profits, an objective magnitude. With the exception of the 'subjective' *vs.* 'objective' natures of the magnitudes, utility theory and production theory up to recent times stressed the essential similarity of these processes. However, modern theory does not view them as exact analogues. The modern concept of utility is that of a preference function which *defines* the consumer's tastes in the sense that, once derived, it allows the prediction of the consumer's choices. It does not purport that those choices were arrived at by the consumer in any extremum search, purposive or intuitive.

Thus, any 'maximization of utility' which we adopt will be merely the specification of one of the data of the model – the consumer's preferences – rather than an actual motivating drive of the consumer. It is even misleading to say that the consumer acts as *if* he were maximizing satisfactions, because the satisfactions cannot exist in any form for maximizing before they have been defined. That is to say, the consumer's 'utility' is maximized *by the economist* to find the alternative the consumer has indicated previously he would choose under the postulated conditions. The measure the model-builder has chosen is maximized because when this is done the consumer's tastes are defined. We may make the assumption in the case of the firm that it will always act to make the difference between revenues and costs a maximum, since both revenues and costs may be defined (at least ideally) independently of the state of mind of the decision-making unit.

Maximization and minimization techniques, therefore, like other components of our vision, are merely methodological conveniences whose value is measured in terms of usefulness in deriving the required interrelationships. Usefulness, in turn, is defined in terms of the insights yielded by the vision into real economic variation. When convenient, we may build choice into the consumer sector of the model without 'maximizing utility': it may be methodologically convenient to assume that under certain conditions the consumer will take a quantity of a good best predicted by other-than-maximization techniques. Whole systems may rest upon consumer choice which abstracts completely from maximization reasoning, as the models of Cassel or Keynes will illustrate.

The methodological convenience of extremum systems in framing models which will yield meaningful theorems goes beyond the initial derivation of interrelationships among the variables. So-called 'second-order conditions' can frequently place restrictions upon the shapes of these interrelationship functions, and these restrictions make an independent contribution to the derivation of such theorems. The appeal to other methods for obtaining the functional interrelationships of the model may well sacrifice these sources of restrictions on the system.

The importance of obtaining such restrictions in the search for operational theorems will be stressed below, but frequently outright assumptions to obtain them may be no more objectionable than assuming an extremum position has been attained and deriving the restrictions from that assumption. For example, if Keynes had obtained his consumption function from a prior maximization analysis he might have been able to derive a marginal propensity to consume of less than unity and more than zero as an implication of the maximum position. As it was, he had to assume in quite arbitrary fashion that it fell within this range of values. Neither method is 'right' or 'sophisticated', 'wrong' or 'naïve': the criterion is one of convenience, and in this simple case there seems little reason to choose one method over the other.

But in general models it is frequently true that such restrictions *can* be obtained from second-order extremum conditions while intuitive methods fail wholly to provide substitute restrictions. For example, as we shall show in chapter 2, the Slutsky-Hicks-Hotelling income-substitution effect analysis allows certain theorems to be framed from a general model because second-order maximization conditions are available to yield the sign of the substitution effect. Thus, we may rise from Pareto's first level of knowledge to the second level with these restrictions, while, in their absence, intuition would afford little justification for one assumption or the other about the relevant determinants' signs.

Equilibrium and Nonequilibrium Solutions

One of the most fruitful of the many economic adoptions from the field of mechanics is the concept of 'economic equilibrium', or a specific solution characterized by a state of balance between opposed forces acting upon economic variables.[16] As we shall employ the term we shall not make it synonomous with 'solution' but rather shall use it to distinguish one type of solution. Thus, for example, we shall not use the term to characterize a steady motion of a variable in a repetitive time path, as might emerge from a specific solution to a dynamic system, as we shall not interpret this type of behavior as reflecting a true equipoise of human motives and natural forces.[17] Therefore, given the existence of the forces described above, we use the term 'equilibrium' as synonomous with 'specific solution' for a static system, but to characterize *one type* of specific solution to a dynamic system, namely, one which attains a position of rest.

Our definition of economic equilibrium is that of Pareto:[18] a specific solution to an economic model describing a state of rest and resulting from the opposition between identifiable desires of men and obstacles to their fulfillment, including among these latter the desires of actors other than any one of the specific firms or consumers under consideration. A model constructed to obtain its solutions on the basis of such an opposition of identifiable forces is called an equilibrium model. Nonequilibrium systems are those whose solutions are not the results of such counteracting forces. For example, a model which predicts that tomorrow's price will be today's price plus a random element is a nonequilibrium system.

In our terminology, therefore, the equilibrium property is not a purely formal one: there must be identifiable and meaningful opposition of motives and/or forces tending toward a net balance as reflected in a position of rest before solutions are acceptable as equilibrium solutions. This is frequently a matter of the model-builder's choice of interpretation. For example, it will be shown in chapter 7 that the Leontief static open input–output model may be interpreted as an equilibrium or nonequilibrium system. The theorems derivable from either system differ considerably from those obtainable from the other, so that the property with which we choose to endow the model is not unimportant.

Since most of the models with which we shall deal are in fact equilibrium models, we shall frequently use the terms 'specific solution' and 'equilibrium' interchangeably where no danger of confusion exists. With this in mind we shall assert an interest in four properties of the 'equilibrium': (*1*) its existence, (*2*) its economic meaningfulness, (*3*) its uniqueness or nonuniqueness, and (*4*) its stability. We shall treat the first three of these properties in great detail in chapter 9, and the last in chapter 8. In this introduction, however, we must indicate the nature of our concern with these concepts.

The Existence of an Equilibrium

It is truistic to say that we are interested in knowing if a solution to a system exists, both in the specific and nonspecific senses of that term. We have indicated in Section III that a model may have value even if we can frame no operational hypotheses from it, although, of course, such usefulness must be of somewhat restricted scope. Similarly, a system can be of value even if it should be impossible to prove the existence of a solution to it. This is not to deny the value of such a proof. Merely to state its implications in a negative fashion – that the data and assumptions of the model will yield a sufficient number of interrelationships among the variables and data which are not mutually inconsistent – is to see that an existence proof is a valuable assurance of potential usefulness for any purpose and may be indispensable if serious efforts to frame fruitful hypotheses are to be made. For example, modern welfare economics must be framed within the confines of a closed system for maximum usefulness, and therefore must concern itself with the properties of a general equilibrium. It is of utmost importance to be assured that the competitive equilibrium which is being described so airily can be shown to exist. In the area of normative economics such proofs would seem to be of extreme importance.[19]

But there is no need to be unbending on the need for this proof, particularly for non-normative uses of models. Quite apart from possible pedagogical uses of the model to fill in the canvas of interdependence, for which an existence proof is only a luxury, it is quite legitimate for the economist to *assume* that the equilibrium exists for acceptable restrictions upon the data of his model which will not disturb the theorems of his model. That such as procedure is less satisfactory intellectually does not establish its illegitimacy nor the nonapplicability of operational theorems derived in the absence of existence-of-equilibrium proofs.

The Economic Meaningfulness of the Equilibrium

Given the existence of an equilibrium solution to a model, it is of some interest to know the nature of it. In particular, negative prices and quantities in general have no economic content, although in certain situations they may have. Therefore, the equilibrium values of the variables, if they exist, must in addition meet any restrictions necessary to guarantee that they do not grossly violate economic reality.

The Uniqueness of the Equilibrium

Still another property of an equilibrium which holds an interest for the analyst is the number of states of the model consistent with a given set of data i.e., how large is the set of specific solutions for a given data set? It is tempting to believe that only one state exists in the real economy for any given constellation of data. But no a *priori* argument can establish that if a solution to the real economy exists it is necessarily unique.[20] We have asserted that to be analytically useful a model must assert hypotheses which restrict the expected variation in the real economy under the stated conditions of the model. It is corollary that the number of solutions to a model consistent with a given data set must be small enough in number to yield useful restrictions, and in many cases this number need not necessarily be unity.

The Stability of the Equilibrium

The stability of an equilibrium solution is concerned with the time path followed by the values of the variables of a model when an equilibrium set of values is displaced by a change in the initial position of the model. Note that the stability analysis is fundamentally one of an equilibrium, i.e. a position of rest, not of a system. And note also that because it is concerned with a time path for values of variables in a model whose initial position has been changed, we are dealing with comparative dynamics. We begin with a dynamic model whose specific solution yields a position of rest, disturb this position of rest by a slight or large displacement which is equivalent to changing the initial position of the model, and find whether the specific solution germane to the new data set will lead to a reattainment of the old position of rest via the same or a new path, whether a new position of rest will be attained, or whether no position of rest will be attained.

A most important point is that a stability analysis is of necessity a dynamic analysis: it can be accomplished only with explicit information about the behavior *through time* of a static or dynamic model in disequilibrium. The values of the variables must at all times be consistent with the data set ruling yet be capable of being in a position of non-rest. From our analysis of statics it is clear that the specific solutions will change instantaneously to equilibrium levels consistent with the nonspecific solutions, and the time paths of these operations will be sharp movements from one equilibrium point to another. In the strictest sense, it is impossible to get the system out of equilibrium since it yields information only about equilibrium states of the model, i.e. positions of rest.

This last point is not true of a dynamic system; among the data of such a system is its initial position, or the values of the variables in some arbitrarily denominated period 0. Therefore, in a dynamic system which yields a position of rest, it is possible to specify the set of values desired as the initial disequilibrium position relative to the first set of data, change the initial position of the system to those values without altering any other characteristic of the system, and compare the subsequent time path of the specific solution with that obtained for the original initial position.

Therefore, a dynamic system must be specified to answer questions essentially dynamic. If the stability properties of an equilibrium solution to a static system are to be investigated, an explicit supplementary dynamic model, specifying the mode of reaction through time to disequilibrium values of the system, must be given, as Samuelson has shown so well.[21] This has not prevented economists of the rank of Walras, Pareto and Marshall from discussing the 'stability' properties of static equilibria, as coming chapters will reveal. They were, of course, implicitly assuming certain dynamic properties for their models in disequilibrium situations which in strict logic could not be brought about. Walras spent pages trying to show, from the slopes of his interrelationship functions, that the time path of his general equilibrium system's values always approached a position of rest corresponding to the specific solution dictated by the given data and assumptions of the static model. Wicksell went so far as to combine the problems of the existence of an equilibrium solution and the stability of an equilibrium, holding that if 'Walrasian stability conditions' (to be discussed below) do not hold at an equilibrium position, it is not truly an equilibrium, but merely a provisional equality of supply and demand.[22]

Of course, it is useless to quarrel over wholly definitional matters, but it is also confusing to move, however deftly, between pure static, quasi-static, dynamic, and pseudo-dynamic models in the manner of Marshall, Walras, and Pareto, or to mix problems concerning the existence of a solution to a static model and of the stability of some dynamic counterpart as Wicksell advocates. It is not to deny these giants of economic analysis their just claims to greatness for insightful analyses to adopt a more puristic approach to the questions of stability. We shall deny the meaningfulness of the term 'stability' to the solution of a wholly static system with no explicitly defined dynamic interrelations among the variables for positions of disequilibrium. Instead of adopting the usual terminology in referring to static 'stability conditions', to be discussed in the following section, we shall term them 'slope conditions' to emphasize the distinction. When they may become true stability conditions is a question we shall pose of our dynamic analysis in chapter 8.

V THE DERIVATION OF OPERATIONAL THEOREMS

We may now return to the question posed at the end of Section III: wherein lie the difficulties of deriving useful theorems from general models? With the discussion of the more basic properties of economic systems completed, it is now possible to illustrate how these properties are employed in the construction and manipulation of the systems. We shall begin the analysis with a nongeneral system – a partial model – and then complicate it by increasing its ambition.

The Partial Model

From an analysis of consumer choice which we shall leave for consideration in chapter 2 and of producers' behavior which will be studied in chapter 3, the respective nonspecific solutions for the consumption of good Y_1 (symbolized X_1) and for the output of good Y_1 (symbolized $\bar{X}_1$) are obtained as functions of the data P_1 and Z, or the price of good Y_1 and consumer's incomes respectively. For consumers

$$X_1 = F_1(P_1, Z) \tag{6.13}$$

and for producers

$$\bar{X}_1 = F_2(P_1) \tag{6.14}$$

Each of these equations represents a set of nonspecific solutions to static, extremum, equilibrium models, each model independent of the others. In 6.13, for every specification of a price and an income level, consumers will desire to purchase the specific equilibrium value X_1 yielded by that function; similarly, for producers, and for a given price, 6.14 yields an equilibrium desire-to-supply. Merely from the nonspecific solutions of 6.13 and 6.14 we have no way of knowing how general or nongeneral were the models for which they are solutions, but from the paucity of the data we might suspect they were quite limited in ambition.

Let us step up the level of ambition of the analysis and attempt to determine P_1; that is, remove P_1 from the status of datum and convert it to a variable. Then, 6.13 and 6.14, which are nonspecific solutions in lesser models, now become the interrelationships among variables and datum (Z) of the grander model. As a condition for the solution of the new model, it will be necessary for consumers and producers simultaneously to be willing to accept the *status quo*. Therefore, equilibrium must be characterized by the condition or interrelationship:

$$X_1^\circ = \bar{X}_1^\circ \equiv X^\circ \tag{6.15}$$

The superscript $^\circ$ indicates that this relationship need characterize the variables only for their solution values in the grand model. This notational procedure has been adopted from Pareto and will be used throughout this book to distinguish relationships holding only in the solution from relationships which may hold out of equilibrium. Thus, in general, for an allowable value of P_1 (given Z) picked at random, 6.13 will dictate a value for X_1 and 6.14 for $\bar{X}_1$, but they will not be equal, and 6.15 will not characterize these values. For simplicity, since the amount demanded in equilibrium must equal the amount supplied, we have defined the symbol X° to denote both equilibrium values.

If our interest lies only in the solution values of X_1, $\bar{X}_1$, and P_1, (i.e., X_1°, $\bar{X}_1^\circ$, P_1°) it is possible to reduce the three-equation system (plus one definitional identity) of 6.13, 6.14, and 6.15 to two equations:

$$X^\circ = F_1(P_1; Z) \tag{6.16.1}$$

$$X^\circ = F_2(P_1) \tag{6.16.2}$$

There are two unknowns in the model, X° and P_1°; one datum, Z; and two interrelationships 6.16.1 and 6.16.2 derived from certain psychological and physical constraints, motivational assumptions not presented explicitly, and conditions 6.15. Further, under certain assumptions, it is possible to solve 6.16 for the nonspecific solutions to the system which would be of the form:

$$X^{\circ} = G_1(Z) \tag{6.17.1}$$

$$P_1^{\circ} = G_2(Z) \tag{6.17.2}$$

In Section IV we have indicated the properties of the equations of 6.17 that hold an interest for us: do they exist, are they meaningful, do they yield unique values, are these equilibrium values stable? But suppose we could answer 'yes' to each of these questions. Of what value is this simplest of economic visions contained in 6.16?

A first point to be made is that such values must be derivable from equations 6.16 in their nonstatistical form; in 999 cases out of 1000 it will prove impossible for one reason or another to fit these equations to realistic data and manipulate the statistical functions. The model-builder's task is to obtain insights into the reality which cannot be poured into the empty containers he has constructed. With this limitation upon his opportunities, the economist may well approach his tools with a good deal of skepticism. If – and this so far by assumption – he can merely assert that for any value of consumers' income stated by an inquiring policymaker there will exist a unique, meaningful, stable price-quantity which will satisfy all actors in the market, he would be wise to expect an 'if so, so what' reaction from the policymaker. To be sure, the body which the model-builder must build upon the skeleton of 6.16 will be a useful representation of economic process for the policymaker: the isolation of the strategic variables involved, the market concept and a description of its operation, the notion of competition, and so forth, have their educating role. But the broadness of the education makes it of limited applicability to some specific problems in which the policy man might be vitally concerned.

For example, suppose the nation became involved in a limited war, and the policymaker anticipated a relatively small rise in consumers' income. Should he expect the quantity of Y_1 exchanged (equals the amount demanded in equilibrium equals the amount supplied in equilibrium) to rise or fall as a result? What of the equilibrium price of Y_1: will it rise or fall? Granted an inability to give quantitative answers to these questions and thus attain Pareto's third level of knowledge, is it possible to deduce from 6.16 together with any other information we may obtain about the system theorems which yield such qualitative predictions? If not, we must fall yet another level to Pareto's first stage and its implied 'broad education'.

One manner of proceeding is to start from an equilibrium solution of 6.16, displace it slightly by a small increase in income, $dZ > 0$, and, by assuming both functions are continuous and possess at least first derivatives, solve for $\frac{dX^\circ}{dZ}$ and $\frac{dP^\circ}{dZ}$.[23] This manner of equilibrium displacement has the disadvantage of yielding theorems for very small changes in the data, and is not applicable for finite changes: it can yield information about the values of the variables in a model only for two solutions which are quite close to each other. Also, its use implicitly assumes that in neither the beginning nor terminal equilibrium is the value of any variable zero. And, moreover, its use depends upon all of the interrelationships of the system being stated in the form of equalities rather than inequalities.

Despite all these limitations, the technique remains a valuable method of obtaining testable theorems from economic models, and we shall rely heavily upon it throughout the book, although we shall not neglect other and more recent techniques for deriving theorems by equilibrium displacement when one or more of the above conditions are not met. By differentiating 6.16 with respect to Z we obtain the following system:

$$\begin{bmatrix} 1 & -F_1^p \\ 1 & -F_2^p \end{bmatrix} \begin{bmatrix} \dfrac{dX^\circ}{dZ} \\ \dfrac{dP^\circ}{dZ} \end{bmatrix} = \begin{bmatrix} F_1^z \\ 0 \end{bmatrix} \qquad (6.18)$$

Note, for the sake of interest, that the matrix of the system is indecomposable, indicating that the changes in the equilibrium quantities and prices are interdependent.

Because this simple system is the first of many equilibrium displacement systems we shall present, it will be useful to spend a little time in interpreting 6.18 in an economic manner. The two equations of 6.18 indicate that if consumer income is changed slightly, there will be two independent paths of influence upon the variables of the system. Along the first path (the demand function) the change in the quantity exchanged from the first to the second equilibrium position will equal the direct impact of the income change on that quantity (F_1^z) plus the indirect impact equal to the change in equilibrium price times the response of the demand function to a small change in price in this neighborhood. Similarly, along the second path (the supply function) the change in equilibrium quantity must equal the direct impact of the change in income on the quantity via the supply curve (in this case zero) plus the indirect impact of the equilibrium price change times the response of quantity supplied to price in this neighborhood of the supply function. The values of the variables change in response to the influences flowing along those paths until a set of values is achieved which lies on *every* path of influence in the system.

The terms F_j^i are the various 'slopes' of the interrelationships in the system. Imagine a three-dimensional space with P along the x-axis, Z along the y-axis, and X_1 or $\bar{X}_1$ along the z-axis, and the demand and supply functions of 6.13 and 6.14 represented as three-dimensional functions in that space. For every value of Z, under our ideal assumptions, these functions will have an intersection at $X°$ and $P°$. Then, choose a Z arbitrarily on the y-axis, and move to the corresponding point ($P°$, $X°$). Now, if we want to get to another equilibrium point $(P°, X°)'$, we can do so only by moving along the demand and supply functions. If $(P°, X°)'$ is very close to $(P°, X°)$, then the slopes of the demand function in the P and Z directions at $(P°, X°)$, or F_1^p and F_1^z, and of the supply function at the same point in the same directions, or F_2^p and 0, must be taken into account in the desired movement.

Let us solve system 6.18 for $\frac{dX°}{dZ}$, to obtain

$$\frac{dX°}{dZ} = F_1^z \left(\frac{F_2^p}{F_2^p - F_1^p} \right) \tag{6.19}$$

We know the direction of impact of income change upon the demand function: dZ is assumed positive, as is F_1^z. Therefore, the direction of change of $X°$ with a slight rise in income depends upon the signs and relative magnitudes of the slopes of the demand and supply functions in the P-direction at $(P°, X°)$. Similarly, if we solve for $\frac{dP°}{dZ}$ we obtain from 6.18

$$\frac{dP°}{dZ} = F_1^z \left[\frac{1}{F_2^p - F_1^p} \right] \tag{6.20}$$

In this case, also, the sign of the change of equilibrium price depends upon the signs and magnitudes of the slopes of the demand and supply functions in the *P*-direction, although if the sign of the difference of the slopes is known the sign of the term can be determined – which is not true of $\frac{dX^\circ}{dZ}$, for which the sign of F_2^p is also needed.

Thus, the ability to answer the questions posed hinges upon knowledge of these slopes of the interrelationship functions – their absolute values, their relative values, their signs, the signs or values of their differences, and so forth. It is impossible to progress further in the derivation of theorems unless restrictions may be placed upon these slopes. In the recognition of this need at this simplest level of model building is the explanation of much of the work that lies ahead for more complicated models. What are the paths to follow to gain this needed information? In our work three such alternatives will be explored:

1. The Postulational Method

It is possible merely to assume information about the slopes in the neighborhood of the equilibrium. For example, we might assume that the slope of the demand function in the *P*-direction is negative and the slope of the supply function in the same direction is positive. On the basis of these assumptions we may evaluate the expressions in 6.19 and 6.20 to obtain unambiguous predictions: $\frac{dX^\circ}{dZ}$ will be positive, $\frac{dP^\circ}{dZ}$ will be positive. These are theorems potentially testable against reality and of some real value to the policymaker. Of course, the assumptions may prove to be incorrect, but at least the theorems derived from the assumptions are potentially capable of validation. A less powerful assumption about these slopes is that $F_2^p - F_1^p$ is positive in the relevant neighborhood. With this information, $\frac{dP^\circ}{dZ}$ is predicted to be positive, but the sign of $\frac{dX^\circ}{dZ}$ is ambiguous in the absence of specific information about the sign of F_2^p.

2. The Extremum Method

In the discussion of extremum models in Section IV it was indicated that their use to derive the needed interrelationships of the model frequently yielded a further dividend to the analysis by facilitating the derivation of operational theorems. We may now illustrate this point. If the demand function was indeed derived by 'maximizing satisfactions', and if in fact maxima have been attained, certain characteristics of the utility surfaces where they touch the budget planes must hold true, for, by assumption, true maxima positions have been reached in the beginning and the terminal equilibrium points. With the implied characteristics of the utility surface in this neighborhood, for example, plus another assumption we may deduce the negative slope of the demand function in the P-direction. Similar maximization analyses permit the deduction of positive P-slopes for the supply function. The methods will be examined in greater detail chapters 2 and 3, but at this point this brief indication suffices.

For the simple model we are now treating, there seems little choice on grounds of analytical nicety between the postulational and extremum methods, when both are possible and yield equivalent information. We need not aspire to deal with models of much greater ambition before the first method, relying as it does upon intuitive guidance to the slope characteristics, becomes much less satisfactory, and, for all practical purposes, unusable.

3. *The Stability Method*

Frequently, particularly for models in whose construction extremum reasoning plays no explicit part, information about these slopes is obtained from an explicit stability analysis. That is, the equilibrium is *assumed* stable, and, from the conditions which must hold for this to be true, information is obtained limiting the values of the slopes. In most 'stability' analysis of partial static models depicting price determination, the reasoning follows one of two routes:

(a) Walrasian slope conditions If price rises above the equilibrium point to a slight extent, and if excess demand $(X_1 - \bar{X}_1)$ becomes negative, competition among the sellers will push price back toward its equilibrium value. Similarly, if price falls below the equilibrium, and if excess demand becomes positive, competition on the demand side will force price upwards toward the equilibrium. For small movements of price around an equilibrium, it is necessary and sufficient for stability of equilibrium that the difference between the slope of the demand curve and supply curve $(F_1^p - F_2^p)$ (which is the slope of the excess demand curve) in the neighborhood of equilibrium be negative.

(b) Marshallian slope conditions Marshall tended to view the market adjustment in terms of prices adjusting to quantity movements, interested as he was in longer-term economic movements. Consequently, he framed his slope conditions on the inverse functions of those we have used in 6.13 and 6.14, defining price as a function of quantities demanded and supplied respectively. Following convention, we should graph P on the z-axis instead of X_1 and $\bar{X}_1$, and these latter on the x-axis. We should then speak of the slopes of these new functions in the X_1 or $\bar{X}_1$, direction, and these slopes would be the reciprocals of F_1^p and F_2^p i.e., $\frac{1}{F_1^p}$ and $\frac{1}{F_2^p}$.

Then, if the quantity on the market fell below the equilibrium quantity, the price at which buyers would take this quantity off the market (demand price) would differ from the price which would just induce suppliers to bring it to the market (supply price). If excess demand price (demand price minus supply price) is positive in this case, the profit motive will lead suppliers to increase their outputs and move back to equilibrium. Also, in the neighborhood of equilibrium, if the quantity on the market rose above the equilibrium level, and if excess demand price were negative, suppliers would be led to reduce output and return to equilibrium. Therefore, it is necessary and sufficient for the stability of an equilibrium for small movements in its neighborhood for the slope of the excess demand price function $\left(F_1^p - F_2^p\right)$ to be negative.

The objection to these types of analyses is not that they are incorrect in a fundamental sense, but rather that they are implicitly expanding an explicit static model by the addition of dynamic relations. This procedure can be misleading in two important respects. First, the very fact that Walras and Marshall arrived at two different sets of slope conditions is an indication that their implicit dynamic analyses were quite different interpretations of the reactions of actors in a market to nonequilibrium states. If these interpretations are allowed to remain implicit, the differences may remain unnoticed and undiscussed. If one were wholly unrealistic, or acceptable only for certain well-defined market structures but not for others, then adoption of the wrong slope-conditions could yield incorrect operational theorems. The Walrasian outlook views demanders and suppliers as acting quickly upon price via competition whenever they cannot buy and sell the desired amounts. The Marshallian outlook puts the major emphasis for adjustment upon suppliers who are viewed as adjusting output to profit opportunities, while consumers tend to be passive. These are two different pictures of reality and they spring from meaningful differences in the construction of the dynamic models.

Second, whenever results are obtained implicitly rather than explicitly, there is a danger when more complicated models are constructed that these *results* will be generalized rather than the *methods*, with error resulting to a greater or lesser degree. In the case of stability analysis this has in fact occurred, in that Hicks's static stability conditions (slope conditions) for general models do not in general correspond to those derived by generalizing the implicit dynamic models of the partial static system.

It is fruitful, therefore, for the matter in hand as well as for the more complicated analysis to follow in chapters 2 and 8, to derive these slope conditions as stability conditions from explicit dynamic models. In the case of Walrasian adjustment, the dynamic interrelationship which must be added to 6.16 is the following:

$$\frac{dP}{dt} \equiv \dot{P} = G(X_1 - \bar{X}_1), \qquad G' > 0,\ G(0) = 0 \tag{6.21}$$

That is, the time rate of change of price in the market is some function of the excess demand in the market such that when excess demand changes, the rate of change of price moves in the same direction, and when excess demand is zero price remains at rest through time. We may simplify 6.21 by the use of Maclaurin's variant of Taylor's series: we retain the linear portion of the expression and neglect all terms with derivatives of higher order than the first.[24] Then 6.21 is approximated by

$$\dot{P} = G(0) + G'(X_1 - \bar{X}_1) = G'(X_1 - \bar{X}_1) \tag{6.22}$$

where G' is the first derivative of the function evaluated at the point of equilibrium. Further, let us define excess demand,

$$E = X_1 - \bar{X}_1 \tag{6.23}$$

so that 6.22 becomes

$$\dot{P} = G'E \tag{6.24}$$

But we may approximate E by Taylor's series. Since

$$E = F_1(P; Z) - F_2(P) = H(P; Z) \tag{6.25}$$

we may approximate E at a point P very close to P° by the following expression, if we again ignore all expressions containing derivatives of H higher than the first order:

$$E = H(P^\circ; Z) + H'(P - P^\circ) = 0 + H'(P - P^\circ) \tag{6.26}$$

where H' is the derivative of H at P°. Substituting from 6.26 we arrive at the final linear approximation of $\dot{P}$ from 6.24:

$$\dot{P} = G'H'P - G'H'P^\circ \tag{6.27}$$

This expression is a first order nonhomogeneous linear differential equation whose solution is

$$P_t = (P - P^\circ)e^{G'E't} + P \tag{6.28}$$

where P_t is the price at time-point t, P is the initial nonequilibrium price, t is the number of time-points away from $t = 0$, and e is the base of natural logarithms. But, from 6.25

$$E' = F_1^P - F_2^P \tag{6.29}$$

and substituting

$$P_t = (P - P^\circ)e^{G'(F_1^P - F_2^P)t} + P^\circ \tag{6.30}$$

Now, from 6.21, $G' > 0$ and, of course, $t > 0$. For stability it is necessary and sufficient that $P_t \rightarrow P^\circ$ as $t \rightarrow \infty$, which will occur if and only if the first term approaches zero. For this to occur, however, it is necessary and sufficient that

$$(F_1^P - F_2^P) < 0 \tag{6.31}$$

Therefore, if we assume the equilibrium of our model to be stable in the sense of Walras, a market reaction function of the type in the dynamic equation 6.21 is implicit, with the set of corrective market adjustments represented by it, and 6.31 is implied.[25]

If the Marshallian market adjustment had been adopted, the dynamic equation counterpart to 6.21 would be

$$\frac{dX}{dt} \equiv \dot{X} = M(P_d - P_s), \qquad M' > 0, M(0) = 0 \tag{6.32}$$

where P_d is demand price and P_s supply price. Approximation by Maclaurin's and Taylor's series and solution of the differential equation would yield

$$X_t = (X - X^\circ)e^{M'\left(\frac{1}{F_1^p} - \frac{1}{F_2^p}\right)t} + X^\circ \tag{6.33}$$

where X is the initial disequilibrium quantity. For stability of the equilibrium it is required that $X_t \rightarrow X^\circ$ as $t \rightarrow \infty$, for which it is necessary and sufficient that

$$\left(\frac{1}{F_1^p} - \frac{1}{F_2^p}\right) < 0 \tag{6.34}$$

Therefore, if stability in the sense of Marshall is assumed, 6.34 is implied.

In the evaluation of 6.19 and 6.20 it may be seen that with the information gained from 6.31 it is possible to predict that $\frac{dP^\circ}{dZ}$ will be positive but it is impossible to obtain a theorem concerning the sign of $\frac{dX^\circ}{dZ}$, since this would require a knowledge of the sign of F_2^p which is not given. On the other hand, 6.34 yields much more ambiguous information. It says that $\frac{dX^\circ}{dZ} < 0$ if the demand curve is negatively sloped in the relevant neighborhood, $\frac{dX^\circ}{dZ} > 0$ if the demand curve is positively sloped in that neighborhood. Also, $\frac{dP^\circ}{dZ}$ will take the sign $(-\ F_1^p\ F_2^p)$, which may be positive or negative.

The General Model

The experiment in the derivation of operational theorems from the partial model was a relatively successful one, for in spite of the triviality of the results, it points the way to the treatment of more realistic variation with more sophisticated models. Specifically, let us introduce the phenomenon of market interdependence into our model by increasing the level of its ambition slightly. We shall merely introduce a second market into the analysis, in which good Y_2 is sold at P_2. From prior maximization analyses we derive the set of demand and supply functions given below:

$$\begin{aligned} X_1 &= F_1(P_1, P_2; Z) \\ \bar{X}_1 &= F_2(P_1, P_2) \\ X_2 &= F_3(P_1, P_2; Z) \\ \bar{X}_2 &= F_4(P_1, P_2) \end{aligned} \tag{6.35}$$

where the additional symbols are simple extensions of those of our partial model. Further, it is required that in equilibrium excess demand be zero in both markets:

$$\begin{aligned} X_1 &= \bar{X}_1 \equiv X_1^\circ \\ X_2 &= \bar{X}_2 \equiv X_2^\circ \end{aligned} \tag{6.36}$$

Substituting from 6.36 into 6.35 yields the simpler system,

$$\begin{aligned} X_1^\circ &= F_1(P_1, P_2; Z) \\ X_1^\circ &= F_2(P_1, P_2) \\ X_2^\circ &= F_3(P_1, P_2; Z) \\ X_2^\circ &= F_4(P_1, P_2) \end{aligned} \tag{6.37}$$

Under assumptions wholly similar to those made about the functions of the simpler model, and beginning from an assumed equilibrium position, it is possible to displace the equilibrium by a very small increase in consumer income. The changes in equilibrium values of the variables will then be given by the following system of equations:

$$\begin{bmatrix} 1 & 0 & -F_1^1 & -F_1^2 \\ 1 & 0 & -F_2^1 & -F_2^2 \\ 0 & 1 & -F_3^1 & -F_3^2 \\ 0 & 1 & -F_4^1 & -F_4^2 \end{bmatrix} \begin{bmatrix} \dfrac{dX_1^\circ}{dZ} \\ \dfrac{dX_2^\circ}{dZ} \\ \dfrac{dP_1^\circ}{dZ} \\ \dfrac{dP_2^\circ}{dZ} \end{bmatrix} = \begin{bmatrix} F_1^z \\ 0 \\ F_3^z \\ 0 \end{bmatrix} \tag{6.38}$$

or

$$[D][dV] = [dK] \tag{6.39}$$

Then, for example,

$$\frac{dX_1^\circ}{dZ} = F_1^z \frac{|D_{11}|}{|D|} + F_3^z \frac{|D_{31}|}{|D|} \tag{6.40}$$

$$\frac{dP_1^\circ}{dZ} = F_1^z \frac{|D_{13}|}{|D|} + F_3^z \frac{|D_{33}|}{|D|} \tag{6.41}$$

where $|D|$ is the determinant formed from $[D]$ and $|D_{st}|$ is the minor of $|D|$ formed from the matrix obtained by crossing out the sth row and t^{th} column of $[D]$. Unless otherwise indicated, throughout the book such expressions as $|M_{st}|$ will denote minors formed in this way. Since the F_1^z and F_3^z terms are assumed to be positive, the signs of the expressions in 6.40 and 6.41 will depend upon the absolute values of F_1^z and F_3^z and the algebraic values of the ratios of determinants.

In order to get some idea of the complexity of these expressions as compared with their counterparts in the partial model, given in 6.19 and 6.20, they are expanded in full below:

$$\frac{dX_1^\circ}{dZ} = F_1^z \left[\frac{1}{1 - \left(\dfrac{F_1^1 F_3^2 + F_1^2 F_4^1 - F_1^2 F_3^1 - F_1^1 F_4^2}{F_2^1 F_3^2 + F_2^2 F_4^1 - F_2^2 F_3^1 - F_2^1 F_4^2} \right)} \right] \tag{6.42}$$

$$+ F_3^z \left[\frac{F_1^1 F_2^2 - F_1^2 F_2^1}{(F_2^1 F_3^2 + F_2^2 F_4^1 - F_2^2 F_3^1 - F_2^1 F_4^2) - (F_1^1 F_3^2 + F_1^2 F_4^1 - F_1^2 F_3^1 - F_1^1 F_4^2)} \right]$$

(43)
$$\frac{dP_1^\circ}{dZ} = F_1^z$$

$$\left[\frac{F_3^2 - F_4^2}{(F_2^1 F_3^2 + F_2^2 F_4^1 - F_2^2 F_3^1 - F_2^1 F_4^2) - (F_1^1 F_3^2 + F_1^2 F_4^1 - F_1^2 F_3^1 - F_1^1 F_4^2)} \right]$$

$$+ F^z_3 \left[\frac{F^2_1 - F^2_2}{(F^1_2 F^2_3 + F^2_2 F^1_4 - F^2_2 F^1_3 - F^1_2 F^2_4) - (F^1_1 F^2_3 + F^2_1 F^1_4 - F^2_1 F^1_3 - F^1_1 F^2_4)}\right]$$

If the reader is impressed with the complexity of these two equations, as compared with the neatness of their counterparts in the partial model, this clumsy method of expression will have served its purpose. It is a basic lesson in the difficulties of manipulating general systems.

By increasing the ambition of the model to the extent of including only one more market, the number of slopes involved in the expressions to be evaluated rose from 3 to 10. No longer will information about the sign of the difference between two slopes yield an operational theorem concerning the sign of $\frac{dP^\circ_1}{dZ}$, or this information plus the sign of one of the slopes suffice to obtain the sign of $\frac{dX^\circ_1}{dZ}$. The information now required includes knowledge about the absolute magnitudes as well as signs of all of these slopes. The problem of deriving theorems in this manner from the model has increased so sharply in difficulty as to have become qualitatively different from that of the partial model.

An increase in the number of equations of a model by one results in an increase in the elements of $[D]$ of $2n + 1$, where n is the previous number of equations. Against this rampant rise in complexity must be counterposed the ability of the economist to extend the methods employed to gain information about these slopes in the partial analysis to the general analysis. To date, as we shall demonstrate in the remainder of the book, this ability has proved quite limited. For obvious reasons less reliance can be put upon the postulational method: who would desire to depend upon his intuition to restrict the signs of the denominator in the terms of 6.42 and 6.43? On the other hand, by appealing to the underlying maximization processes of this model, discoveries made by Slutsky, Hotelling, and Hicks and Allen do allow some restrictions to be placed upon some of the determinants in 6.40 and 6.41, but only seldom are they sufficient to yield completely unambiguous signs for the derivatives. Lastly, if one is willing to accept 'static stability' analysis, other restrictions are possible, while if one insists upon dynamic stability conditions less information is forthcoming in general.

The lesson is a simple one. An increase in the ambition of a vision implies the removal of phenomena from the data category and their placement in the variable category. This requires an increase in the number of equations and a much larger increase in the number of 'slope terms' which are the elements in the expressions that must be evaluated if operational theorems are to be forthcoming from the model. In the absence of specific statistical information about these slopes, the usefulness of models declines very quickly as more variables are added. With the present ability to restrict these slopes by plausible assumption one may conclude with reasonable confidence: *either the economist has a simple theory or he has no theory at all!*

VI THE CLOSED GENERAL MODEL

If these lessons are apparent in the analysis of a general model with so little ambition, the limited analytical applicability of the closed model – the 'general equilibrium model' of economic theory – with its ultimate pretensions will come as no shock. As Leontief writes:

> The comprehensive general equilibrium theories developed by the Lausanne school are essentially grand classificatory taxonomic devices. They define and stratify the basic types of economic phenomena and describe their mutual interrelationships in so general a form that very few, if any, operational propositions concerning the measurable properties of specific economic systems can actually be derived from them. Neither Walras nor Pareto made any stipulations concerning the possible empirical implementation of their explanatory schemes. The fact that at least one of the two – Pareto, with his well-known statistical 'law of income distribution' – gave much thought to the role of measurement in economics seems to indicate that such reticence was not accidental. The theory of general interdependence – no less than the classical 'theory of value' to which it was generically related despite the many differences – was not intended by its creators to become a tool of factual analysis. . . .[26]

As we have emphasized in Section III, the more ambitious general model is most useful in its portrayal of interdependence, as Professor Leontief stresses for the case of the closed general model. The phenomenon of interdependence had been pointed out by John Stuart Mill in his statement of the demand schedule:

> the quantity demanded is not a fixed quantity, even at the same time and place; it varies according to the value; if the thing is cheap, there is usually a demand for more of it than when it is dear. The demand, therefore, partly depends on the value. But it was before laid down that the value depends on the demand. From this contradiction how shall we extricate ourselves? How solve the paradox, of two things, each depending on the other?[27]

But the clearest vision of interdependence as a phenomenon making the economy a coherent whole, and the one which influenced most the thinking of Walras, was that of Cournot. His mathematical background facilitated an understanding of the simultaneity of determination of the values of variables and forestalled him from viewing it as paradox, as Mill had. At the same time it led him to see the complexity of general analysis:

> the economic system is a whole of which all parts are connected and react on each other. . . . It would seem, therefore, as if, for a complete and rigorous solution of the problems relative to some parts of the economic system, it were indispensable to take the entire system into consideration. But this would surpass the powers of mathematical analysis and of our practical methods of calculation, even if the values of all the constants could be assigned to them numerically.[28]

Although both Walras and Pareto stressed the need for general economic analysis, the interdependence of economic variables, and the mutuality of such interdependence in the sense of Section II, Pareto stressed interdependence more consistently than Walras. Even though no one can dissociate the name of Walras from the contribution that his general equilibrium model made in *depicting* mutual interdependence, nor deny that the latter lay at the heart of the former, Walras was quite ready to become a Marshallian pragmatist:

> Theoretically, all the unknowns of the economic problem depend upon all the equations of economic equilibrium. Nevertheless, even from the static, theoretical viewpoint, it is indeed permissible to consider certain of these unknowns as more especially dependent upon the equations which were introduced with them into the problem in order to determine them. . . .[29]

Pareto, more rigid in his insistence upon a consistent recognition of mutual interdependence, took Walras to task for lapses from the outlook in the first sentence of the above quotation. He viewed the *whole* purpose of employing mathematics in economics to originate in the need to build models capable of coping with that phenomenon.[30] Indeed, some contemporary economists regard the concept of full economic equilibrium in a mutually interdependent system as a Paretian contribution rather than a Walrasian one.[31] And, unlike Walras, who always sought to convey the impression that the equations of the system had some potential for statistical fitting, Pareto was willing to base the value of the closed economic model on its 'broader' lessons of interdependence:

> if the conditions that we have just enumerated are not able to serve us practically in numerical calculations of quantity and price, they are the unique method known at present to arrive at a notion of the manner in which prices and quantities vary, or more exactly, in a general fashion, to know how economic equilibrium is produced.[32]

The desirability of deriving greater insights into economic reality from the closed model than those implicit at this 'first level' of knowledge was made quite clear by Walras. He believed it to be a responsibility of the theorist to go beyond the mere derivation of models in order to strive for operational theorems of the type we sought in the less ambitious systems of 6.42 and 6.43. He termed these 'laws of variation' and of them he wrote:

> The theorist has the right to assume that the underlying price determinants are invariant over the period he has chosen to use in his formulation of the law of equilibrium prices. But, once this formulation has been completed, it is his duty to remember that the forces that underlie prices are by their nature variable, and consequently he must formulate the law of variation of equilibrium prices. . . .[33]

Walras made little progress down this tortuous route. The real beginning of the derivation of theorems from general systems in a workmanlike manner dates from Pareto. We shall study Pareto's initiating efforts in chapter 2, but, from our analysis of Section V, we should not be surprised if his success was somewhat limited. Walras and Pareto left the theory of general equilibrium resting on the first level of Pareto's classification of knowledge, with clearly perceived goals on the second level, but with only the beginnings of efforts to achieve them.

It was this lack of achievement in general equilibrium theory which led U. Ricci to remark that 'The whole construction gives the effect of an enchanted palace which delights the fantasy but does not help to solve problems of housing. Or, to drop the metaphor, the theory remains abstract and intangible.'[34] And Hicks would write in the 1930s:

> Now the reason for this sterility of the Walrasian system is largely, I believe, that he did not go on to work out the laws of change for his system of General Equilibrium. He could tell what conditions must be satisfied by the prices established with given resources and given preferences; but he did not explain what would happen if tastes or resources changed.[35]

And, as a last example of this criticism, we may quote from Samuelson on the general aims of analysis:

> It is hardly enough, however, to show that under certain conditions we can name enough relations (equations) to determine the values of our unknowns. It is important that our analysis be developed in such terms that we are aided in determining how our variables change qualitatively or quantitatively with changes in explicit data. . . .[36]

It is against this body of criticism and the orientation spelled out at such length in this chapter that the contributions of Slutsky, Hotelling, Hicks, Keynes, Lange, Samuelson, Patinkin and others will be developed and viewed. What progress has been made in increasing the analytical power of general systems by working within the neo-classical framework of Walras-Pareto-Fisher-Wicksell to facilitate the derivation of operational theorems? Even the study of minor progress is of great value, for the vision of interdependence is too valuable to be thrown up before the most exhaustive efforts to derive information about the relevant slopes have been expended. We shall devote part II to a succession of general models in the neo-classical tradition, building successively and cumulatively through exchange models (chapter 2), production models (chapter 3), investment models (chapter 4), and money models (chapter 5). We shall end chapter 5 with a closed general model incorporating each of these aspects of a realistic economy.

In chapter 7 an attempt will be made to extend this neo-classical analysis by building in an explicit spatial dimension; in chapter 9 restrictions upon the data and the assumptions will be imposed which will assure that certain of these neo-classical systems will have solutions, and the conditions for uniqueness of solutions for some of these systems will be explored. In chapter 8 general dynamic systems will be constructed with a primary goal being that of seeing what true dynamic stability analysis can contribute to the derivation of operational theorems from such systems as those contained in part II.

The number of these general systems from which we can derive any operational theorems is relatively small, and the empirical content of these hypotheses quite disappointing in scope and restrictiveness. Having attained the high ground, however, it will be possible to discern three lower-lying alternative paths of analytical promise:

(a) The Simplification of the Interrelationships of the General Model One route is that of retaining the generalness of the model by approximating the interrelationships in it with simple functions whose slopes may be fitted by simple statistical techniques. In this way not only may operational hypotheses be forthcoming but they may achieve Pareto's third level of knowledge: quantitative predictions. The rather recent techniques of linear economics are relevant under this heading, and we shall study the Leontief static input–output system in chapter 6, the dynamic system in chapter 8, linear programming or activity analysis in chapters 3, 6, and 7, and von Neumann's growth model in chapter 8.

(b) The Elimination of Variables from the Model by Aggregation Yet another path open to the model-builder in his strivings for meaningful systems is to reduce the number of equations in his model by aggregation techniques. Instead of n goods in the consumer's action space let there be one: 'consumption goods'. By such Gordian-knot approaches, the matrices of the systems can be drastically reduced in the number of slope-elements they contain, and, as we have seen, this is a large leap forward toward the theorems sought. Except in the trivial sense that every model is aggregated to some extent, we shall deal with only one aggregate general system in this book: the Keynesian model. It will be viewed as a contribution to

the stream of neo-classical general equilibrium theory and will be analyzed in chapter 5.

(c) The Elimination of Variables from the model by Converting Them to Data Lastly, we may tread the path of partial analysis, turning our backs (however reluctantly) upon the mutual interdependence of economic variables. Certainly we should be following a most persuasive leader:

> It was not till the seventeenth century that the physical sciences appreciated the full importance of the fact that when several causes act together and mutually affect one another, then each cause produces two classes of effects: those which are direct, and those which result indirectly from the influence exerted by it on other causes: for indeed these direct and indirect effects are apt to become so intricately interwoven that they can by no means be disentangled. So far the results are negative: they seem to indicate that the task of following out and understanding the combined action of several causes, which are in various degrees mutually interdependent, is beyond the power of human faculty. But a way out of the difficulty was found, chiefly under the guidance of Leibnitz and Newton. An epoch-making process of reasoning showed that, though the indirect effects might grow cumulatively, and ere long become considerable, yet at first they would be very small indeed relatively to the direct effects. Hence it was concluded that a study of the tendency to change, resulting from each of several disturbing causes, might be made the starting point for a broad study of the influence of several factors acting together. This principle is the foundation of the victory of analytical methods in many fields of science. . . .[37]

It is but a short step from this position to the assertion of the negligibility of the indirect effects and a full-blown partial analysis.[38] This is equivalent to substituting zeros for the off-diagonal terms in the matrix of a general system depicting an equilibrium displacement, or to eliminating interdependence of variables from the original interrelationships by placing most of the variables among the data and holding them constant.

Certainly this has proved to be a productive analytical method – indeed, to date, few have been the economic laws or hypotheses which have not been derived from its use. So great is its predominance in Anglo-American economics that we need not consider it in this work. Rather, we shall somewhat idealistically follow less well-known paths of a great deal less analytical promise, but with the compensating reward of leading to heights from which new perspectives and a panoramic view of the whole economic terrain may be attained.

Notes

1. The importance of the 'vision' with which an investigator approaches a set of phenomena, and its essentially prescientific character, has been stressed by J. A. Schumpeter in his later writings. His concept of the vision is a different one than the one used above in that his includes only the phenomena chosen for study and the *initial* view of their interrelationships. See J. Schumpeter [30], pp. 345–59, especially pp. 350–1, and J. Schumpeter [31], pp. 41–3, 561–3.
2. Cf. J. Schumpeter [30], p. 356.
3. Operationalism, as enunciated by P. W. Bridgman, stresses (*a*) the identity of the scientific concept and the set of physical or mental operations which must be specified to determine it, and (*b*) the exclusive concern of science (social as well as physical) with questions which are operationally meaningful in the sense that a set of operations may be specified to obtain an answer. See P. W. Bridgman [2], pp. 1–32.

   We do not presume to judge the issues in the debate on operationalism, but merely assert that the outlook, as presented to some extent in P. Samuelson [26], pp. 3–20, is a valuable guide in framing the vision, interpreting such concepts as 'utility', and testing the usefulness of a model's theorems.

   Our use of the terms 'operationally meaningful', 'meaningful', and 'operational theorem', in the final analysis, is less a blind adoption of Bridgman's outlook and more the purely practical desire to avoid the term 'testable theorem', since this implies a *current* ability to verify hypotheses and does not connote our acceptance of methods which yield theorems only potentially testable. We seek, that is, to avoid such propositions as Cotton Mather's hypothesis: 'That there is a Devil is a thing Doubted by none but such as are under the influence of the Devil' – which may be true but is not subject to validation by scientific techniques. The operational outlook seems a convenient way to avoid dealing with such questions.
4. M. Friedman and L. Savage [10], p. 465.
5. Cf. W. Evans [9], pp. 60–2.
6. This may be confirmed by solving for, say $\frac{dX_1}{dQ_1}$ by Cramer's method. It will be seen that in both numerator and denominator determinants all products which include F_3^1, F_3^2, or $F_1^{Q_1}$ are uniformly zero.
7. H. Chenery and P. Clark [5], pp. 202–11, 230. However, studies of the

United States, Japanese, Italian and Norwegian economies discussed in these pages disclose broad patterns of triangularity.

8. J. Schumpeter [29], vol. I, pp. 41–2.
9. For an interesting discussion of the relevance of indirect evidence see M. Friedman and L. Savage [10], pp. 466–7.
10. V. Pareto [20], vol. II, p. 6, and A. Osorio [18], pp. 158–60.
11. See, for example, Cassel's identification of statics with the explanation of how the data determine equilibrium values and dynamics with equilibrium displacements. G. Cassel [4], pp. 153–4.

    This same view is expressed explicitly and implicitly by Walras, in [33], pp. 319 and 377–81. In Walras, however, dynamics has the cast of a *medley* of comparative statics adjustments occurring with time lags. See also H. Moore [17], pp. 21–3, for a repetition of this Walrasian outlook.

    To J. B. Clark, statics dealt with economic systems whose factor quantities (among other data) were given and invariable, while dynamics coped with the laws of change of factors, technology and wants. Robbins has shown a possibility of this interpretation in J. S. Mill's writings as well. See L. Robbins [24], pp. 202–4.

    Pareto, while holding views on the nature of statics which appear confused from the modern view, did depart from his contemporaries in defining two types of dynamics, the first of which is clearly comparative statics in modern terminology and the second of which is true dynamics: V. Pareto [21], p. 147.

    Thus, in classical and neo-classical economics, there was a widespread tendency to consider comparative statics a form of dynamics, with the implied treatment of statics discussed above.
12. A dynamic system may have as solutions more than one path converging to the same or different positions of rest, if indeed one or more such paths do converge to such limiting positions.
13. As E. Schneider writes: 'It is essential to understand that in modern theory "statics" and "dynamics" refer to a particular *mode of treatment or type of analysis* of the phenomena observed, while the adjectives "stationary" and "changing" describe the actual economic phenomena. *A static or dynamic theory is a particular kind of explanation of economic phenomena; and, indeed, stationary and changing phenomena can be submitted either to a static or to a dynamic analysis*' E. Schneider [28], p. 185. Quoted by permission.

    The neo-classical theorist who came closest to this distinction between static-dynamic vs. stationary-nonstationary systems and solutions respectively was Pareto, as pointed out in note 11. His distinctions were not clear however: for example, he identifies a state of rest in a static system with a stationary state in a dynamic system. In effect, he defines static equilibrium in dynamic terms. See V. Pareto [21], pp. 153–4.

    Not all modern theorists would acquiesce in our interpretation of statics and dynamics. For example, M. Allais writes: 'An economic system may be said to be static when all of its elements – production, consumption, prices, and so forth – do not introduce time. In the contrary case, the system may be said to be dynamic.' M. Allais [1], p. 28.

14. Cf. P. Samuelson [26], pp. 311–33.
15. This statement holds even when the static model includes expectations of prices in the future among the data. In Week 0 a specific solution will include the magnitudes of variables in the future *n* weeks. If the same set of expectations and other data is repeated in Week 1, the same specific solution will emerge in Week 1, and so forth.
16. Cf. H. Moore [17], pp. 19–23. For a discussion of the relationship of equilibrium in the social sciences to that in the physical sciences, and especially the analogous roles of motive and force in the respective areas, see F. Knight [13], pp. 161–85, and especially p. 163. To Knight's way of thinking, economics is properly concerned with equilibrium not as a pure state of rest but as a state of a 'process' in which the data are not truly constant but moving more slowly than the variables and enforcing a kind of moving tendency to a state of rest. See pp. 169–70. The relevance of this outlook for his capital theory will be made clear in chapter 4.
17. Cf. J. Sargan [27], p. 382.
18. V. Pareto [21], pp. 150, 155.
19. R. Dorfman, P. Samuelson and R. Solow, in R. Dorfman [7], pp. 350–1, take a more positive position on the need for an existence proof than the one developed here. They assert, among other things, that 'a system of equations whose assumptions do not guarantee the existence of a solution may fail to be a useful idealization of reality', and again that it is necessary to establish existence of equilibrium to be sure a model 'really captures what is really important about economic reality'.

    They themselves point out that the existence question hinges on a property of the model, not reality. Merely to assert that the complicated dynamic system which describes reality functions and therefore an evolutionary path exists as a solution does not imply that a general static system approximating it grossly must have a solution under all conditions. Indeed, the assumptions concerning continuity of functions and correspondences which must be made to guarantee the existence of equilibrium in such static model approximations to reality may be more accurately described as a retreat from reality, not a guarantee that it has been approached.
20. Indeed, as we shall show in chapter 9, the intuitive plausibility of the restrictions upon consumer preference functions for unique solutions to static approximations to reality is something less than compelling.
21. P. Samuelson [26], pp. 260–9. These dynamic supplementary models may be made to appear 'timeless' by assuming (*1*) the directions of movements of variables are known and (*2*) the rates of movement are so rapid as to be essentially instantaneous.
22. K. Wicksell [34], p. 87.
23. Since only one price exists in this simple system, we may omit the subscript 1 for simplicity in the following operations.
24. Less pretentiously, and more honestly, we shall linearize 6.21.
25. For an attempt to relate the dynamic market process to iterative processes for solving equations, see R. Goodwin [11].
26. W. Leontief [14], pp. 41–2. Quoted by permission.
27. J. S. Mill, [16], book III, ch. 2, section 3.

28. A. Cournot [6], p. 127. Quoted by permission of Miss Anne Ashley.
29. L. Walras [32], p. 289. Quoted by permission.
30. V. Pareto [21], p. 246, for the objection to Walras, and pp. 42–3, 146, 247, 557n, for the view on the role of mathematics in economics. See also V. Pareto [22], pp. 60–2.
31. See M. Brodsky [3], pp. 135–47, and L. Einaudi [8], pp. 14–16.
32. Pareto [21], p. 234. See also p. 665.
33. L. Walras [33], p. 146. Quoted by permission of Richard Irwin, Inc., George Allen & Unwin, Ltd., and the American Economic Association.
34. Quoted in H. Moore [17], p. 28. The article has been translated and published in U. Ricci, [23], and the quotation may be found on p. 20.
35. J. Hicks [12], p. 61. Quoted by permission. This also seems to be the burden of the charge by F. Oulès that the extreme concentration upon equilibrium conditions by both Walras and Pareto led them to neglect certain 'interdependences' in the system. See F. Oulès [19]. So many charges are made, however, many of them badly taken (e.g. Walras's and Pareto's adoption of the mathematical method *interfered* with their handling of interdependence systems), that it is difficult to discern the relative weight assigned each by Oulès.
36. P. Samuelson [26], p. 7. Quoted by permission. See also pp. 257–8.
37. A. Marshall [15], pp. 677–8. Quoted by permission.
38. H. Moore [17], pp. 24–5.

References

[1] Allais, M., *Economie et Intérêt*, Librairie des Publications Officielles, Paris (1947).
[2] Bridgman, P., *The Logic of Modern Physics*, Macmillan, New York (1927).
[3] Brodsky, M. and P. Rocher, *L'Economie Politique Mathématique*, R. Pichon et R. Durand-Auziers, Paris (1949).
[4] Cassel, G., *Theory of Social Economy*, Harcourt, Brace, New York (1924).
[5] Chenery, H., and P. Clark, *Interindustry Economics*, John Wiley & Sons, New York (1959).
[6] Cournot, A., *Researches into the Mathematical Principles of the Theory of Wealth*, Macmillan, New York (1927).
[7] Dorfman, R., P. Samuelson and R. Solow, *Linear Programming and Economic Analysis*, McGraw-Hill, New York (1958).
[8] Einaudi, L., 'Fifty Years of Italian Economic Thought: 1896–1946 Reminiscences', *International Economic Papers*, no. 5, Macmillan, New York (1955).
[9] Evans, W., 'Input–Output Computations', in T. Barna (ed.), *The Structural Interdependence of the Economy*, John Wiley & Sons, New York (1955).
[10] Friedman, M., and L. Savage, 'The Utility Analysis of Choices Involving Risk', *Journal of Political Economy*, 56 (1948), pp. 57–96.

[11] Goodwin, R., 'Iteration, Automatic Computers and Economic Dynamics', *Metroeconomica*, 3(1951), pp. 1–7.
[12] Hicks, J., 'Wages and Interest: The Dynamic Problem', *Economic Journal*, 45(1935), pp. 456–68.
[13] Knight, F. H., *Ethics and Competition*, Harper, New York (1935).
[14] Leontief, W. 'Input–Output Analysis and the General Equilibrium', in T. Barna (ed.), *The Structural Interdependence of the Economy*, John Wiley & Sons, New York (1955).
[15] Marshall, A., *Industry and Trade*, Macmillan, London (1919).
[16] Mill, J. S., *Principles of Political Economy*, D. Appleton, New York (1901).
[17] Moore, H., *Synthetic Economics*, Macmillan, New York (1929).
[18] Osorio, A., *Théorie de l'Echange*, Giard, Paris (1913).
[19] Oulès, F., 'Les Insuffisances Théoriques de la Doctrine Economique de la Première Ecole de Lausanne', *Metroeconomica*, 2 (1950), pp. 20–43, 134–71.
[20] Pareto, V., *Cours d'Economie Politique*, Rouge, Lausanne (1897).
[21] Pareto, V., *Manuel d'Economie Politique*, V. Giard & E. Brière, Paris (1909).
[22] Pareto, V., 'Mathematical Economics', in *International Economic Papers*, no. 5, Macmillan, New York (1955).
[23] Ricci, U., 'Pareto and Pure Economics', *Review of Economic Studies*, 1 (1933–4), pp. 3–21.
[24] Robbins, L., *An Essay on the Nature and Significance of Economic Science*, Macmillan, London (1935).
[25] Samuelson, P., 'Dynamics, Statics and the Stationary State', *Review of Economics and Statistics*, 25 (1943), pp. 58–68.
[26] Samuelson, P., *Foundations of Economic Analysis*, Harvard University Press, Cambridge, Mass. (1947).
[27] Sargan, J., 'The Instability of the Leontief Dynamic Model, *Econometrica*, 26 (1958), pp. 381–92.
[28] Schneider, E., *Pricing and Equilibrium*, Allen & Unwin, London (1952).
[29] Schumpeter, J., *Business Cycles*, I, McGraw-Hill, New York (1939).
[30] Schumpeter, J., 'Science and Ideology', *American Economic Review*, 39 (1949), pp. 345–59.
[31] Schumpeter, J., *History of Economic Analysis*, Oxford University Press, New York (1954).
[32] Walras, L., *Eléments d'Economie Politique Pure*, F. Rouge, Lausanne, 2nd edn (1889).
[33] Walras, L., *Elements of Pure Economics*, Allen & Unwin, London (1954).
[34] Wicksell, K., *Value, Capital and Rent*, Rinehart, New York (1954).

7 Oligopoly under Rivalrous Consonance: An Exploration of Phantom Objective Functions and some Algorithmic Considerations*

In economic theorizing economic agents are generally depicted as optimizing objective functions based on narrow self-interest, with global or social variable values emerging from the interaction of the individual optimization procedure. An interesting question is whether there exists a 'phantom' objective function whose optimization would yield the same solutions as the set of interacting optimizing decisions does. For example, to cite a famous case, in a purely competitive capitalistic general equilibrium model, is there a social welfare function whose optimization would yield the same solutions for the model as those obtained from solving the whole set of first-order conditions of the conventional model?

More practically, in the economic modelling of interactive, interdependent decision-making, occasions arise when the first-order necessary conditions for an optimum may acquire a behavioral meaning rather independent of the objective function that formally yields them. Consumers, for example, may confront decisions about spending increments of money and optimize through marginal adjustments that are never related to a utility function. Indeed, if a set of such functions is believed to represent the revealed motivations and

* Published originally in *Rivista Internazionale di Scienze Economiche e Commerciali*, 32(1985), pp. 393–411, and reproduced with permission.

actions of the agents, it may be difficult or impossible to integrate back to an objective function whose differentiation gives them as optimizing conditions. And, as in the case of the consumer, even when such an objective function can be found it may be meaningless in behavioral terms, being a wholly formal or 'phantom' objective function whose only existential relevance inheres in its partial derivatives.[1]

The problem arises also in the modelling of partial and general oligopolistic equilibrium, and this paper will develop its theoretical nature by using simple examples in that area.[2] However, I believe that the problem will be seen to exist in other areas of economic theory and the social sciences, where agents, through bounded rationality or the structure of the total framework within which they move, follow intuitively appealing modes of behaviour that may or may not be projected by definable objective functions.

This paper is a rather formal and ground-clearing exercise in the problems inherent in seeking an overarching objective function in an oligopolistic context. My purpose in the paper is to develop a general interpretation of the problem in oligopoly terms (section II) and to indicate the nature of an algorithm that permits its solution for specified structures (section III). I have included an analysis of the straightforward necessary conditions of constrained optimization as a brief preparatory step for the crippled optimization presentation (section I). Because this may be a useful pedagogical approach to interpreting in economic terms the standard Lagrangean conditions of economic equilibrium, it may possess some independent interest for economists. A brief conclusion will be found in section IV.

I have termed the field that includes the definition of the oligopoly problem, the theoretical solution and its interpretation, and the policy questions that inhere in the problem 'rivalrous consonance'. That title conveys the notion that mature oligopolistic industries contain a mixture of rivalry and co-operation that must be explicitly recognized in their modelling. The algorithmic means of reaching solutions to such models I have called 'crippled optimization', because each rival must restrain or 'cripple' his own search for maximum profits by a concern for competitors' profits. In a transcendent, global sense, however, a constrained optimum is sought on a single implied objective function. It is frequently non-derivable and is a behaviorally meaningless 'phantom' objective function. But if capable of derivation, the phantom function would be most useful for direct solutions to what in its absence must be interpreted and solved iteratively as interacting optimizations of sets of objective functions. Also, its isolation may yield insights into the nature of the interactive process itself.

I AN INTERPRETATION OF A CONSTRAINED OPTIMUM

Consider the following classical joint-profit maximizing differentiated duopoly model, complicated somewhat by constraints. Two firms seek to maximize their joint profits noncollusively subject to *maximum* market share constraints, where their sales functions are approximated linearly by

$$d_i = a_i - b_{ii}p_i + b_{ij}p_j,\ i, j = 1, 2,\ i \neq j. \tag{7.1}$$

We define m_i as the maximum share of industry output acceptable to firm i in view of its worry about anti-trust prosecution, p_i as price, and c_i as (constant) average (and marginal) cost. We assume that each firm sets its price and supplies sufficient output to satisfy the amount demanded, and further that both firms are 'whole-hearted' joint-profit maximizers (subject to meeting their market share constraints). That is, each firm treats a dollar of its rival's profits as equivalent to a dollar of its own, and in its price decisions acts in this light.[3] Finally, to assure the existence of a feasible region and a solution in which both firms produce, we assume that $m_i, m_j\ \varepsilon\ ]0, 1[$, and that $m_1 + m_j \geq 1$.

The initial model of duopolistic equilibrium is to find prices $P^0 = [p_1^0, p_2^0]$ that

$$\max Z = \sum_{i=1}^{2} (p_i - c_i)\, d_i \tag{7.2}$$

subject to

$$L_{i1} \equiv d_i - (m_i/(1 - m_i))\, d_j \leq 0,\ i = 1, 2,\ i \neq j, \tag{7.3}$$

and

$$L_{i2} \equiv p_i \geqslant 0,\ i = 1, 2, \tag{7.4}$$

where d_i is defined in 7.1.

To solve the problem we form the Lagrangean

$$L\,(P,\lambda) = Z + \sum_{i=1}^{2} \lambda_{i1}\, L_{i1} - \sum_{i=1}^{2} \lambda_{i2}\, L_{i2}. \tag{7.5}$$

Given second-order differentiability of the objective and constraint functions, the necessary conditions for a local maximum are the following:

$$\nabla Z + \sum_{i=1}^{2} \lambda_{i1} \nabla L_{i1} - \sum_{i=1}^{2} \lambda_{i2} \nabla L_{i2} = 0 \tag{7.6.1}$$

$$\lambda_{ij}\, L_{ij} = 0 \tag{7.6.2}$$

$$\lambda_{ij} \geq 0 \tag{7.6.3}$$

$$L_{ij} \leq 0 \tag{7.6.4}$$

$$Y^T\, \nabla^2 L\,(P^0, \lambda^0)\, Y \leq 0, \tag{7.6.5}$$

where ∇ denotes a gradient vector, Y is any vector orthogonal to all the constraints that bind as equalities at the solution $[P^0, \lambda^0]$, and $\nabla^2 L$ is the Hessian of L evaluated at the same point. The linearity of the constraints guarantees the existence of a finite λ (equivalently, the Kuhn-Tucker Constraint Qualification holds at any local solution).[4]

Consider ∇Z in detail:

$$\nabla Z = \begin{Bmatrix} \{a_1 - 2b_{11}\, p_1 + b_{12}\, p_2 + b_{11}\, c_1\} + \{b_{21}\, (p_2 - c_2\} \\ \{a_2 - 2b_{22}\, p_2 + b_{21}\, p_1 + b_{22}\, c_2\} + \{b_{12}\, (p_1 - c_1\} \end{Bmatrix}$$

$$= \begin{Bmatrix} \{OC_1\} + \{RC_1\} \\ \{OC_2\} + \{RC_2\} \end{Bmatrix}, \tag{7.7}$$

where OC is the 'own-profit' component of the firm's marginal conditions and RC is the 'rival's-profit' component. Assume for the moment that $\lambda = 0$: then 7.6.1 and 7.6.2 state that for an effectively unconstrained joint profit maximum each firm must set price such that at the margin a slight change in its price will affect its own profits at a rate that is equal but opposite in sign to the impact on its rival's profit. It is in this sense that each firm values a dollar gain to itself as equivalent to a dollar gain for its rival.

The linearity of the constraints (which is not required in any of the analysis to follow) implies that

$$\nabla^2 L = \nabla^2 Z = \begin{bmatrix} -2b_{11} & b_{12} + b_{21} \\ b_{12} + b_{21} & -2b_{21} \end{bmatrix}. \tag{7.8}$$

We shall assume that $(b_{12} + b_{21})$ is sufficiently small relative to $4b_{11}\, b_{22}$ to make the matrix negative definite. This condition is stronger than necessary to satisfy 7.6.5, and, moreover, is sufficient to make the problem strictly convex and therefore yield a unique global maximum. This will permit a crisper presentation, but our analysis could be performed within the restrictions of a search for a local optimum only.

We have assumed that in all optima $P^0 > 0$, so that $\lambda_{i2} = 0, i = 1, 2$. Henceforth, we will deal only with λ_{i1} and suppress the second subscript. When $\lambda_{11} = 0$ is assumed we suppress both subscripts so that $\lambda = \lambda_{12}$.

On Figure 7.1 we graph the two types of constrained joint-profit boundary maximum with $P > 0$.[5] Case 1 depicts an optimum on the edge of the polygon (in n-space, polytope) where one ($n - 1$ or fewer) constraint (s) bind at the optimum P^0. For a feasible region defined by nonlinear constraints this corresponds to a point of support of the objective function and one ($n - 1$ or fewer) constraint (s). In Case 2 the optimum occurs at an intersection of two (n) constraints, or in our case of linear constraints, a vertex of the polygon (polytope).

Case 1: Solutions

For Case 1, from conditions 7.6.1 and 7.6.2, we write

$$\begin{aligned} &\nabla Z - \sum_i \lambda_j \nabla L_i = 0 \\ &\lambda_j L_i = 0. \end{aligned} \tag{7.9}$$

We have, in a geometrical depiction of these conditions on Figure 7.2, shifted the origin of axes to the solution P^0 for consideration of perturbations in its neighborhood, and have graphed ∇Z and ∇L_2 against them.

We show now that 7.9 requires for Case 1 that ∇Z and ∇L_2 be collinear (in n-space, that ∇Z and ∇L for binding constraints be coplanar or co-hyperplanar). Any allowable perturbation along L_2 in the neighborhood of P^0 must meet the condition that

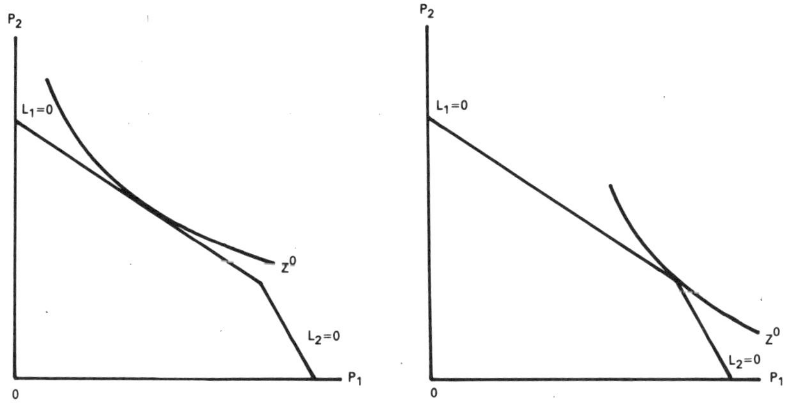

(*a*) Case 1. *One Constraint Binds* (*b*) Case 2. *Both Constraints Bind*

Figure 7.1 The initial duopoly equilibrium

$$L_2^1\, dp_1 + L_2^2\, dp_2 = 0, \tag{7.10}$$

where $L_j^i = \partial L_j / \partial p_i$, or

$$dp_2 / dp_1 = -\, L_2^1 / L_2^2, \tag{7.11}$$

which is the slope of L_2. Being normal to L_2 at P^0, $\nabla\, L_2$ has a slope which is the negative of the inverse of L_2's slope, as depicted in Figure 7.2.

The corresponding movement along Z in that neighborhood is

$$Z^1\, dp_1 = +\, Z^2\, dp_2 = dZ, \tag{7.12}$$

and for $dZ = 0$,

$$Z^2 / Z^1 = L_2^2 / L_2^1. \tag{7.13}$$

Hence, $\nabla\, Z$ and $\nabla\, L_2$ must have the same direction at the optimum.

Define

$$\lambda = Z^1 / L_2^1, \tag{7.14}$$

so that 7.13 may be written

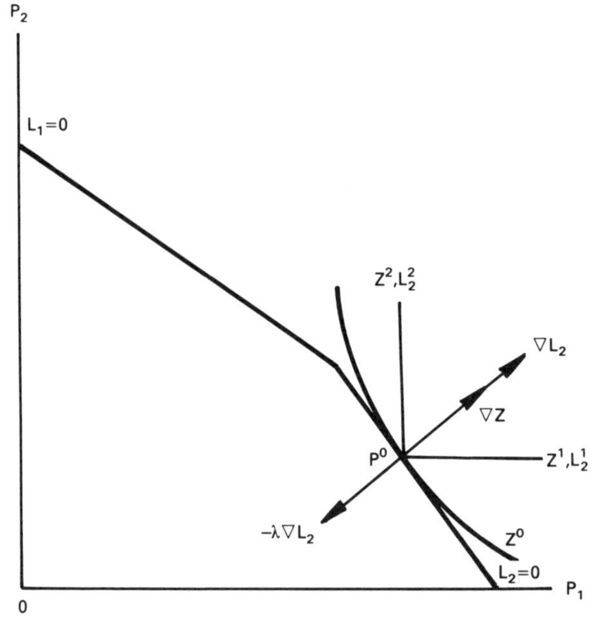

Figure 7.2 A geometric interpretation of Case 1 equilibrium

$$Z^2 = \lambda L_2^2, \tag{7.15.1}$$

and from 7.14

$$Z^1 = \lambda L_2^1, \tag{7.15.2}$$

Then 7.15.1 and 7.15.2 are 7.9.1 in Case 1, and λ may be seen to be the 'carrier' of the allowable tradeoffs in quantities enforced by the constraint. Suppose, for example, at P^0 we choose $dp_2 > 0$, so that the gross movement in objective function value is $Z^2\ dp_2$. From 7.15.2 we are told that if P^0 is indeed a local optimum, this must equal $\lambda L_2^2 dp_2$. Then, from 7.15.1,

$$\lambda L_2^2\ dp_2 = (Z^1/L_2^1)\ L_2^2\ dp_2,$$

and from 7.11,

$$Z^1\ (L_2^2/L_2^1)\ dp_2 = Z^1\ (-dp_1/dp_2)\ dp_2 = -\ Z^1\ dp_1,$$

where $-dp_1$ is the reduction in p_1^0 enforced by L_2. Hence,

$$Z^2\, dp_2 = -Z^1\, dp_1 \tag{7.16}$$

or the gross gain in objective function value arising from $dp_2 > 0$ must be exactly balanced by a gross loss due to a necessary $dp_1 < 0$.[6]

Further, λ performs this 'carrier' function by defining the common marginal benefit/marginal cost ratio that must rule for all dp_i, or the marginal movement in objective function value per marginal unit of constraint capacity used up or released by any change in p_i.[7] In addition to defining this ratio for dp_i, therefore, λ enforces their equality for both variables, so that no net increase in Z can occur by shifting a unit of constraint capacity to permit an increase in one variable at the expense of another.

We can, therefore, interpret λ in two equivalent ways: (*1*) as the carrier of an equal but opposite force upon Z when one variable is increased by enforcing a decrease upon the other, and (*2*) as the common marginal-benefit-per-unit-of-capacity-used magnitude for small movements of the variables.

We may present the first of these interpretations on Figure 7.2 in a vector space. ∇Z contains the rates at which Z increases for small positive changes in P. But we have seen that these applied to vectors dP must lead to equal and opposite forces produced by the tradeoff that the inflexible capacity of the constraint enforces between the variable increments. The ∇L_2 vector maps the constraint capacity absorptions required by small positive changes in P. By multiplying each component of $-\nabla L_2$ by λ, because the latter enforces negative offsetting movements, it is ensured that net movements in Z are reduced to 0, as show in Figure 7.2. Hence, $\nabla Z + (-\lambda \nabla L_2) = 0$, or the resultant of the two vectors is the null vector.

Implied by this interpretation is the well-known argument about movement in the direction of the gradient.[8] The most rapid rate of climb on the objective function, in the neighborhood of P^0 and per unit of Euclidean distance, is in the direction in which ∇Z points, in the sense that of all $[dp_1, dp_2]$ that lie on the unit circle about P^0 as center, $dp_2/dp_1 = Z^2/Z^1$ is the ratio of increments that maximizes the rate of ascent. Moreover, the negative of this direction maximizes the rate of descent. Therefore, we may view the forces of unconstrained optimization as pushing to escape the feasible region in the most efficient manner (as defined above) by attempting to move along ∇Z in Figure 7.2. But these forces put strains upon the constraints to which they are unable to adjust by moving outward. Hence these forces must be overcome by enforcing resisting pressures through variable trade-off. If we wish to minimize these resistance forces we should move in the direction opposite to ∇L_2, and when we do λ enforces the minimum trade-offs $[-dp_1, dp_2]$ necessary to keep the solution P^0 in the feasible region without forcing it into the interior.

Finally, let us note another well-known result already mentioned in note 7, for from 15.1 and it is clear that

$$Z^i = \lambda L_2^i = \partial Z / \partial L_2 \cdot L_2^i, \tag{7.17}$$

so that λ, when the constraint binds as an equality, reflects the marginal objective function value of the constraint capacity.

Case 2: Solutions

We turn now to the consideration of conditions 7.6.1 and 7.6.2 for Case 2, which we have presented geometrically in Figure 7.3. The solution occurs at P^0 with both constraints binding[9] so that ∇Z cannot be collinear with both ∇L_1 and ∇L_2, and in general will have a direction equal to neither. Obviously, since $\nabla L_1 \neq \nabla L_2$, ∇Z can equal ΔL_1's direction at one extreme or ∇L_2's direction at

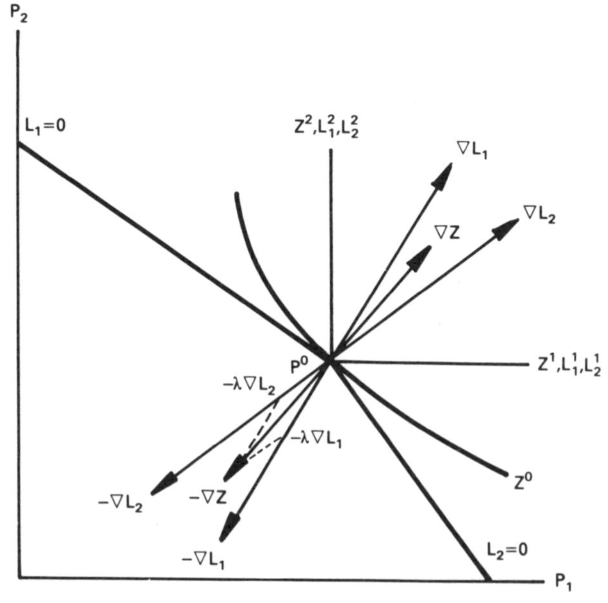

Figure 7.3 A geometric interpretation of Case 2 equilibrium

the other, and in general will lie in the interior of the convex polyhedral cone generated by the constraint gradients at the shifted origin P^0. Hence ∇Z will form acute angles with ∇L_1 and ∇L_2 if both L_1 and L_2 are negatively sloped. In Case 2, therefore, the constraint gradients form a basis for the space of partial derivatives, and ∇Z may be expressed on that basis with non-negative weights.[10] This is shown in Figure 7.3 with λ_1 and λ_2 the applied weights.

Algebraically, our interpretation is similar to that of Case 1, and can be speedily made. Condition 7.6.1 becomes

$$Z^1 - \lambda_{11} L_1^1 - \lambda_{21} L_1^2 = 0$$
$$Z^2 - \lambda_{11} L_1^2 - \lambda_{21} L_2^2 = 0. \qquad (7.18)$$

With two constraints binding there are two capacities that inhibit movement upward on the objective function, and therefore a dual requirement to carry forward substitution trade-offs along the constraints, or, equivalently, to inject the opportunity costs of using constraint capacity into alternative usages. At an optimum, the gross benefit obtained from any $dp_i > 0$ must be exactly offset by the sum of the opportunity costs forgone by the need to reduce p_j in order to free constraint capacities.

Geometrically, ∇Z represents the rates at which objective function value responds to variable differentials in the neighborhood of the optimum, and ∇L_1 and ∇L_2 carry the responses of capacity usage to them. From 7.18 it can be seen that every element in ∇Z must be responded to in equal force but opposite direction by the sum of the opportunity costs along all binding constraints, and, of course, each unit of every constraint capacity must have the same opportunity cost whether it is associated with one variable or another to eliminate profitable exchanges. Both of these conditions imply that ∇Z must be expressed as a resultant of $-\nabla L_1$ and $-\nabla L_2$ weighted by opportunity costs that can never be negative. This is depicted in Figure 7.3.

And, finally, we may again interpret λ_1 and λ_2 as the minimum opportunity costs that are needed to restrain the unconstrained maximization forces, when moving upward in the most efficient manner, by inflicting movement downward due to constraint capacity usage.

With the completion of this elementary presentation of the necessary conditions for a constrained optimum, we are ready to study the less well-behaved case of the crippled optimum in rivalrous consonance. Note, however, in our interpretation, that there exists a 'global' objective function whose first-order conditions describe exactly the duopolists' motivations. Neither of them may actually have it in their consciousness in adjusting marginally on the basis of whether dp_i raises their profit more than it reduces firm *j*'s. Alternatively, each may be consciously maximizing a joint profit function. In either case there exists a single function over the decision space whose optimization yields the solution.

II CRIPPLED OPTIMIZATION IN RIVALROUS CONSONANCE: THEORETICAL BASES

We return to our duopoly model, now, to assume more realistically that the firms' interest in the welfares of their rivals is somewhat less than total, and perhaps the minimum necessary to ward off disastrous price wars. Therefore, firm *i* regards its rival's loss of \$1 or gain of \$1 as worth only $\theta_i \varepsilon [0, 1]$ of its own loss or gain. If $\theta_i = 1$, for $i = 1, 2$, the case of whole-hearted joint-profit maximization discussed in section I exists; if $\theta_i = 0$ for both rivals the firms act with Cournot-intense myopia, ignoring the impacts of their price decisions upon rivals. Interior values of θ_i define varying degrees of rivalrous-co-operative behavior – a rivalrous consonance of interests of greater or lesser degree which we believe offers more realistic approximations to behavior.

Let us go directly to the marginal conditions of 7.6.1, assuming $\theta = [\theta_1, \theta_2]$, are determined by independent empirical study, and replace ∇Z in 7.7 by

$$\nabla Z^* = \begin{Bmatrix} \{(a_1 - 2b_{11} p_1 + b_{12} p_2 + b_{11} c_1) + \theta_1 b_{21} (p_2 - c_2)\} \\ \{(a_2 - 2b_{22} p_2 + b_{21} p_1 + b_{22} c_2) + \theta_2 b_{12} (p_1 - c_1)\} \end{Bmatrix}$$

$$= \begin{Bmatrix} \{OC_1\} + \{\theta_1 RC_1\} \\ \{OC_2\} + \{\theta_2 RC_2\} \end{Bmatrix}, \tag{7.19}$$

We assume the behavior of the rivals is best approximated by their adjustment of marginal profits through price policy to equal rivals' marginal losses discounted at the relevant θ_i factor, and subject to the firms' constraints.[11] The algorithm in section III simulates that iterative approach to the steady state solution.

The first question that arises is whether the integral of the Pfaffian

$$dZ^* = \nabla Z^* \cdot dP = Mdp_1 + Ndp_2 \tag{7.20}$$

exists as an exact equation, and, if it does not, whether some integrating factor can be found to obtain a transformation of it. If it can be obtained in either of the above senses, Z^* can be substituted for Z in 7.2 and 7.5, and the formal analysis of Case 1 and Case 2 behavior can be repeated (although the general counterpart of 7.19 may imply a nonconcave function and the convexity of the problem, where it existed, may be sacrificed). However, even if this is possible, Z^* is wholly formal, without behavioral content, a 'phantom' objective function of our creation. We will still be interested in the relationship of P^{*0}, the constrained optimum for the new problem, to the separate crippled objective functions. Nonetheless, finding the expression of Z^* is a step forward in interpreting the solution of the problem.

But this is not an easy problem. In our simple case, $\partial M/\partial p_2 \neq \partial N/\partial p_1$, so that the integrability conditions are not met for Z^*. It follows that Z^* is not an exact equation, and that we must search for an integrating factor that will transform M and N to forms that permit integration of Z^*. This has been possible for our example, but consideration of the complexity of the result will convince one that in general, for more complex problems this will be difficult or impossible.[12]

Define:

$$
\begin{aligned}
A_i &= a_i + b_{ii}\, c_i - \theta_i\, b_{ji}\, c_j \\
D_i &= b_{ij} + b_{ji} \\
u &= p_1 - \frac{(2A_1\, b_{22} + A_2\, D_1)}{(4b_{11}\, b_{22} - D_1\, D_2)} = p_1 - h \\
v &= p_2 - \frac{(2A_2\, b_{11} + A_1\, D_2)}{(4b_{11}\, b_{22} - D_1\, D_2)} = p_2 - k \qquad (7.21) \\
w &= v/u \\
X &= b_{11} - (D_1 + D_2)\, w + 2b_{22}w_2 = r + sw + tw^2 \\
q &= 16b_{11}b_{22} - (D_1 + D_2)^2 \text{ (assumed} > 0)
\end{aligned}
$$

Then

$$
Z^* = \ln u + .5 \ln X - \frac{2D_2 + s}{\sqrt{q}} \tan^{-1} \frac{2tw + s}{\sqrt{q}} + K \qquad (7.22)
$$

where K is a constant of integration. We may substitute for u, v, and w in 7.22 to translate Z^* into a function of P, and the contour lines of the phantom function may be used in the presentation to follow. Note from 7.19 that given our previous assumptions about the relative magnitudes of $4b_{11}\,b_{22}$ and $b_{12} + b_{21}$, and given θ defined on the unit square, Z^* retains the strict concavity of Z.

Case 1: Crippled Optimization

In Figure 7.4 we have drawn the Case 1 optimization using Z^* as objective function with conditions 7.6.1 and 7.6.2:

$$\nabla\, Z^* - \lambda^*\, \nabla\, L_1 = 0 \tag{7.23.1}$$

$$\lambda^*\, L_1 = 0 \tag{7.23.2}$$

But 7.23.1 may be written

$$\nabla\, \bar{Z} - \lambda^*\, \nabla\, L_1 = \nabla\, a, \tag{7.24}$$

where $\nabla\, \bar{Z}$ is evaluated at the new optimum P^{*0} and where

$$\nabla\, a^T = [(1 - \theta_1)\, b_{21}\, (p_2 - c_2),\ (1 - \theta_2)\, b_{12}\, (p_1 - c_1)]. \tag{7.25}$$

The λ^* valuation of L_1's capacity takes into account as an opportunity cost only those portions of the rivals' components remaining in $\nabla\, Z^*$. Thus, P^{*0} is an equilibrium where these fractional RC_i, when added to OC_i, are equated to a common marginal opportunity cost as seen by the individual firms and neglecting some or all profit externalities of their decisions.

In the solution $\nabla\, Z^*$ and $\nabla\, L_i$ stand in the standard Case 1 relations as discussed in Section I and illustrated in Figure 7.2. But our present interest resides in the relation of this solution to the crippled objective function contour that goes through it ($\bar{Z}$), as well as to the whole-hearted joint profit maximum (all θ_i, = 1) at P^0. $\nabla\, \bar{Z}$ differs from $\nabla\, Z^*$ by the (non-negative) vector $\nabla\, a$, both of which vectors are drawn in relation to $\nabla\, Z^*$ on Figure 7.4. Obviously, no single λ-variable can be applied to $\nabla\, L_1$ to yield $\nabla\, \bar{Z}$, which is not collinear with it. Therefore, we must search for two different variables such that

$$Z^1 = \pi_1\, L_1^1$$

$$Z^2 = \pi_2\, L_1^2. \tag{7.26}$$

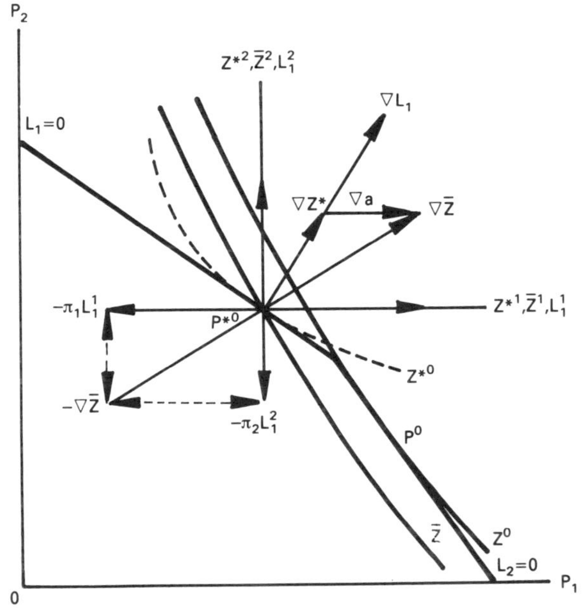

Figure 7.4 A geometric interpretation of Case 1 crippled optimization

The inequality of the π implies that the marginal opportunity costs of associating the constraint capacity with variables p_1 and p_2 are different, which on the face of it means that, were the optimization not crippled, it would be possible by moving in the positive direction of that p_i whose π_i is largest to obtain higher levels of Z in the neighborhood of P^{*0}. Indeed, it would be possible to escape the neighborhood and to move to P^0, where a standard Case 1 solution holds for Z. The institutional crippling, therefore, enforces the loss in objective function value of $Z^0 - \bar{Z}$.

To assess the costs of crippling in the neighborhood of P^{*0} more exactly, we may write from 7.24 and 7.26

$$\pi_1 - \lambda^* = a_1/L_1^1$$
$$\pi_2 - \lambda^* = a_2/L_1^2. \quad (7.27)$$

We have seen that λ^* assesses myopic marginal opportunity costs of constraint capacity when $\theta_i \cdot RC_i$ are included as marginal objective function payoffs to capacity usage. The π_i values – which reflect the opportunity cost from the viewpoint of genuine, whole-hearted joint maximization – add to these 'crippled' opportunity costs the neglected portions of marginal objective function payoffs to yield 'corrected' opportunity costs of L_1 capacity. We may construct an index of crippling by computing

$$\omega = \frac{\sum_i (1 - \theta_i) \cdot RC_i}{\sum_i RC_i} = \frac{\sum_i L_1^i (\pi_i - \lambda^*)}{\sum_i RC_i}$$

$$= \frac{\nabla a \cdot S}{\sum_i RC_i}, \tag{7.28}$$

where S is the sum vector [1, 1].

Note, finally, for Case 1, that π_1 and π_2 are non-negative, and could be negative only if ∇L_1 and ∇Z^* had negative slopes, which indicates satiation in the respective variable dimensions. These are pathological cases and we shall not discuss them.

Case 2: Crippled Optimization

The crippled optimization where P^{*0} occurs at the point where both constraints bind is presented in Figure 7.5. In this case the peculiarity of the solution as it relates to the crippled objective function is not the departure of the corrected opportunity cost values for each capacity from equality, but their potential negativity. That is, the use of a constraint capacity to full employment under the aegis of decentralized egoism may imply a forgoing of potential joint profit that could be recouped by movement to a Case 1 solution. For 7.26 may now be written

$$\nabla \bar{Z} - \pi_1 \nabla L_1 - \pi_2 \nabla L_2 = 0. \tag{7.29}$$

A basis of ∇L_i vectors will always exist to form a basis for $\nabla \bar{Z}$ in Case 2, but the latter may well lie outside the convex polyhedral cone formed by this basis, implying that one of the corrected constraint capacity opportunity costs, π_i, must be negative. As drawn on Figure 7.5 it is π_1 which is negative, inferring that a small movement into the interior of the set defined by L_1 in the neighborhood of P^{*0} would make $\pi_1 = 0$ and move the solution to a higher contour of the Z function. The crippling that forces the full-employment of both constraint capacities (both firms at their maximum acceptable market

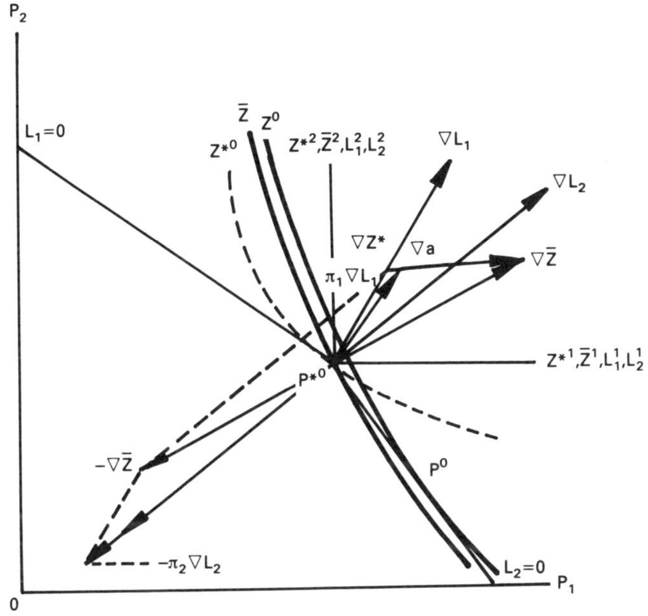

Figure 7.5 A geometric interpretation of Case 2 crippled optimization

shares) forces p_1 to be held too low and p_2 too high from the viewpoint of joint profits. As in Case 1, the sum of the differences in myopic and corrected opportunity costs, taking both constraint capacities into account, equals the sacrifices in Z made necessary by the crippling:

$$(\pi_1 - \lambda_1^*) \nabla L_1 + (\pi_2 - \lambda_2^*) \nabla L_2 = \nabla a. \tag{7.30}$$

And ω, the index of crippling, may once more be written

$$\omega = \frac{\nabla a \cdot S}{\sum_i RC_i} \tag{7.31}$$

III AN ALGORITHMIC SOLUTION FOR CRIPPLED OPTIMIZATION

In this simple case a transcendent objective function of some complexity can be derived. But as a final consideration we ask how it is possible to solve for a crippled optimization equilibrium when Z^*, the phantom objective function, cannot be obtained by integrating the altered first-order conditions. We have developed an algorithm for doing this by altering Fiacco-McCormick's SUMT (Sequential Unconstrained Minimization Technique) algorithm, and we will describe it briefly.[13]

SUMT is a penalty function algorithm which solves a succession of unconstrained subproblems which have the formal structure of a Lagrangean function, and whose solutions converge in the limit to a local optimum to the original problem, or, for convex problems, approach the global optimum to a controlled degree of approximation – a control achievable through the existence and use of a dual.

The structure of SUMT is peculiarly well-suited to a crippled optimization, since it literally solves equations of the form 7.6.1 in its original mode and for our purposes can be adapted to solve problems of the form 7.24 or its Case 2 counterpart. Hence the crucial needs of the algorithm at every step are the gradients $\nabla Z^* = \nabla Z - \nabla a$ and the ∇L_i, all of which are provided in separate routines. The crippled objective function, Z, is defined in the subroutine that calls for the specification of objective and constraint functions, but when partial derivatives are called by the program it branches into a subroutine that specifies the $\nabla - \nabla a$ gradient for the objective function. Hence, the program approaches P^{*0} from the interior of the feasible region, operating wholly upon the known gradient of an unknown objective function. The Lagrangean variables λ^* are yielded in the solution, but the π variables must be solved for from 7.27 or 7.30.

IV CONCLUSION

A frequent criticism of formal economic analysis is that although agents may know their objective functions (e.g. profit functions), they are not perceptive or sophisticated enough to seek optima through marginal conditions. Our point of departure is quite the opposite: important cases arise when behavior is most fruitfully analyzed by studying 'common sense' or institutionally formalized marginal behavior when the objective function has no easily defined substantive meaningfulness. This arises, for example, when firms are broadly seeking to obtain a constrained profit maximum in a market with high interdependence, and have learned to 'muddle through' to a mutually acceptable solution by taking account of their impacts upon their rivals' fortunes at differential discounts, perhaps related to the potential relative ability to retaliate or to market strength.

Crippled optimization permits the analyst to seek out the implications of such behavior upon social variables, and may, as in the case of oligopoly analysis, permit a range of new models to be investigated as improvements upon those existing in the field.

Notes

1. The utility function is really a creation of the economist with the goal of predicting the consumer's behavior by optimization techniques. It is a representation of a consumer preference pre-ordering, obtained by answers to a binary relation, and as such has behavioral significance though only a phantom existence.
2. For the full analysis of the oligopolistic equilibria referred to see Kuenne [3], [5] and especially Kuenne [6] and [7] where rivalrous consonance and crippled optimization techniques are used in the analysis of general oligopolistic equilibrium.
3. Again, this paper deemphasizes the economic justification of oligopoly models of this basic form to focus upon crippled-optimization issues. We do not assert that this is a realistic model, but rather that it is a convenient means of presenting the techniques. We shall approach reality more closely in Section 2, however.
4. See Kuhn and Tucker [8].
5. To conserve space we ignore, besides end-point solutions on the boundaries of the feasible region, optima that occur in its interior.
6. We are employing the model to interpret the equilibrium of a noncollusive, non-cartellized industry. Hence, conditions (7.6) are viewed as the implicit conditions that are attained by both firms acting to maximize profits taking into account their impacts upon rivals' profits and subject to remaining on or within their market share constraints. In this paper we are not greatly concerned with the dynamics by which the firms will be led to conditions (7.6), and will speak, therefore, as if they explicitly consider trade-offs along the constraints and λ variables in their actions. This device is a mere shortcut to avoid a full development of the mores and folkways of the industry that evolve to permit the industry equilibrium to be attained. That development is recognized to be a necessary and important complement to the model in a full-blooded application to an economic analysis, but our goals in this present paper are more limited.
7. Constraint capacity in the present case is the excess by which actual sales of firm 2 are permitted to rise above maximum desired sales (m_2 (d_1 + d_2)). We have set this equal to zero (i.e. the *restraint* of the *constraint* is zero). If we permitted this to become slightly positive, L_2 would shift outward to a tangency with a higher Z-contour, and $\lambda = \partial Z/\partial L_2$ is the marginal rate of benefit per marginal increment of restraint.
8. The fact local movement in the direction of the gradient yields the highest rate of ascent or descent for the Euclidean metric is *well-known*

in mathematical programming. See, for example, Hadley ([2], pp. 79–81), for a proof.

9. In our simple model, of course, such an intersection of the constraints will be possible if and only if $m_1 + m_2 = 1$.
10. In general, Case 1 is distinguished from Case 2 by the fact that a basis of ∇L binding at the solution does not span the space, and hence ∇Z must be coplanar with the reduced number of vectors.
11. Perhaps of even greater interest to the oligopoly theorist is the ability to introduce marginal conjectural variation in 7.19. Firm i may feel that a marginal change in p_i leads its rival to change its price by $\alpha_i \, dp_i$, $\alpha_i \lesseqgtr 0$. This leads to an obvious marginal profit term which may be included in 7.19. Nonmarginal conjectural variation behavior can frequently be incorporated in constraints in the model.
12. The integrating factor was obtained in the following manner. M and N are linear forms in p_1 and p_2, but are nonhomogeneous. Convert dZ^* to a homogeneous equation of degree one in the variables by transformation of the variables and a shift of their origin to the intersection and M and N when these latter are equated to zero and treated as linear equations. This new Pfaffian is solved easily because its variables are separable, and is integrated by parts.
13. See Fiacco and McCormick [1]. For a brief discussion of the unaltered techniques with special reference to oligopoly needs, see the appendix to Kuenne [4].

References

[1] Fiacco, Anthony V. and Garth McCormick, *Nonlinear Programming: Sequential Unconstrained Minimization Technique*, John Wiley & Sons, New York (1968).

[2] Hadley, George, *Nonlinear and Dynamic Programming*, Addison-Wesley, Reading, Mass. (1964).

[3] Kuenne, Robert E., 'Toward the Incorporation of Product Differentiation in Economic Theory: The Uses of Inexact Measurement', *Applied Economics*, 6 (1974,) pp. 225–73.

[4] Kuenne, Robert E., 'Toward a Usable General Theory of Oligopoly', *De Economist*, 1974 (122), pp. 491–502.

[5] Kuenne, Robert E., 'Towards an Operational General Equilibrium Theory with Oligopoly: Some Experimental Results and Conjectures', *Kyklos*, 27 (1974), pp. 792–820.

[6] Kuenne, Robert E., 'General Oligopolistic Equilibrium: A Crippled Optimization Approach', in *Pioneering Economics: International Essays in Honour of Giovanni Demaria*, Cedam, Padova, 1978, pp. 537–77.

[7] Kuenne, Robert E., 'Duopoly Reaction Functions Under Crippled Optimization Regimes', *Oxford Economic Papers*, 32 (1980), pp. 224–40.

[8] Kuhn, Harold W. and Albert W. Tucker, 'Nonlinear Programming', in J. Neyman (ed.), *Proceeding of the Second Berkeley Symposium on Mathematical Statistics and Probability*, University of California Press, Berkeley (1951), pp. 481–93.

8 The Rivalrous Consonance Framework for Oligopoly Analysis*

I THE INADEQUACIES OF EXISTING THEORIES

Oligopoly theory occupies a peculiar position in economic analysis. On the one hand, lavish theoretical attention is bestowed upon it in the form of game-theoretic and bargaining models, entry-barrier theory, and core theory.[1] Yet the ghosts of Cournot, Bertrand, Edgeworth, and Stackelberg process and equilibrium analysis continue to parade through the pages of textbooks, frequently updated to relate them to Nash or core equilibrium concepts.

On the other hand, it is difficult to rid one's self of the belief that, despite the intensity of analytical effort, this body of analysis stands apart in antiseptic isolation from the major fields of economists' microeconomic efforts. General equilibrium theory remains almost wholly untouched by it, despite the unchallenged fact that modern industrial economies are permeated with such market structures.[2] Despite conjectures that recent inflation has a microeconomic component in the increased readiness of oligopolistic industries to grant higher wages to labor and pass them on in higher prices, no comparative statics theorems bearing upon such problems have been derived from interdependence theory.

Oligopoly theory has proved unrewarding for several reasons. First, it has aimed at the most general results, seeking to derive theorems of universal applicability to oligopolistic behavior. By its nature, however, each oligopolistic industry is importantly distinctive, with pricing, output, selling cost, innovation and efficiency outcomes that are heavily impacted by personalities, industry history,

* Published originally as Ch. 1, Robert E. Kuenne, *Rivalrous Consonance: A Theory of General Oligopolistic Equilibrium*, North-Holland, Amsterdam, 1986, pp. 1–18, and reproduced with permission of Elsevier Science Publishers. Cross-references to chapters refer to the original publication.

and the nature of the product. Existing theoretical approaches are resistant to close tailoring to specific industry analysis and therefore misleading in their treatment of oligopolistic industries in universals.

Second, classical and game-theoretical frameworks are both tied closely to 'conjectural' strategies of the 'if I, then he' variety. Admittedly, intuitive and experimental support exists for a component of such attitudes in oligopolistic behavior, but, unfortunately, the insight has not proved to be capable of translation in empirical modeling, nor is it an exclusive component. In illustration, although consumer preferences are not, practically speaking, capable of specification, they can be made empirically effective in demand functions. Analogous instruments have been difficult to formulate in other than overly simple modes (e.g. Cournot) and impossible to isolate empirically.

Third, conjectural schemata are strongly wedded to single-objective modeling, generally the maximization of profit or expected utility. They cannot capture the richness or the plurality of objectives sought by a typical oligopoly.

In recent years the theorist has received welcome guidance from the valuable empirical work done by such investigators as Kaplan, Dirlam, and Lanzillotti [16] in the United States, Fog [9] in Denmark, and Wärneryd in Sweden.[3] From this work has emerged a major theoretical guidepost: the modern oligopolistic firm follows a multiple-objective strategy. In very few cases can any simply defined profit maximization be discerned; rather, to an almost exclusive degree, the goals sought are a set of mutually constraining and conflicting aims, imperfectly conceived and articulated, uncertainly co-ordinated, and decentralized in administrative origin and achievement.[4]

The set of goals frequently sought includes some subset of the following;

1. A target rate of return on dedicated capital;[5]
2. Maintenance or expansion of market share;
3. Maintenance of a profit margin over a standard cost figure;
4. Stabilization of prices in the industry;
5. Conformance to current antitrust trends;
6. Meeting competitors' price leads;
7. Minimization of customer inconvenience;
8. Exclusion of potential industry entrants;[6]
9. Enhancing the welfare of managers.

Another major conclusion emerges implicitly from this recent literature: the preconditions to make formal game-theoretic approaches promising as a general analytical framework for the pricing decision do not exist. The modern corporation is a pluralistic organization with multiple, poorly co-ordinated, and to some extent, competing, loci of decision-making.[7] Within such groups, decisions are arrived at in heuristic, informal or nebulous manners.[8] Of course, it is possible to use game theory or conjectural frameworks as a formal encapsulation of the informal procedures of rivalrous corporations, but those models are not likely candidates to serve this metaphoric role. Their plausible application requires a clear-cut demonstration of a well-defined locus of decision-making authority with a comprehensive canvas of possibilities, grasp of strategic alternatives, concern with rivals' potential reactions, and constant preoccupation with firm survival or the payoffs implied by the paraphernalia of such theories.

It is difficult, therefore, to accept the war-of-survival paradigm typified by Shubik's early work [25], or, indeed, most of the classical or modern approaches which stress only the individual-firm oriented one-dimensional 'reaction function' in one form or another.[9] Rather, the realistic image of rivalrous competition is one which includes competing intrafirm goals and elements of an 'organic' subordination of the individual unit's welfare somewhat to the welfare of the whole. That is, there is a definite commitment of the rivalrous firm to the continued peaceful survival of the industry – a 'rivalrous consonance of interests' – even if that implies the forgoing of immediate opportunities by the firm to seek greater benefit through conflict.

In summary, misplaced theoretical ambition, nonobservable variables, and inability to cope with the goal multiplicity and subtlety of the firm in oligopoly, as well as the industry's communitarian aspects have led to a dissatisfaction among economists with existing formal frameworks in this market structure. In their simplicity and generality they serve as often to mislead as to illuminate. We agree with Telser [28] who observes.

> The formal models often have great mathematical elegance and are fun to play with and teach, but at least from this author's vantage point their existence has had a negligible effect on applied research and policy analysis in the industrial organization area. General policy prescriptions and uncertainties regarding concentration, price rigidities, barriers to entry, market definition, tacit coordination, etc.,

seem to exist independently of the formal modeling tradition (pp. 213–14).

II SOME DESIDERATA

It would be incorrect to react to such criticism by rejecting formal theory as a means of gaining structural insights into realistic oligopolies. Formal analysis exists to supplement the limitations of the analyst's unaided mind in manipulating more than a small number of interdependent variables. Interdependence of firms' decisions is the very essence of the oligopolistic problem, and we are unlikely to probe very deeply into the implications of hypotheses concerning a specific oligopoly without formal models.

However, we agree with Koopmans's assertion[10] that the role of the theorist is to follow the empiricist, not to lead him. Theoretical advance in the social sciences lies in the direction of incorporating, enlightening, and co-ordinating empirical discoveries. The unrealistic must yield to the realistic postulate unless the simplifying abstraction from reality can be shown to yield more fruitful results through a gain in the ability to manipulate the theory to derive useful insights. Hence it is always unfortunate for a field when empirical investigator and theorist fail to communicate in continuous fashion, or when the theorists' frameworks are not applicable to the empiricist's tasks.

To be useful, therefore, we shall insist that new theoretical departures conform to a set of requirements that reflect the criticisms in Section I above.

First, the theory must be operational. Its variables must be observable or derivable from real-world data by practically available – not merely conceptually feasible – methods. To the extent valid theoretical concepts are not so translatable, summary or surrogate variables or attributes must be incorporated to capture the impacts.

Second, the theory must accord with general equilibrium frameworks in two senses. Within the industry, it must be tailored to treat the interdependent decision-making of the major rivals, retaining their individual identities to incorporate the essence of the oligopoly problem. Moreover, it must be capable of fitting into a larger-scale model that features the interdependence among industries and customers. Otherwise the framework will be incapable of contributing to the analysis of such problems as the transmission of inflationary impacts.

Third, it must be flexible enough to incorporate the richness of realistic variety in rivals' goods (primary and secondary), industry power structures, and attitudes toward competition and cooperation. We reject the notion of a universally applicable theory of oligopolistic behavior, and embrace the need for less ambitious theoretical analyses tailored specifically to given industries.

Within this goal we are most anxious to temper the tendencies to extremes of assumptions concerning rivalrous behavior that the rigid classical and game-theory models foster. Conjectural models tend by their nature to incorporate a reactive passivity of firms' expectations to rivals' actions. This is true even for more recent approaches that stress the necessity for consistent conjectures in equilibrium.[11] Conjectures tend – in more sophisticated models than Cournot's – to be altered in a time-phased, adaptive fashion in the light of rivals' recent behavior. Such viewpoints also ignore the active signalling to rivals that occurs continuously in realistic mature oligopolistic communities sensitized by years of coexistence. Firms engage actively in the shaping of their rivals' expectations, calming fears of destabilizing the price structure, assuring rivals of a willingness to remain within production ceilings, or, at the other extreme, warning them of contingent intentions to match or outdo rivals' aggressive actions.

Game-theoretic approaches with bargaining and coalition formation implant the analysis in an environment of conflict, ranging from the war-of-survival context to narrow self-interest defined in one-dimensional units of benefit. The competitive overwhelms the co-operative, the latter becoming of interest only if narrow self-interest is served.

Mature oligopolistic industries, in which major rivals have coexisted for extensive periods of time, are much more peaceful *communities* than such frameworks picture. They are blends of the rivalrous and the co-operative, with coexistence generally frankly accepted, and, indeed, with some positive and motivating concern for the image or welfare of the industry. Each rival is solicitous of the impacts of his decisions upon the welfares of each of his major rivals. A great source of this concern, of course, is the fear that if that impact is too painful the rival or rivals will respond with punishing reaction. The desire for a peaceful, stable industry seems a general one in mature industries. 'Good citizenship' in the industry implies a recognition of a rivalrous consonance of long-term interests by all participants. Rivalry coexists with strivings for harmonistic existence, in different proportions depending on the industry; live-and-let-live is a characteristic of communitarian life.

In our view, any hope for successful formal analysis of oligopolistic decision-making must be based upon the incorporation of these potentials and characteristics. If this work is successful in constructing such analytical frameworks, they will be used to obtain insights into the structure and functioning of the industry. By parametric displacement, it should be possible to illuminate the distribution of benefits and costs that results from the industry's power structure, the consequences of increasing or reducing co-operative tendencies, and the price-output impacts of cost increases.

III RIVALROUS CONSONANCE AND CRIPPLED OPTIMIZATION

To analyze mature oligopoly in a rivalrous consonance context a nonlinear programming approach to firm and industry behavior is formulated in which each rival strives to optimize an objective hobbled by the need to consider other firms' objectives as well as secondary goals of its own. We have termed this framework, together with the algorithm used to solve the system, 'crippled optimization'.

Because this approach is flexibly tailored to specific industries' cases, it is necessary to present it in exemplifying forms. A specific application of the approach has been made to the Organization of Petroleum Exporting Countries (OPEC) as it functioned in the period 1973–81 (Kuenne [19]). In this book, however, prototypical but hypothetical cases will be used to formulate the approach in a more controlled manner than realistic research applications permit.

Model 1 – Crippled Profit Maximization: Single Industry

Let us hypothesize an industry whose firms typically seek to maximize profits within the context of recognized rivalry as a primary objective, subject to the simultaneous striving for a set of secondary objectives. Assume an oligopolistic industry with n firms and associated differentiated products. Rival i (producing product i) strives as a primary objective to maximize 'extended profits', E_i:

$$\operatorname*{Max}_{p_i} E_i = \sum_{j=1}^{n} \theta_{ij}(p_j x_j - C_j x_j), \tag{8.1}$$

where we define:

1. p_i: price of rival i's product;
2. x_i: amount demand by consumers of product i;
3. C_i: average cost, assumed constant, of product i;
4. θ_{ij}: binary consonance factors for firm i relevant to firm j,

$$\theta_{ii} \equiv 1.$$

Note that the firm is assumed – realistically – to set price rather than quantity of its product. The amount demanded follows from the consumer demand or sales function it faces, which is assumed to be specified linearly:

$$x_i - a_i + b_{ii}p_i \sum_{j \neq i} b_{ij}p_j - \mathbf{b_{iq}} \cdot \mathbf{q} = 0, \qquad (8.2)$$

where **q** is a vector of variables other than prices of the industry's products affecting sales to consumers.

The Binary Consonance Factors

The distinctive term in 8.1 and the factor that distinguishes the rivalrous consonance model is the presence in rival i's objective function of the profits of all other rivals, each 'discounted' by a coefficient, θ_{ij}, where $\theta_{ii} \equiv 1$. The *binary consonance factor*, θ_{ij}, is an element of the $n \times n$ consonance matrix θ which defines the industry power structure as well as its specific position on the rivalry-cooperation spectrum.

Specifically, θ_{ij} is the dollar equivalent in rival *i*'s profits at which it values one dollar of rival *j*'s profit or loss. Hence, if $\theta_{ij} = .25$, rival *i* treats \$1 of profit or loss by rival *j* as the equivalent of a 25¢ gain or loss to itself. Rival *i* has $n-1$ (θ_{ii} is trivially unity) such factors to define its relevant attitudes toward *each* of its rivals. Those coefficients are the behaviorally effective net resultants of rival *i*'s beliefs concerning the ability of rival *j* to retaliate against it; the effectiveness of rival *j*'s management; the claims rival *j* has upon its solicitude as a member of the industry; the importance for industry welfare of keeping rival *j* viable (perhaps for antitrust reasons); and the degree of deference owed rival *j* on the basis of the mores and folkways of the industry.

Obviously, θ_{ij} need not equal θ_{ji}, and indeed in terms of power structure definition the pair would generally be expected to be inversely related. In any event, they are defined in the consciousness of the relevant rivals' decision-makers and therefore need bear no consistent relation to each other. Further, because the elements of θ are binary in their statement, that matrix is a detailed specification of the behaviorally relevant attributes of the industry.

Of interest, both theoretically and in the empirical determination of the rivalrous consonance structure, is the *aggregate consonance factor*,

$$\theta_i = \sum_j \theta_{ij},$$

an element of the vector θ_i. It may be interpreted as firm i's valuation in own-profit dollars of $n-1$ dollars of profit spread uniformly over its rivals. In a macro sense, it indexes the firm's overall position in the consonance structure of the industry, as determined by its market position, its attitudes towards live-and-let-live strategies, and its feelings of confidence as an industry participant.

The binary consonance factors are related to the aggregate factor by the power structure factors, $k_{ij} \varepsilon \mathbf{K}$, where

$$\theta_{ij} = k_{ij}\theta_i, \sum_j k_{ij} = 1.$$

For the industry as a whole the power structure is defined by **K**, a matrix whose row vectors are the $\mathbf{K_i}$. The k_{ij} describe, for firm i, the specific manner in which its industry solicitude is distributed over *specific* rivals. Hence, **K** is a concise and quantitative description of the *relative* power position of each firm in the industry. The binary consonance matrix, θ, converts these positions to absolute terms. The theoretical presentations to follow work largely with θ, but the techniques for the numerical definition of **K** and the θ_i as steps toward the isolation of θ will be discussed in chapter 7.

The binary consonance factors incorporate – but are more information-inclusive than – the 'conjectures' so dear to the analytical hearts of classical theorists and their modern followers. Moreover, the θ_{ij} are determinate in realistic industrial settings which conjectures are not, and therefore are operational concepts.[12]

In a fully developed, long-run theory of oligopoly in rivalrous consonance the θ_{ij} would be endogenous to the model, determined by economic, psychological, political and social variables too complicated in nature to contemplate comfortably. In chapter 6 θ_i is permitted to become variable (**K** fixed) and determined by own-profit maximization considerations in a long-run analysis which attains a generalized Stackelberg solution. Elsewhere, however, θ is treated as fixed in a short-run analysis, and the θ_{ij} are viewed as behavioral parameters that summarize a complex of market and nonmarket forces acting to shape rival i's conduct toward rival j.

When the θ_{ij} are lifted out of the context of constrained optimization and n-firm oligopoly, they may be seen to be related to prior work of other theorists. Within the confines of game-theoretic analysis of unconstrained duopoly, such coefficients have been employed by Robert L. Bishop [5], James W. Friedman [10] – [12], A. Hoggatt [13] – [14], and R. M. Cyert and M. H. De Groot [7] – [8].

Goal-Defining Constraints

Firm i's goal profile can be completed by defining a set of constraints that specify other 'secondary' goals. These will vary with firm and industry, and the following set will be used as merely illustrative of such secondary goals, although they are believed to have some general applicability in American industry.
First, output must be within the firm's capacity, B_i, to produce:

$$C^1 \quad : \quad x_i - B_i \leq 0. \tag{8.3}$$

Second, the firm insists upon retaining a minimum level of sales, M_i, set perhaps by the desire to achieve an expected target market proportion:[13]

$$C^2 \quad : \quad M_i - x_i \leq 0. \tag{8.4}$$

Third, the firm sets upper and lower bounds, T_i^+ and T_i^-, on period-by-period price changes, specified as proportions of previous period price, $\bar{p}_i$:

$$C^3 \quad : \quad p_i/\bar{p}_i - T_i^+ \leq 0, \tag{8.5}$$

$$C^4 \quad : \quad T_i^- - p_i/\bar{p}_i \leq 0. \tag{8.6}$$

Such limits are frequently used to prevent consumer antagonism, anti-trust attention, delayed purchases because of customers' expectations of further price falls, and so forth.

Lastly, price is bounded away from a specific lower bound, which we will in further analysis assume is zero, but which may not be:

$$C^5 \quad : \quad N_i - p_i \geq 0. \tag{8.7}$$

In this book we will term B_i, M_i, T_i^+, T_i^-, and N_i, as well as parameters of similar functions, the *restraints* of the relevant constraints.

Mathematical Properties

Crippled optimization models are designed to be numerically specified by econometric or other techniques and to be solved by appropriate algorithms. The solution technique used in our work – a penalty function nonlinear programming algorithm – will be discussed in some detail in chapter 3.3. However, as in the case of most such programming algorithms, convergence of the component problem solutions to global optima and the existence of a dual problem are dependent upon the model's 'convexity'. At this point, therefore, Model 1 will be examined briefly for this quality.

Effectively, by substitution of expression 8.2 for x_i in 8.1, 8.3, and 8.4, and of similar expressions for x_j in the like models for firms $j \neq i$ (which models are solved simultaneously as submodels in Model 1), the p_j, $j = 1, 2, \ldots, n$, are the only endogenous variables. Further, since the solution algorithm will solve each firm's constrained maximization problem sequentially, assuming all other firms' prices are held constant at current levels, in the firm's optimization only its own price is variable. Hence, if Model 1 is convex in own-price only all of the subproblems will be convex problems.

Sufficient conditions that each firm's subproblems be convex in the iterations are that (*1*) its objective function be concave in own-price, and (2) its constraints be convex in own price. Clearly,

$$d^2E_i/dp_i^2 = -2b_{ii} < 0, \tag{8.8}$$

since $b_{ii} > 0$, and therefore E_i is (strictly) concave in p_i. Also, constraints 8.3 and 8.5 – 7 (with the x_j eliminated by substitution) are linear in own-price and therefore trivially convex. Also, constraint 8.4 is convex in p_i. Thus, the feasible region will be convex. Therefore, the solutions to Model 1's subproblems (to be explained further in chapter 3) will be global extended-profit maxima.

Model 2 – Crippled Profit-Maximization: General Equilibrium

As an illustration of the important potential for crippled optimization modelling to serve as a means of coping with oligopoly in operational general equilibrium systems, let us extend Model 1 into such a context by a few simple adaptations.

Assume that the entire economy contains t firms divided into oligopolistic industries, and that *each* of the firms possesses a constraint set of the type illustrated in 8.3–8.7. The θ matrix will be a $t \times t$ matrix with zero elements for those pairs of firms which have neither rivalrous relations (as in Model 1) nor customer–supplier relationships. That is, we now include the solicitude of a supplier firm for its industrial customers among the consonance factors of the model.

Consumer demands, x_i, remain as defined in 8.2, where firms j continue to be interpreted as rivals within firm i's industry. However, we now include an industrial demand component in total demand,

$$z_i - \textstyle\sum_c v_{ic} z_c = x_i \tag{8.9}$$

where we define

1. z_i: total – consumer plus intermediate – demand for product i;
2. v_{ic}: units of product i used in the production of a unit of good c.

We will assume a Leontief technology in which the v_{ic} are fixed technical coefficients, that each firm uses one activity to produce its product, and that each activity produces only one product.

Then we may redefine C_i, unit cost, to be

$$C_i - f_i - \textstyle\sum_s v_{si} p_s = 0, \tag{8.10}$$

where

1. f_i : unit factor cost of firm i's product, assumed constant;
2. p_s : price of a supplier firm's product.

It is then necessary to rewrite firm i's objective function; letting m_j denote firm j's own-profit margin, we define

$$\underset{p_i}{\text{Max}}\ E_i = \sum_{j=1}^{q} \theta_{ij}(p_j - C_j) z_j = \sum_{j=1}^{q} \theta_{ij} m_j z_j, \tag{8.11}$$

and to constrain it with demand functions 8.2 and 8.9, unit cost definition 8.10, and goal functions of the form 8.3–8.7. These include the constraints

$$p_i \geq N_i. \tag{8.12}$$

Sufficient conditions for the concavity of E_i become a bit more complicated. The Leontief economy can be depicted as

$$\begin{bmatrix} z_1 \\ z_2 \\ \cdot \\ \cdot \\ \cdot \\ z_t \end{bmatrix} = \begin{bmatrix} 1-v_{11} & -v_{12} & \cdots & -v_{1t} \\ -v_{21} & 1-v_{22} & \cdots & -v_{2t} \\ \cdot & \cdot & \cdots & \cdot \\ \cdot & \cdot & \cdots & \cdot \\ \cdot & \cdot & \cdots & \cdot \\ -v_{t1} & -v_{t2} & \cdots & 1-v_{tt} \end{bmatrix}^{-1} \begin{bmatrix} x_1 \\ x_2 \\ \cdot \\ \cdot \\ \cdot \\ x_t \end{bmatrix} \tag{8.13}$$

or

$$\mathbf{Z} = \mathbf{A}^{-1}\,\mathbf{X} = \mathbf{BX}. \tag{8.14}$$

Then, for example,

$$[dz_1/dp_1, dz_2/dp_1, \ldots, dz_t/dp_1]' = \mathbf{B}[-b_{11}, b_{21}, \ldots, b_{t1}]'. \tag{8.15}$$

E_i is concave in own-price if and only if

$$d^2E_i/dp_i^2 = 2[(1-v_{ii})dz_i/dp_i - \sum_{j \neq i} \theta_{ij}v_{ij}dz_j/dp_i] \leqslant 0. \tag{8.16}$$

Since $\mathbf{B}$ in 8.14 is non-negative if the Leontief technology is viable, the sign of dz_j/dp_i will depend upon the relative sizes of the elements in the jth row of $\mathbf{B}$ and the vector of demand function coefficients in 8.15. If consumers treat all goods as gross substitutes, with $dz_i/dp_i < 0$, $dz_j/dp_i \geqslant 0$, then, with $\theta_{ij} \in [0, 1]$, the extended profit function will be concave. In general, if the θ_{ij} are small, so the firms do not take rivals' and industrial customers' welfares much into account in their decision making, the probability of 8.16 holding is also raised, since own-price effects should then dominate.

Model 2, therefore, should yield global optima to each firm's optimization in each iteration of the algorithm. The model incorporates broad decision interdependence among sectors of the economy in manners that are operational. It is viewed, therefore, as a promising beginning to the inbuilding of oligopolistic pricing into general equilibrium models that are realistic and capable of empirical specification.

Model 3 – Crippled Target-Rate-of-Return Optimization – Single Industry

As another illustration of the flexibility of the rivalrous consonance framework, suppose the firms in an industry attempt as a primary objective to attain a target rate of return, $\bar{r}_i$, treating shortfalls or overages of the actual rate of return, r_i, as equally undesirable. Moreover, they once more feel constrained to take into account the power structure of the industry as they perceive it. Assume that they strive for a set of subordinate objectives which is, for purposes of this example, the same as the set 8.3–8.6 in Model 1. Finally, assume each firm has a dedicated capital, K_i, devoted to the production of product i. For simplicity, industrial demand will be ignored as in Model 1.

The matrix, θ, now performs the same function as it did in Models 1 and 2 but must be given a different interpretation. The consonance factor θ_{ij} is simply a weight that reflects the degree of importance rival i accords the attainment of rival j's goal relative to the importance ($\theta_{ii} \equiv 1$) it attaches to attaining its own goal. The factor becomes a pure scaling value without ready interpretation as own-profit translator.

The firm's objective function becomes

$$\text{Min } Z_i = \sum_{j=1}^{n} \theta_{ij}|\bar{r}_j - r_j| \tag{8.17}$$

subject to the definitional constraints

$$r_j - m_j x_j/k_j = 0, \tag{8.18}$$

the definition of the sales function 8.2, the subordinate goal set 8.3–8.6, and the lower bound constraints 8.7.

The objective function 8.17 is difficult to manipulate mathematically, and, therefore, a more conventional function that retains the symmetry of deviations on both sides of $\bar{r}_j$ and which is convex in p_j is preferable. One acceptable form is simply

$$\text{Min } Z_i = \sum_{j=1}^{n} \theta_{ij}(\bar{r}_j - r_j)^2, \tag{8.19}$$

but to punish deviations more severely we adopt an exponential form:

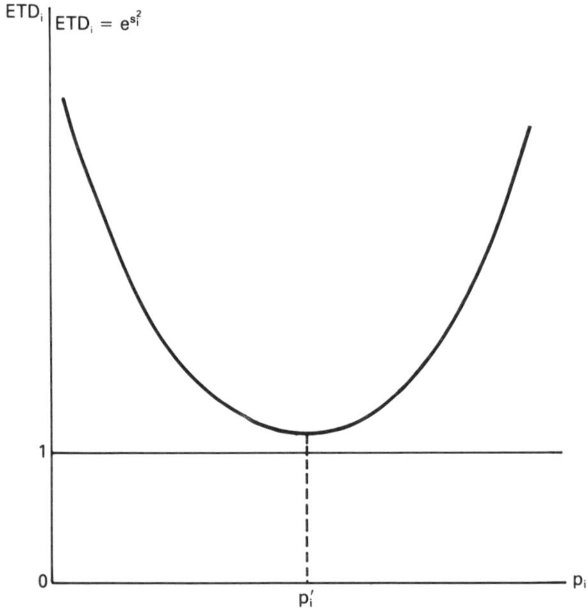

Figure 8.1a Target deviation function for isolated firm i when $r_i < \overline{r}_i$

$$\text{Min } ETD_i = \sum_{j=1}^{n} \theta_{ij} e^{s_j^2}, \tag{8.20}$$

where ETD_i is firm i's *extended target deviation* and $s_i = \overline{r}_i - r_i$ is its *target deviation*. Convexity of this function requires that

$$d^2ETD_i/dp_i^2 = 2\left[\left[((b_{ii}m_i - x_i)/K_i)^2(1 + 2s_i^2) + 2b_{ii}s_i/K_i\right]e^{s_i^2} + \sum_{j \neq i} \theta_{ij}\left[(-b_{ji}m_j/K_j)^2(1 + 2s_j^2)\right]e^{s_j^2}\right] \tag{8.21}$$

The summation term is obviously positive if any rival is over- or underachieving its target rate. As an adverse case, therefore, assume for $j \neq i$, $\theta_{ii} = 0$. The sign of 8.21 then hinges upon the sign of s_i. If $s_i \geq 0$, the condition of 8.21 holds. When $s_i < 0$ it will hold only in neighborhoods of $s_i = 0$, and hence ETD_i will not be convex outside of these neighborhoods. In Figure 8.1a we illustrate ETD_i for the case where r_i is everywhere greater than or equal to r_i, so that the firm is effectively maximizing profits and convexity rules. On the other hand, Figure 8.1b illustrates the nonconvexities that occur when r_i

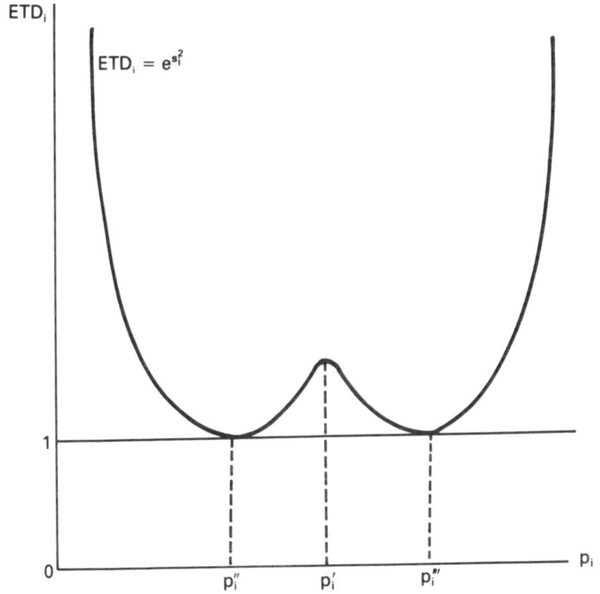

Figure 8.1b Target deviation function for isolated firm i when $r_i > \overline{r}_i$

lies above $\overline{r}_i$ in domains of p_i. Obviously, in seeking an unconstrained minimum of 8.19 we run the danger of becoming hung up on a local maximum cusp such as exists at p_i' of Figure 8.1b rather than achieving the minimum at one of the points of tangency with the horizontal line at p_i'' or p_i''' where $r_i = \overline{r}_i$. Hence, the model will not always be well behaved.

Our reason for striving for a convex programming problem is wholly one of analytical expediency. If a given realistic model specification for an industry is not convex, that fact must be accepted, for it reflects the reality that the industry's decision making leads to the attainment of 'good' solutions rather than necessarily globally optimal results. Human decision-making, in all likelihood, is more frequently than not of this type. However, although in reality if a firm attains a poor local optimum it will move to improve it, our rigid models are not so flexible in their goal striving. Hence, aiming for convex nonlinear programming formulations even at the expense of some distortion of reality is to achieve some protection against poor local optima, not an assertion that firms always strive for the very best attainable from their goal set and market setting. If the convexification cannot be obtained without great distortion of reality, it is preferable to embrace the nonconvexity as an important feature of the industry's operation.

Hybrid Models and Extensions

In this book we will remain exclusively with variants of Models 1, 2 and 3 in our illustrations and analyses, with major emphasis on Model 2. However, that should not obscure the great diversity of approach that is possible within the framework. In this section some of this versatility will be illustrated in brief fashion.

It is possible to mix modes of behavior for different firms. Some firms may have as a dominant interest the maximization of profits while others strive toward target rates of return. One can simply use the $e^{s_j^2}$ formulation for all firms j, setting $\bar{r}_j$ to realistic values for rivals who strive for target rates of return and formally defining $\bar{r}_j$ equal to unrealistically high magnitudes for firms that maximize profits, for this will effectively specify that behavior. Hence, $e^{s_j^2}$ is the general functional form that subsumes both types of behavior.

Other motivations are easily formulated. For example, if the industry operates with a dominant firm or firms as leaders, those firms may be depicted as maximizing profits or seeking target rates, with or without positive consonance factors for the like goals of their dominant rival compeers, if any. Followers i may be depicted as having objective functions of the form

$$\text{Min } Z_i = \sum_k \theta_{ik} e^{(p_i - \alpha_{ik} p_k)}, \tag{8.22}$$

subject to follower i's other goals as specified in constraints, where firms k are the leaders and θ_{ik} their absolute importance in firm i's view of power structure. Hence, follower i seeks to get as close to the leaders' prices as industry convention permits, subject perhaps to a minimum output or market share constraint, among others. The function 8.22 is globally convex in p_i which is instrumental for achieving the convexity discussed in Section III.

Simon's 'satisficing' or bounded rationality behavior[14] may be specified in several manners. The parameters which the firm attempts to satisfice are simply forced to lie between constraints that specify upper and lower bounds. Profits, or market share, or sales, or a group of such goal variables can be constrained to lie in such intervals. The objective function may then be taken to contain one goal with perhaps a primary importance for the firm in question in addition to the sum of rivals' objectives as discounted by consonance factors.

The firm's objective function might be specified to minimize firm i's price, for example, taking into account rivals' rates of return:

$$\text{Min } Z_i = p_i + \sum_{j \neq i} \theta_{ij} e^{s_j^2},$$

where the units of i's product for which p_i is stated can be varied to give greater or lesser impact of its minimization of Z_i. If rivals' terms are taken to be profits, presumably p_i will be insignificant in comparison to other-profit impacts on Z_i, and p_i might be eliminated entirely. In conditions of satisficing it presumably is a rather formal inclusion in any event.

Satisficing formulations are likely to be nonconvex programming problems where constraints cannot be kept linear or, even if linear, will occur in conjugate pairs that are the negatives of one another. For example, a satisficing goal for profits would be written as

$$\begin{aligned} m_i x_i &\geq U_i \\ -m_i x_i &\geq -L_i, \end{aligned}$$

where U_i and L_i are upper and lower bounds respectively. If the first constraint in the pair is convex, the second will be concave, and vice versa. Hence, one will violate the sufficiency conditions for a convex program. But the necessity of accepting nonconvexity in general as a condition of the real world applies with even greater force in industries in which satisficing prevails, for 'fuzzy' solutions are of the essence of such formulations.

IV SIMULATIVE THEORIZING WITH RIVALROUS CONSONANCE MODELS

General economic interdependence models, functionally and numerically unspecified, have yielded few insights into real phenomena. Indeed, nothing is so well established both in theory and in practice than the proposition that they can *at best* yield few unambiguous theorems,[15] and that in order to reap that meager harvest the most constraining assumptions must be included. Perhaps the best manner of illustrating this failure is to point to the flight of general equilibrium theorists to point set topology away from classic calculus formulations. That movement has reinforced the attachment to pure competition, and has all but made the notion of general equilibrium synonomous with existence and uniqueness of equilibrium proofs. It offers a classic example of the over-theoretization of a field, and the detached sterility that is the fate of a behavioral science that divorces itself from actual behavior. Small wonder that the focus of modern effort is the narcissistic one of constructing models largely for the purpose of proving they have solutions, however unattainable or uninteresting they may be.

Is it not time that the theory community switches its efforts away from the search for the grail of the theorem of universal applicability within the confines of a fantasy world toward the derivation of insights of relevance to specific real-world situations? Should not the generalizations about oligopoly behavior, for example, arise as consistent tendencies in the movements of prices and quantities in models specifically tailored to industries? Is not the lessening of ambition in the face of a rich and complex economic reality that can never be fully understood a need that must be grasped if economic theory is to progress beyond the unattainable fantasies of its youth?

In this spirit I have urged the development of a low-brow theoretical technique called 'simulative theorizing' (Kuenne [20]). It is concerned with the use of sensitivity or post-optimality analysis of numerically specified large-scale models to derive insights into the structural and functional characteristics of large-scale systems. In the study of oligopoly, of course, such work is directed to the obtaining of propositions about the implications of the power structure, primary and secondary objectives of rivals, cost changes, and so forth, of specific industries.

What is truly needed is a new 'econometrics of simulative parameter displacement' that would develop optimal techniques for designing and using specified models for such purposes. Available techniques at the present are most primitive and *ad hoc*. Nonetheless, even simple experiments with such oligopoly models are useful to demonstrate the extremely complicated causal pathways that exist and, therefore, the rather formidable tasks that are given ordinary least squares regression in tracing them. Such manipulations will be illustrated for a prototypical Model 2 in chapter 5. It was also employed to manipulate the GENESYS model of OPEC in [19].

Rivalrous consonance models are well suited to the purposes of simulative theorizing. Their parameters are selected for both relevance to industry behavior and empirical derivability. Their theoretical structure is simple in basic concept, being composed of functions that depict primary or secondary objectives and definitional relations. They are optimizing models and therefore generally more straightforward in solution than Monte Carlo or expected value simulation models. And, because of their relatively simple structure and optimizing behavior, the search for the driving factors of a solution or the sources of a counterintuitive result is simpler than for more complex models with diffuse motivational modes, as, for example, in pure simulation analyses.

With these features, therefore, the rivalrous consonance model of oligopoly reflects a particular view of the role of theoretical models. Their function is neither to yield propositions of universal validity on the one hand, nor to predict 'the' solution for a specific set of parameters describing a single present or future environment. The first, as argued above, do not exist in realistic oligopoly other than as empirical central tendencies derivable from many specific industry studies. The second goal is that of forecasting, and generally is useful only for a degree of quantitative accuracy that is simply not attainable given the limited degree of detail that can be included in theoretical constructions and the limitations of econometric techniques in deriving parameter values.

Rather, the usefulness of theoretical models is to yield primarily qualitative theorems about the structure and functioning of economic entities – theorems that ideally can be tested against the evidence of reality. That usefulness depends upon the practical ability to manipulate the models by parameter displacement to derive insights into the relevant phenomena. Those insights will be deduced of necessity in numerical terms, given the limitations of purely qualitative models, but they should be translated into qualitative or, at best, order-of-magnitude propositions. These structural/functional theorems of imprecise magnitude will be pursued in the theoretical and empirical work that follows. Their deduction constitutes the substance of simulative theorizing.

V THE STRUCTURE OF THE ANALYSIS

The presentation of rivalrous consonance theory is structured at three levels of analysis. The first level is the determination of oligopolistic prices when each firm assumes its rivals' prices are fixed in choosing its optimal price. Models 1, 2 and 3 are analyzed at this simple level in chapter 2. In chapter 3 the analysis turns to solution and parametric displacement techniques for the programming formulations of rivalrous consonance at this first level.

The second analytical level concerns itself with the classical short-run oligopoly problem of strategy choice when rivals' policy variables (prices in the present book) are simultaneously determined. Chapter 4 deals at length with the theoretical derivation of theorems under constrained and unconstrained conditions. Then, given the limitations of unspecified models, chapter 5 presents an extensive illustration of the techniques and power of simulative theorizing at this second level of analysis, using a fictitious Model 2 for its operations. Methods, concepts, and relevant questions are stressed in furtherance of the basic goals of the book.

The third level of analysis is long run, in that it frees the θ_{ij} to react under the eroding forces of self-interest. This analysis is conducted in chapter 6, and derives a point of Nash equilibrium which is suggested as a gravitational price vector for rivalrous consonance in mature oligopolies over time.

Finally, in chapter 7, the theoretical and practical problems of isolating $\mathbf{K}$ and θ_i (hence, θ) are faced, given their crucial importance in rivalrous consonance applications. Methods are developed and tested against the simulative theorizing results of chapter 5 to suggest their feasibility and the worthwhileness of further experimentation.

Notes

1. For surveys of this recent literature see Lester G. Telser [28]; James Friedman [11]; Tun Thin [29]; and William Baumol, John C. Panzar, and Robert D. Willig [4].
2. One recent attempt to integrate oligopoly theory into a general equilibrium framework is Thomas Marschak and Reinhard Selten [22]. Unfortunately, however, it is a very formal analysis which, in the current tradition of general equilibrium theory, equates the latter with proofs of existence of equilibrium. Negishi's [23] classic earlier attempt is really the generalization of monopoly rather than oligopoly in such an interdependent framework.
3. For a general summary of Wärneryd's work, see Johnsen [15], ch. 3.
4. Arthur A. Thompson [30], ch. 9, contains an excellent summary of the multi-objective nature of the firm's decision-making. Carl Kaysen [17] has been a particularly strong advocate of the view that a variety of goals may characterize firms' behaviors.

5. Kaplan, Dirlam and Lanzillotti [16] found this goal the most frequently stressed by the large firms in their study (p. 130), but it is typical of the poorly co-ordinated decision-making of the corporations that they found it to be of concern at the higher levels of management, the lower rungs of officialdom stressing profit margins on sales (p. 17).
6. This goal is emphasized by Bain [3], Andrews [1], Sylos-Labini [27], and Baumol, Panzar and Willig [4], among others.
7. For example, cf. Kaplan, Dirlam and Lanzillotti [16]: 'The officials who discussed policy with the interviewers found it difficult to analyze pricing as a separate activity distinct from others entering into company policy. Pricing decisions were made or influenced by executives from production, development, engineering, accounting, finance, sales, and public relations divisions. They were usually considered part of the general strategy for achieving a broadly defined goal' (p. 3). And: 'Indeed, top management frequently did not take these decisions and was surprisingly ignorant of the manner in which they were derived – that is, prices' (p. 5). An excellent presentation of the firm from the viewpoint of organization theory can be found in Oliver Williamson [32], especially 36–7.
8. To quote from Kaplan, Dirlam and Lanzillotti [16] again: 'Economists, legislators, and the public generally would like to see pricing decisions by big companies analyzed in logical fashion, with historical comparisons of competitors' prices, cost factors, and profit margins given consistent and quantitative weight in detailed memoranda of officials involved. Unfortunately for those who would insist on fully ordered business behavior, such strategic memoranda summarizing the considerations at an important conference leading to a price decision are rarely found. Perhaps the presumed formal conference was never held. Even where the people doing the pricing tended to have certain staff information placed before them while making up their minds, whether and just how that information was taken into consideration often remained obscure. There was no document tracing the steps by which the staff information could be said to account for the price decision. Repeatedly, reference was made to the "art" or "feel" of pricing rather than observance of a formula' (p. 5).
9. Among modern treatments, Tun Thin's [29] typifies this 'monolithic' approach. He identifies two possible motives of the oligopolist: either (*1*) to seek the maximum payoff for himself regardless of the impact on his rivals, or (*2*) to wage a price war to kill them off even at the cost of his own short-run profits. This type of 'vision' of oligopolistic behavior does not gain support from empirical study.
10. [18], p. (143).
11. See, for example, T. Bresnahan, [6].
12. In R. E. Kuenne [21] θ has been derived for the OPEC cartel in the 1973–81 time frame. And, again, further techniques for the derivation of power structure and binary consonance factors are presented in chapter 7.
13. The constraint can be written explicitly as a market share constraint in the following manner:

$$N_i - x_i \,/ \sum_{j \neq i} x_j \leqslant 0,$$

where $N_i = M_i / (1 - M_i)$, and where M_i is minimum acceptance market share. Unfortunately this constraint will not be convex, and hence destroys the convexity of the feasible region and the problem.

14. Herbert Simon [26].
15. See, for example, the implications of Samuelson's theorem of conjugate pairs that the profession has resolutely ignored except in the specific case of the theorem of the second best. P. Samuelson, [24], p. 19, and G. C. Archibald [2].

References

[1] Andrews, P. W. S., *Manufacturing Business*, Longmans, London (1949).

[2] Archibald, G., 'The Qualitative Content of Maximizing Models', *Journal of Political Economy*, 73 (1965), pp. 27–36.

[3] Bain, J., *Barriers to Competition*, Harvard University Press, Cambridge, Mass. (1956).

[4] Baumol, W., J. Panzar and R. Willig, *Contestable Markets and The Theory of Industry Structure*, Harcourt, Brace, Jovanovich, New York (1982).

[5] Bishop, R., 'Game-Theoretic Analyses of Bargaining', *Quarterly Journal of Economics*, 77 (1963), pp. 559–602.

[6] Bresnahan, T., 'Duopoly Models with Consistent Conjectures', *American Economic Review*, 71 (1981), pp. 934–45.

[7] Cyert, R. M. and M. H. DeGroot, 'Multiperiod Decision Models With Alternating Choice as a Solution to the Duopoly Problem', *Quarterly Journal of Economics*, 84 (1970), pp. 410–29.

[8] Cyert, R. M. and M. H. DeGroot, 'An Analysis of Cooperation and Learning in a Duopoly Context', *American Economic Review*, 63 (1973), pp. 14–37.

[9] Fog, B. *Industrial Pricing Policies: An Analysis of Pricing Policies of Danish Manufacturers*, North-Holland, Amsterdam (1960).

[10] Friedman, J. W. 'Individual Behavior in Oligopolistic Markets: An Experimental Study', *Yale Economic Essays*, 3 (1963), pp. 359–417.

[11] Friedman, J. W., 'Reaction Functions and the Theory of Duopoly', *Review of Economic Studies*, 35 (1968), pp. 257–72.

[12] Friedman, J. W., 'On Experimental Research in Oligopoly', *Review of Economic Studies*, 36 (1969), pp. 319–415.

[13] Hoggatt, A., 'Measuring the Cooperativeness of Behavior in Quantity Variation Duopoly Games', *Behavioral Science*, 12 (1967), pp. 109–21.

[14] Hoggatt, A., 'Response of Paid Student Subjects to Differential Behavior of Robots in Bifurcated Duopoly Games', *Review of Economic Studies*, 36 (1969), pp. 417–32.

[15] Johnsen, E., *Studies in Multiobjective Decision Models*, Studentlitteratur, Lund (1968).

[16] Kaplan, A., J. D. Dirlam and R. Lanzillotti, *Pricing in Big Business: A Case Approach*, Brookings, Washington, DC (1958).

[17] Kaysen C., 'The Corporation: How Much Power? What Scope?', ch. 5 in Edward S. Mason, *The Corporation in Modern Society*, Harvard University Press, Cambridge, Mass. (1959), pp. 85–105.
[18] Koopmans, T., *Three Essays on the State of Economic Science*, McGraw-Hill, New York (1957).
[19] Kuenne, R. E., 'The GENESYS Model of OPEC, 1974–1980: Structural Insights from a Non-Forecasting Model', *Energy Economics*, 4 (1982) pp. 146–58.
[20] Kuenne, R. E., 'Towards an Operational General Equilibrium Theory with Oligopoly: Some Experimental Results and Conjectures', *Kyklos*, 27 (1974), pp. 471–502.
[21] Kuenne, R. E., 'Rivalrous Consonance and the Power Structure of OPEC', *Kyklos*, 32 (1979), pp. 19–35.
[22] Marschak, T. and R. Selten, *General Equilibrium with Price-Making Firms*, Springer-Verlag, Berlin (1974).
[23] Negishi, T., 'Monopolistic Competition and General Equilibrium', *Review of Economic Studies*, 28 (1961), pp. 196–201.
[24] Samuelson, P., *Foundations of Economic Analysis*, Harvard University Press, Cambridge, Mass. (1947).
[25] Shubik, M., *Strategy and Market Structure*, John Wiley & Sons, New York (1959).
[26] Simon, H., 'Theories of Decision-Making in Economics and Behavioral Science', *American Economic Review*, 49 (1959), pp. 253–83.
[27] Sylos-Labini, P., *Oligopoly and Technical Progress*, Harvard University Press, Cambridge, Mass. (1969).
[28] Telser, L., *Competition, Collusion, and Game Theory*, Macmillan, London (1972).
[29] Thin, T., *Theory of Markets*, Harvard University Press, Cambridge, Mass. (1960).
[30] Thompson, A. A., Jr, *Economics of the Firm: Theory and Practice*, Prentice-Hall, Englewood Cliffs, NJ (1973).
[31] Wärneryd, W., see [15], ch. 3.
[32] Williamson, O., *The Economics of Discretionary Behavior: Managerial Objectives in a Theory of the Firm*, Prentice-Hall, Englewood Cliffs, NJ (1965).

Part II

Spatial Interdependence

Introduction to Part II

Among the most challenging problems in interdependence theory and modelling are those that arise in determining optimal locations for firms, consumers and public facilities in two-dimensional space. Discontinuities are rife – in some cases the agent should move to a distant point or should not, marginal changes being of no help; distance metrics give rise to complicated extremum conditions which are difficult to solve by closed analytic methods; problems may involve sequences of movements with large numbers of potential destinations possible at every choice node, all of which must be examined by combinatorial methods; and optimal moves may be complicated by all the uncertainties of oligopolistic rivalry, since location is a major form of non-price competition. Movements through time are generally one-directional, one-dimensional, and occur in a continuous manner; movements in its twin dimension do not offer these simplifications. Small wonder, therefore, that two characteristics are frequent in locational analysis: on the one hand, it has seen pioneering applications of mathematical elegance in economics, including the calculus of variations, the theory of functional equations, spatial differential equations, and real and topological analysis. On the other hand it must frequently rely upon heuristics and complete enumeration algorithms for solutions.

The five papers in this Part – two of them co-authored with mathematicians, without whose skills it would have been impossible to have written them – are concerned with spatial problems in two dimensions. The first four are operational in character, seeking to obtain exact or approximate solutions to difficult problems of location or movement in space. The last is a theoretical effort to apply Poisson probability distributions to spatial markets where demands for goods or services are only occasional for individual buyers (e.g. repair or emergency services).

Chapter 9 was co-authored with Harold W. Kuhn and addresses the 'generalized Weber problem', which remained without direct solution techniques for about fifty years. The problem can be stated simply: suppose a firm must supply products to *n* noncollinear markets with possibly different amounts demanded. Given transport rates per unit of weight and distance, and transport costs proportionate to weight and distance, where is the minimum transport-cost point for the firm to locate? The problem as stated by Weber and attacked by other analysts contained $n = 3$ points, so that the convex hull of those points was a triangle. Various geometrical solution methods were worked out for this case, but they are not applicable to cases where $n > 3$. Hence, the 'generalized' problem is for one where $n > 3$.

The generalized problem has a long history of mathematical interest – a fact of which we were unaware when we began this paper. It was made relevant to spatial economists by Alfred Weber in 1909 in its triangular form. The only known methods for its solution were the geometrical techniques referred to above or by physical analogue, the latter using Varignon's frame to find the resultant of forces set in train by suspending weights (proportionate to the relative importance of the markets) attached to strings over pulleys.

The algorithm presented here – a type of gradient method – was rediscovered independently by two other researchers: W. Miehle in 1958 and Leon Cooper in 1964. However, upon further search, it was discussed in one form or another by three other persons much earlier, with results published in journals not readily accessible to economists. Those sources are listed in the paper. This article has had the greatest role in publicizing the algorithm to spatial economists and also contains the most thorough proofs of existence and convergence.

Chapter 10, co-authored with Richard M. Soland, is an extension of the problem solved in Chapter 9, but a simple extension it is not. Suppose within the convex hull of the markets (generalized to 'sinks') one wished to locate $m > 1$ production points ('sources') in such manner as to allocate sinks to sources and locate sources in order to minimize transport costs. The distinctive feature of the 'multisource Weber problem', therefore, as compared with the 'single source' problem of Chapter 9, is the number of sources specified parametrically.

The multisource problem is vastly more complicated. Where, in the single source problem, the objective function was strictly convex over the interior of the convex hull and piecewise linear over the edges, hence everywhere convex, the multisource objective function meets none of the criteria for well-behaved functions in minimization problems. Where in the single source problem the iterative solutions rolled downhill speedily to the optimum from arbitrary starting points, in the new problem they are likely to be captured by local minima cusps.

Since transport costs are linear in weight and distance and no capacity constraints on shipments are placed upon the sources, all of a sink's shipments will originate from a single source in the optimal solution. Hence, shipments from a source to a sink will be a 0,1-problem, which suggests a combinatorial method of solution. MULTIWEB, a branch-and-bound algorithm is designed to obtain the global minimum. Because MULTIWEB requires large amounts of computer storage (at least in terms of computer capacity at the time the paper was written), it may not be feasible to use for large problems. An approximative heuristic method that yields local minima is also presented: CROSSCUT. The latter has been suggested by others, and we have had good success with it. Three different metrics are presented for distance measurement in both algorithms: Euclidean, approximate great circle, and metropolitan.

The branch-and-bound algorithm is presented in detail, computational experience is detailed, and the use of MULTIWEB and CROSSCUT in tandem to obtain increasingly better answers for large problems is discussed. With the storage capacity of today's computer, no doubt much larger problems can be solved than was true when the article was written. Even when the optimum solution for large problems was not obtained, MULTIWEB exits with very good approximations to it. It remains, to the best of my knowledge, the only known method of obtaining optimal solutions to the multi-source Weber problem.

Chapters 11 and 12 also form a related pair. Both approach a 'dynamic' (a better term would be 'space-sequenced') problem involving the definition of routes by delivery vehicles located at given sources through a set of sinks in a minimum-transport-cost manner. This is a problem of extreme complexity, and in Chapter 11 I urge that good approximative solutions be sought rather than exact solutions which were infeasible because of demands on computer time, or, indeed, the effort required to specify the problem for solution. The method I employ consists in finding a feasible solution to the problem, then generating feasible solutions efficiently and moving to revise the solution at each step of the sequence as improved feasible solutions emerge. The movement from one best feasible solution found to another is continued until a given amount of computer time is expended or until the successive improvements between such solutions become acceptably small. An approximate solution to a 7-vehicle, 13 source problem – which in terms of complexity is very large, as will be seen in Chapter 12 – is found by use of the method. Although not known at the time Chapter 11 was written, the approximate solution is probably the optimum, again as revealed by the different techniques of Chapter 12. In the course of the paper I suggested that branch-and-bound methods be used to generate feasible solutions more efficiently.

I wrote Chapter 12 four years later, by which time I had become more knowledgeable about the analytical power of branch-and-bound combinatorial techniques, largely through my work with Richard M. Soland in Chapter 10. This resulted in the design of DYNCOM for the *exact* solution of such space-sequenced problems as those dealt with in Chapter 11, at least for moderately sized problems, and good approximate solutions for large problems. The paper also served to introduce branch-and-bound programming to economists and regional scientists, as at that time (and indeed presently) it was not a well-known technique in either field.

The problems of Chapter 11 are revisited with DYNCOM, and new problems are devised to demonstrate its effectiveness and limitations. DYNCOM also includes options not available in the earlier method, such as recycling vehicles to their original starting sources after completion of their travels (although no conceptual problem prevents inclusion of this requirement) and the opportunity to use the three different metrics mentioned in the introduction to Chapter 10 or to simply read in the matrix of distances between sources and sinks as well as inter-sink distances.

A series of five problems is addressed with DYNCOM, including the 7-source, 13-sink problem addressed in Chapter 11. Node storage as a limitational factor constrains rapidly as problem size increases, and after 60 minutes of IBM/91 computer time (a near-paleontological time span) and storage of 100 000 nodes, the algorithm had found no better solution than that arrived at by the heuristic of Chapter 11. Hence the conjecture that that solution is in fact optimal. This experience highlights an important observation first made by Leon Cooper in dealing with such spatial problems: the objective function tends to be shallow-bottomed, so that the algorithm rolls down to a good solution quickly, but then spends huge amounts of time to find feasible solutions that provide very little improvement.

DYNCOM is a more systematic manner of searching for feasible solutions than the earlier method and, of course, has the advantage of providing a proof that the optimal solution will be reached after a finite (possibly very large) number of operations. It also prints out successively better feasible solutions as it moves down an inverted tree graph. Also, the ability to specify a percentage tolerance factor for the distance from the optimal solution that is acceptable can reduce time, and with it necessary node storage, substantially. Present day computer speed and memory capacity should provide much-improved performance of DYNCOM in solving large problems.

Further, in conformance with my stubborn views about the need for economists to strive for the operational and practical, and abandon an excessive concern for the 'optimal' when 'very good' is perfectly acceptable, I urge that such methods be used. This is especially advisable in light of the probable accuracy of the data they are analyzing. Thus in both chapters I have stressed the explicit complexities that bedevil such spatial problems to demonstrate the frequent futility and nonnecessity of searching for the 'very best'.

Chapter 13 addresses a spatial problem of a quite different sort. It analyzes a demand regime in two-dimensional space which follows a Poisson distribution. The product or service sold in this environment is demanded when a consumer's normal activity is disturbed at long intervals and restoration of the status quo requires immediate satisfaction of that demand, although some price elasticity exists. The demand is, therefore, for any single consumer a 'rare event', falling under the Poisson rule in time, and, for a group of consumers spaced over a market area, under same rule over space for any fixed time period. The demand for durable household good repairs or for emergency health services are good examples, as noted earlier in the introduction. The analysis seeks the relationship between firms' spatial behaviour and their price-output decisions. Monopolistic and competitive market structures are examined and both fixed and variable market areas are studied, when all firms are assumed to have rising costs.

In such an environment the monopolist's price will rise with the size of market area, but by far less than proportionately to the increase in market size. That is, a firm which increases its market area will raise price to reduce the density of demand. When transportation costs rise, the monopoly will lower price to increase demand density in the area, although price reductions are less proportionately than service area reduction. Hence the analysis leads us to expect that a monopoly selling such services in the face of a large price increase in petrol would reduce the radius of its market area and lower the price of the service simultaneously in order to increase demand intensity.

In competition as firms enter to service a common fixed market area, each firm's demand density falls initially, but it is led to reduce price to increase density, and by more than what density would be for the same market area under monopoly. Interesting results are obtained when only transportation costs are raised in this competitive environment. Market areas increase as a function of the number of firms over an initial domain, but then begin to fall as more firms enter.

The analysis can be adapted to analyze situations of this sort when transportation costs are borne by the customer, e.g. when physicians compete in a fixed market area for patients who pay the costs of transportation from their homes to physicians' offices. A rise in such costs should reduce demand density for a fixed number of physicians and lead them to reduce prices of their services. An increase in the number of physicians should have the same effect. If this does not occur – as some studies suggest – one is led to conjecture about the existence of rivalrous consonance in an oligopoly structure.

The spatial analysis of Part II serves well, I believe, to evoke the extreme complexity of interdependence in this frequently neglected dimension. Problems whose statements possess a surface simplicity are frequently seen to be incapable of analytical solutions and must be solved by iterative algorithms or investigated with heuristics. Other areas of microeconomics may equal spatial economics in these regards, but I know of none that surpasses it. It offers a range of challenges to the theorist of elementary realistic importance, and deserves a great deal more attention than it receives.

9 An Efficient Algorithm for the Numerical Solution of the Generalized Weber Problem in Spatial Economics*

Spatial economics in the past has been hampered in its progress by certain peculiarities which are inherent in the functional relationships among its variables. The frequency of corner solutions in its extremum problems, a common tendency for the constraints upon these problems to be stated as inequalities, and the importance of discontinuous functions in the field, are examples of these peculiarities. It is not fortuitous that the origins and development of programming techniques are so closely tied to the problems of spatial economics, and these techniques have resulted in an advance of breakthrough proportions in obtaining solutions to formerly insoluble spatial problems.

Another, and related, peculiarity of spatial economics, with both advantageous and disadvantageous implications, is the applicability of geometric methods of solution to many of its problems. The use of Euclidean (straight-line) distances in a problem insures that the resulting equations will be at least of the second order and frequently will involve the additional complication of the radical sign. An inability to solve such equation systems directly by algebraic means, and the difficulties involved in applying known numerical analytic

* Co-authored with Harold W. Kuhn, Professor of Mathematical Economics, Princeton University. Published originally in the *Journal of Regional Science*, 4 (1962), pp. 21–33, and reproduced with permission. The algorithm to be presented was developed by Kuenne as a research associate of the Econometric Research Program, Princeton University, and by Kuhn as a recipient of a grant from the National Science Foundation.

techniques to them, led to a dependence upon geometric methods for solutions. Of course, the ability to fall back upon them was fortunate in one sense, but it may well have played a part in the failure to search out usable numerical techniques of more specific applicability. Consequently, in such fundamental areas of the field as the location of a minimum transport-cost point, we find ourselves today limited to the primitive techniques involving plane geometry or physical analogue methods.

This paper presents an efficient algorithm for the derivation of a minimum transport-cost point for any number of elements, (i.e. raw materials and products) and for any system of weights in a generalized Weber problem. Our description of the algorithm as 'efficient' rests wholly upon an empirical basis: we have solved about 15 such problems involving 3 to 24 elements, and in each case the iterative solutions converged to very close approximations to the solution within 7 iterations. The calculations are well within the capabilities of the desk calculator for large numbers of elements, and the acceptability of the approximation of any iteration can be readily checked. At the present writing we have not been able to prove that the straightforward method will always yield a converging sequence of approximations, but we have developed a modified method which is proved to converge to the solution, and the latter may be used with but a slight increase in computational tedium.

Although the problem is of importance in contexts other than plant location,[1] as we shall illustrate by example, we have chosen to present it in the format of the generalized Weber problem, Section I contains a brief statement of this problem as Weber formulated it, as well as a brief outline of the existing methods of solutions. Section II presents the generalized Weber problem and derives a set of relevant theorems for the analysis to follow. The algorithm is then presented in Section III and is illustrated with two examples. Lastly, in Section IV we present some concluding remarks concerning the algorithm.

I THE WEBER PROBLEM

In his pioneering work, *Über den Standort der Industrien*, Alfred Weber [13] reduced the problem of locating a firm under the regime of pure transport orientation[2] to that of determining the point(s) of minimum transportation costs. In his formal presentation Weber limited his analysis to the case of the firm which produced a product at one site with two localized raw materials from given, distinct points, for sale in a single, given market point, when the three points were not collinear. Throughout this paper we retain the assumption of non-collinearity of all points, to avoid the trivial case of linear points.

In Weber's presentation, therefore, there were three elements j – one product and two non-ubiquitous inputs – and we shall adopt the convention that $j = 1$ symbolizes the product. The technological conditions were those of fixed input coefficients per unit of product, independent of scale of output. The unit of output was defined in terms of weight, $u_1 = 1$, and the fixed input coefficients $u_j \neq 1$, were defined in weight units per unit weight of product. Transport rates, r_j, $j = 1, 2, 3$, were prices of moving the respective elements per unit weight-distance; distances were defined for Euclidean 2-space; and total costs of transporting any element were proportional to weight and distance only.

We define

$$w_j = u_j r_j \tag{9.1}$$

which Weber termed 'ideal weights', or weights of elements adjusted for differential transport rates. We seek a point $[p]$, such that

$$\Phi = \sum_j w_j s_{pj} \tag{9.2}$$

is a minimum, where Φ is total transport costs, and s_{pj} is the distance from $[p]$ to $[m_j]$, the site of element j.[3]

Weber's 'locational triangle' is illustrated in Figure 9.1, with vertices at $[m_j]$, $j = 1, 2, 3$, and with the weights w_j written in parentheses at the relevant vertices. Clearly and trivially, the triangle forms the convex hull of the points $[m_j]$, and to economize effort we shall proceed to the generalization of Weber's problem in which more than three points form the convex hull. The results of our analysis of the generalized problem may be applied directly to this simpler case.

Before analyzing the generalized Weber problem, we may note that several methods for finding $[p]$ for the three-element Weber problem exist in the literature. For the special case where all w_j are equal, the Weber problem becomes in fact a classic geometry problem posed and solved by Jacob Steiner. He showed that if any angle of the triangle equalled or exceeded 120°, $[p]$ was to be found at the vertex of the obtuse angle, while if all angles were less than 120°, $[p]$

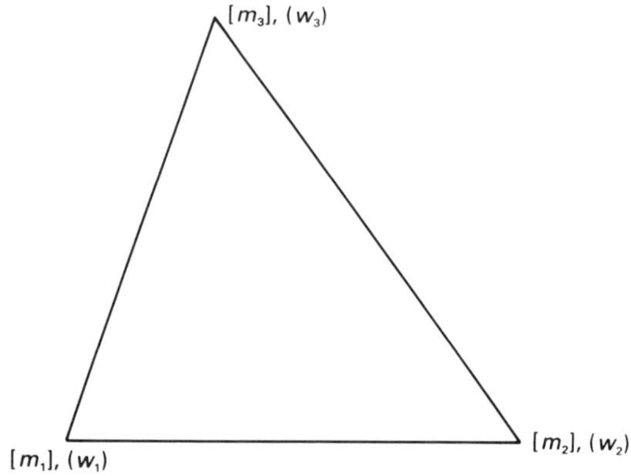

Figure 9.1 The Weberian locational triangle

was to be found in the interior of the convex hull. Steiner proved that [*p*] would be, in the latter case, at that point at which each side of the triangle subtends an angle of 120°, as illustrated in Figure 9.2.[4]

To solve the problem when the w_j are not equal, four methods have been developed, all of which depend upon geometrical or physical analogue constructions. The four methods, which we shall merely list, are:

1. *The Use of the Weight Triangle When it Exists*. In conjunction with the locational triangle, the weight triangle permits a geometrical determination of the resultant of forces. Weber employed this technique, and it was analyzed more deeply by Dean [3], p. 19.[5]
2. *The Use of Launhardt's Pole Principle*. This method was discussed by Launhardt [7] and Palander [10], ch. 7.
3. *The Use of Isodapanes*. Although Weber invented the isodapane – a line of equal total transport costs – he employed it only for the analysis of forces tending to draw the firm away from [*p*]. Its development as a tool for finding [*p*] we owe to Palander and Hoover [5], pp. 42–52.
4. *The Use of Varignon's Frame*. The use of this physical model was suggested by Georg Pick in his Mathematical Appendix to Weber's book, and the essential idea can be found in more modern treatments of force systems.[6]

Steiner's results are, of course, of limited applicability, while the four general methods of solution are cumbersome.

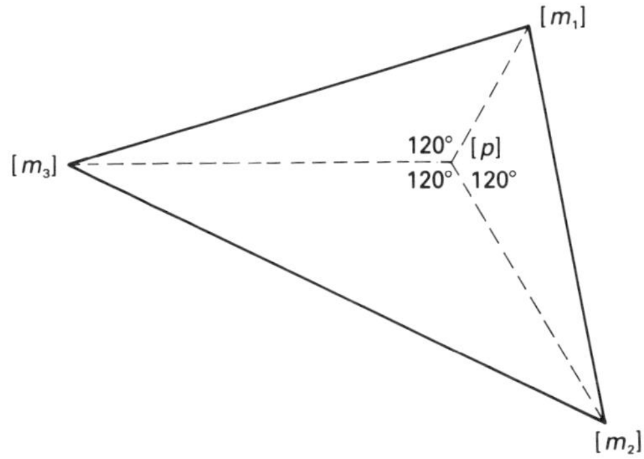

Figure 9.2 Steiner's problem: interior solution

II THE GENERALIZED WEBER PROBLEM[7]

We generalize the Weber problem simply by allowing n, the number of elements j and the number of points in space associated with them, to be greater than 3. The locational triangle then becomes a locational polygon with an $[m_j]$, $j = 1, \ldots n$, at each vertex. The polygon may or may not be convex. Therefore, we form the convex hull of the n distinct points, and symbolize the set of points $[a]$ in the convex hull as H and the set of points not in the convex hull as $[b] \notin H$. We seek to minimize 9.2 for a set of $w_j > 0$ and $s_{pj} = \sqrt{(x_j - x_p)^2 + (y_j - y_p)^2}$ by finding the coordinates (x_p, y_p) of $[p]$.

Theorem 1. If the n distinct points $[m_j]$ are not collinear, $\Phi([p])$ is a strictly convex function. If the points are collinear, $\Phi([p])$ is piecewise linear on the line through them and strictly convex elsewhere.

Proof. From the geometric interpretation of s_{pj} as in Euclidean space, it is obvious that $w_j s_{pj}$ is a convex function of $[p]$. Indeed, except on the half lines ending at $[m_j]$, it is strictly convex, and on each of these half lines the function is linear and therefore trivially convex. Since the sum of (strictly) convex functions is (strictly) convex, the theorem is proved.

Corollary. If the n distinct $[m_j]$ are not collinear, $\Phi([p])$ has a unique minimum.

We proceed now to the proof that this minimum must lie at a point $[p] \in H$, restricting ourselves to the case of noncollinear $[m_j]$. For a variety of reasons, either mathematical or physical and deriving from

Varignon's frame, it is natural to consider the gradient of Φ (i.e., the resultant of forces in the physical model). To this end, the gradient of Φ at any point $[p'] \neq [m_j]$ for all j is defined as

$$[R_{p'}] = \left[\sum_j \frac{w_j}{s_{p'j}}(x_j - x_{p'}), \quad \sum_j \frac{w_j}{s_{p'j}}(y_j - y_{p'}) \right] \tag{9.3}$$

The gradient is not defined at any vertex $[m_j]$, but, by physical analogy, we may set

$$[R_k] = \left[\sum_{j \neq k} \frac{w_j}{s_{kj}}(x_j - x_k), \quad \sum_{j \neq k} \frac{w_j}{s_{kj}}(y_j - y_k) \right] \tag{9.4}$$

and extend the definition of $[R_{p'}]$ by defining

$$[R_{m_k}] = \max\left(1 - \frac{w_k}{\|[R_k]\|}, 0\right)[R_k], \qquad k = 1, \ldots, n \tag{9.5}$$

where $\|[R_k]\|$ denotes the modulus of $[R_k]$ and is compared with w_k. If $w_k > \|[R_k]\|$, then $[R_k] = [0]$; otherwise, we have assigned a 'gradient' of magnitude $\|[R_k]\| - w_k$ in the direction of $[R_k]$.

The theorem we seek follows as a corollary from the following theorem:

Theorem 2. The point $[p]$ *minimizes* $\Phi([p])$ *if and only if* $[R_p] = [0]$.

Proof. If $[p] \neq [m_j]$ for all j, the convexity and differentiability of Φ imply that the first-order condition $[R_p] = [0]$ is both necessary and sufficient for a minimum.

If $[p] = [m_k]$, where k is any j, we consider a perturbation from $[p]$ to $[p^*] = [x_p + ta, y_p + tb]$, where $\sqrt{a^2 + b^2} = 1$. Then direct calculation yields

$$\frac{d\Phi([m_k])}{dt} = w_k - [R_k] \cdot [a, b] \tag{9.6}$$

and hence the direction of greatest decrease is $[a, b] = [R_k]/\|[R_k]\|$. Clearly $[m_k]$ yields a local minimum if and only if $(w_k - \|[R_k]\|) \geqq 0$, which is equivalent to $[R_{m_k}] = [0]$. Again, the convexity of Φ implies that $[R_{m_k}] = [0]$ is both necessary and sufficient for a global minimum.

We may now show that $[p] \in H$, and that if $[p] \neq [m_j]$ for any j, it does not lie on the boundary of H and therefore is an interior point.

Corollary. If $[p]$ minimizes $\Phi([p])$, $[p] \in H$.

Proof. For $[p] \neq [m_j]$ for any j, $[R_p] = [0]$ gives $[p]$ as a weighted sum of the $[m_j]$ with positive weights that sum to 1. Otherwise, $[p] = [m_j]$ for some j and is trivially an element of H.

Corollary. If $[p]$ minimizes $\Phi([p])$ and $[p] \neq [m_j]$ for any j, $[p]$ is not on the boundary of H.

Proof. This is a consequence of the fact that $[R_p] = [0]$ yields an expression for $[p]$ as a *positive* weighted sum of the $[m_j]$.

The vertex solution will occur under well-defined conditions:

Theorem 3. The minimum cost point $[p]$ will equal $[m_k]$ if and only if

$$w_k^2 \geqq \left(\sum_{j \neq k} w_j \cos \theta_j \right)^2 + \left(\sum_{j \neq k} w_j \sin \theta_j \right)^2 \tag{9.7}$$

Proof. By *Theorem 2* and equation 9.5 above, $[m_k]$ minimizes Φ if and only if $w_k \geqq \|[R_k]\|$. Since

$$\cos \theta_j = \frac{x_j - x_k}{s_{kj}}, \quad \sin \theta_j = \frac{y_j - y_k}{s_{kj}} \tag{9.8}$$

the theorem follows immediately:

Corollary. A sufficient condition for the inequality to hold in equation 9.7 is that $w_k \geqq \sum_{j \neq k} w_j$.

Proof. This corollary follows from *Theorem 3* and the elementary inequality

$$\sum_{j \neq k} w_j \geqq \sqrt{\left(\sum_{j \neq k} w_j \cos \theta_j \right)^2 + \left(\sum_{j \neq k} w_j \sin \theta_j \right)^2} \tag{9.9}$$

with the equality holding if and only if the θ_j are equal for all $j \neq k$.

The solution to Steiner's problem for $n = 4$ has been obtained and is shown to lie at the intersection of the diagonals of the polygon (when convex).[8] Methods 1 and 2 in Section I, involving the use of the weight triangle and Launhardt's pole principle, cannot be generalized to cases where $n > 3$. The isodapane method can be employed, but since isotims or isovectors must be drawn about all points, and a succession of isodapanes constructed until the highest level of isodapane is obtained, the method is tedious. Lastly, Method 4, involving the use of Varignon's frame, is also applicable, but it too may involve a good deal of labor.

III AN ALGORITHM FOR THE SOLUTION OF THE GENERALIZED WEBER PROBLEM

We have developed an algorithm for the solution of the generalized Weber problem which defines an iterative technique that, in our experience, has not failed to yield $[p]$ after six or seven iterations. The technique is applied to the first-order minimization conditions on Φ, which we may obtain by taking the partial derivatives of equation 9.2 with respect to x_p and y_p, and setting these expressions equal to 0:

$$\frac{\partial \Phi}{\partial x_p} = \sum_j \frac{w_j}{s_{pj}}(x_j - x_p) = 0 \tag{9.10.1}$$

$$\frac{\partial \Phi}{\partial y_p} = \sum_j \frac{w_j}{s_{pj}}(y_j - y_p) = 0 \tag{9.10.2}$$

To illustrate the procedure from this point we shall employ it to find $[p]$ in Steiner's problem for $n = 4$.

Steiner's Problem, $n = 4$.

Assume the $[m_j]$ to lie at [0,0], [1,0], [0,1], and [5,8] respectively, and that $w_1 = w_2 = w_3 = w_4 = 1$. The polygon is graphed in Figure 9.3. We shall proceed in step-by-step fashion to the determination of $[p]$.

Step 1. We derive an initial estimate of $[p] = [x_p, y_p]$ by minimizing the sum of the *squares* of the distances of each $[m_j]$ from $[p]$, when these squared distances are weighted by the w_j (in the present case equal to unity for all j). This is, of course, equivalent to finding the center of gravity of the weights w_j. We define the function

$$\Phi^* = \sum_j (w_j s_{pj}^{2\,*}) = \sum_j (w_j[(x_j - x_p^*)^2 + (y_j - y_p^*)^2]) \tag{9.11}$$

To minimize this, necessary conditions are

$$\frac{\partial \Phi^*}{\partial x_p} = \sum_j w_j(x_j - x_p^*) = 0$$
$$\frac{\partial \Phi^*}{\partial y_p} = \sum_j w_j(y_j - y_p^*) = 0 \tag{9.12}$$

and, therefore,

$$x_p^* = \frac{\sum_j (w_j x_j)}{\sum_j w_j}$$
$$y_p^* = \frac{\sum_j (w_j y_j)}{\sum_j w_j} \tag{9.13}$$

Since the point $[x_p^*, y_p^*]$ is the one which would be obtained at the intersection of a straight line passed through the scatter of the $[m_j]$ to minimize the sum of the weighted squares of the deviations in the y-direction, and of a similar line minimizing the sum of the squares of deviations in the x-direction weighted in the same manner, we shall term $[p^*]$ the 'weighted least-squares point'.

In Table 9.1, after listing the data of the problem in columns 1 through 4, we have computed the weighted co-ordinates needed for solving the equations 9.13 in columns 5 and 6, and substituting the entries in the Σ-row for columns 4, 5, and 6 into the equations, we obtain $[x_p^*, y_p^*] = [1.50, 2.25]$. These values become $[p^1] = [x_p^1, y_p^1]$, our first estimate of $[p]$.

Step 2. It will be convenient to rewrite the equations in 9.10 as follows:

$$x_p = \frac{\sum_j (w_j^0 x_j)}{\sum_j w_j^0}$$
$$y_p = \frac{\sum_j (w_j^0 y_j)}{\sum_j w_j^0} \tag{9.14}$$

Table 9.1 The solution to Steiner's problem for n=4

m_j	x_j	y_j	w_j	w_jx_j	w_jy_j	x_j-x^1	y_j-y^1	s_{p^1j}	w_j^{01}	$w_j^{01}x_j$	$w_j^{01}y_j$	x_j-x^2	y_j-y^2	s_{p^2j}	w_j^{02}	$w_j^{02}x_j$	$w_j^{02}y_j$
1	2	3	4	5	6	7	8	9	10	11	12	13	14	15	16	17	18
1	0	0	1	0	0	–1.5	–2.25	2.70	0.37	0	0	–0.81	–1.17	1.42	0.70	0	0
2	1	0	1	1	0	–0.5	–2.25	2.30	0.43	0.43	0	0.19	–1.17	1.19	0.84	0.84	0
3	0	1	1	0	1	–1.5	–1.25	1.95	0.51	0	0.51	–0.81	–0.17	0.83	1.20	0	1.20
4	5	8	1	5	8	3.5	5.75	6.73	0.15	0.75	1.20	4.19	6.83	8.01	0.12	0.60	0.96
	Σ		4	6	9	$[p^1]$=[1.5, 2.25]			1.46	1.18	1.71	$[p^2]$=[0.81, 1.17]			2.86	1.44	2.16

where

$$w_j^0 = \frac{w_j}{s_{pj}} \tag{9.15}$$

For s_{pj} in equations 9.14 and 9.15 we substitute s_{p^1j}, obtained by using $[p^1]$ in the definition of distance:

$$s_{p^1j} = [(x_j - x_p^1)^2 + (y_j - y_p^1)^2]^{1/2} \tag{9.16}$$

These calculations are found in columns 7, 8, and 9 of Table 9.1.

Step 3. Compute the w_j^{01} by dividing the values in column 4 by the values in column 9, row by row. These results are entered in column 10 and summed.

Step 4. Compute $w_j^{01}x_j$ and $w_j^{01}y_j$ by multiplying column 2 by column 10 and column 3 by column 10 respectively, row by row. These are entered in columns 11 and 12, together with the sum for each column.

Step 5. Substitute the values in the Σ-row for columns 10, 11, and 12, into (14) to get $[p^2] = [x_p^2, y_p^2]$, the second estimate of $[p]$. For our example, $[p^2] = [0.81, 1.17]$.

Step 6. Repeat steps 1 through 5 using $[p^2]$ as the initial estimate of $[p]$ and derive $[p^3]$. Then repeat steps 1 through 5 using $[p^3]$ as the initial estimate of $[p]$, and so forth, until, on the sth iteration, $x_p^s - x_p^{s-1}$ and $y_p^s - y_p^{s-1}$ are negligible.

Step 7. $[p^s]$ is a close approximation to $[p]$, except when $[p]$ is a corner value. The closeness is subject to check by substitution of x_p^s and by y_p^s into the equations of 9.10 to see how close the results are to zero.

In Table 9.1 we have carried out one more iteration to illustrate the method. In Table 9.2 we have summarized the results obtained for 7 iterations. The last column, $s_{i,\ i-1}$, gives the distances between successive $[p^i]$.

We substitute the estimates $x_p = 0.38$, $y_p = 0.61$ into the equations of (9.10) and obtain a residual of $0.01 \simeq 0$ for 9.10.1 and of $-0.014 \simeq 0$ for 9.10.2. From Figure 9.3 it is seen that our estimate is at the intersection of the diagonals, whose equations are $y = 1.6\,x$ and $y = 1 - x$.

Note the efficiency of the method. We deliberately chose an example in which one vertex was at an extreme position compared

Table 9.2 Results of the iterations in the solution to Steiner's problem

Iterations (i)	$\Sigma\, w_j^{0i}$	x_p^i	y_p^i	$s_{i,\, i-1}$
1	1.46	1.50	2.25	–
2	2.86	0.81	1.17	1.28
3	4.21	0.50	0.76	0.51
4	4.45	0.40	0.68	0.13
5	4.49	0.38	0.63	0.05
6	4.49	0.38	0.62	0.01
7	4.48	0.38	0.61	0.01

with the other three, for this will result in a poor first estimate of [p]. That is, the least-squares method will exaggerate the importance of an extreme item by virtue of the nature of squaring. Indeed, [p*] is a poor estimate of [p]. Yet the solution was obtained in 4 more iterations, while a good approximation to it was obtained as early as the fourth iteration. The speed of the convergence is well attested by the $s_{i,\, i-1}$.

A Population Center Problem

Let us turn now to a generalized Weber problem where the w_j are not the same. Further, to illustrate the applicability of the Weber problem to other areas we shall step out of the sphere of spatial economics. Suppose we have n cities with populations w_j, $j = 1, \ldots, n$. Let us seek that point [p] from which the sum of the distances to these cities weighted by the populations is a minimum. We may view [p] as a focal point for population in the region containing the n cities – a point which may be useful for many purposes.

In the field of spatial economics, [p] may be interpreted in another way. Suppose the n sites [m_j] were population clusters in a large city and we sought to locate a shopping center. If we assumed that the attraction of the shopping center upon each cluster varied directly as the population of the cluster and inversely as the distance, then [p] will be the point of maximum interaction between populations and shopping center. Of course, the w_j may be interpreted as any number of other variables which may be important in the problem: income, families, automobiles, and so forth.

For our example we have chosen the 24 cities in the Russian Ukraine which rank among the 100 most populous cities in the

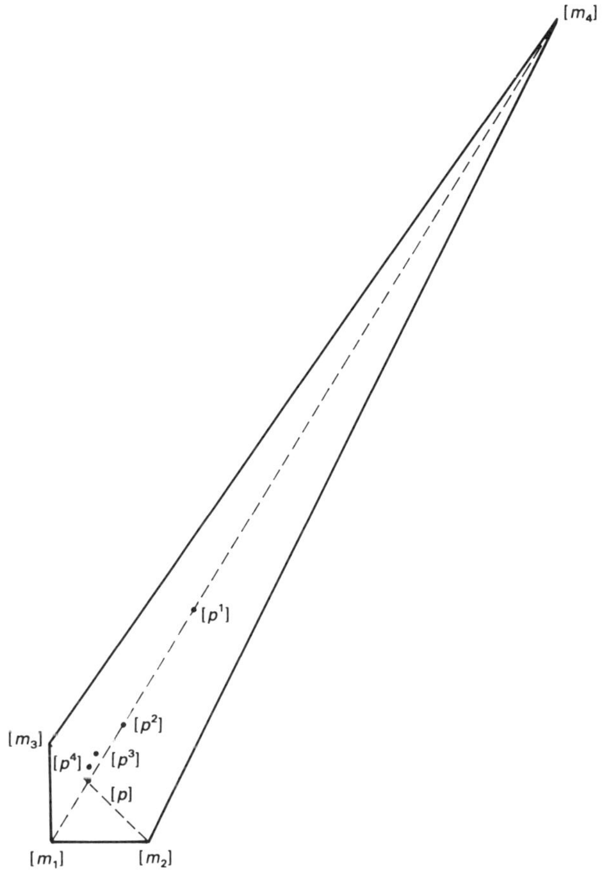

Figure 9.3 Solution to Steiner's problem, $n = 4$

U.S.S.R.[9] The cities are listed in Table 9.3 with their weights w_j equal to the proportion each has of the population of the 100 largest Russian cities. The x_j and y_j are degrees of north latitude and east longitude respectively correct to the nearest full degree.[10]

Table 9.4 lists the results of the iterations for the problem. The residual errors obtained by substituting 47.60 for x_p in 9.10.1 and 35.32 for y_p in 9.10.2 are $-0.00082 \simeq 0$ and $0.01516 \simeq 0$ respectively.

Two details remain to complete the description. The first deals with vertex solutions. Although Theorem 3 provides a necessary and sufficient condition for a vertex solution, and therefore may be used to test all vertices before beginning the algorithm proper, it is tedious

Table 9.3 Locations and weights for major Ukraine cities

City	x_j	y_j	w_j	*City*	x_j	y_j	y_j
Minsk	54	28	0.012	Zhdanov	47	38	0.007
Lvov	50	24	0.010	Stalino	48	37	0.016
Kishinev	47	28	0.005	Makeyevka	48	38	0.008
Odessa	46	31	0.014	Gorlovka	48	38	0.007
Nikolaev	47	32	0.005	Taganrog	47	39	0.005
Kherson	47	33	0.004	Krasnodar	45	39	0.007
Sevastapol	45	34	0.004	Rostov	47	40	0.014
Simferopol	45	34	0.004	Shakhty	48	40	0.004
Krivoi Rog	46	34	0.009	Kadiyevka	49	39	0.004
Dneprodzerhinsk	48	35	0.004	Kharkhov	41	37	0.021
Dnepropetrovsk	48	35	0.015	Kiev	50	31	0.026
Zaporozhe	48	36	0.010	Gomel	52	31	0.004

Table 9.4 Results of the iterations in the solution to the USSR problem

Iterations (i)	$\Sigma\, w_j^{0i}$	x_p^i	y_p^i	$s_{i,\, i-1}$
1	0.0782	47.48	34.36	–
2	0.0914	47.51	35.74	0.381
3	0.0957	47.64	35.51	0.264
4	0.0963	47.62	35.35	0.161
5	0.0965	47.60	35.32	0.036

to do so. Instead, we suggest only a preliminary check of the condition of the Corollary to Theorem 3 for weight dominance. If this test does not produce a dominant vertex, but the algorithm constructs a sequence which appears to be converging toward a vertex, the more complicated condition 9.9 should be checked for that vertex.

Finally, a last qualification must be added to insure convergence of the algorithm. A simple condition, which is sufficient for convergence but which we have been unable to prove in general is that $\Phi([p^{i+1}]) < \Phi([p^i])$ for each step of the algorithm: that is, that the objective function yield monotonically diminishing values with successive iterations. In our experience with the algorithm transport costs have *always* fallen with each iteration, leading us to suspect that cases in which this convergence does not occur will be rare indeed.

However, a slight modification of the procedure in the event that

$\Phi([p^{i+1}]) \geqq \Phi([p^i])$ is possible which will lead to this convergence. It is simply to cut the step length in half until this criterion is met, and then to continue with the stepwise procedure. Suppose, for example, that $\Phi([p^3]) \geqq \Phi([p^2])$ for the problem whose results are summarized in Table 9.4. The step in the x-direction which resulted in this misadventure was $47.64 - 47.51 = 0.13$, and the step in the y-direction was $35.51 - 35.74 = -0.23$. The modified procedure then dictates that the x-step be cut to 0.065 and the y-step to -0.115, so that the next iteration is begun with $[p^{3'}] = [47.445, 35.625]$, and that this procedure be continued until the value of $\Phi([p^{i+1}]) < \Phi([p^i])$. This does not complicate the computation excessively, since the distances, which require most of the computational time in determining the values of the objective function, must be computed for each iteration anyway.

Again, we do not know of an example in which this modified procedure is necessary, but the rigorous proof of convergence requires it in our present state of knowledge of the problem.

IV CONCLUDING REMARKS

The algorithm proposed in Section III may be regarded as a transformation which maps each trial point into a new trial point. Informally, if $[p]$ is not a vertex, the transformation T: $[p] \rightarrow [p'] = T([p])$, where $[p'] = (x'_p, y'_p]$ is the solution of

$$[R_{p'}] = \left[\sum_j \frac{w_j}{s_{pj}}(x_j - x'), \quad \sum_j \frac{w_j}{s_{pj}}(y_j - y') \right] = 0 \tag{9.17}$$

For the sake of continuity, the definition of T must be extended to the vertices by $T([x_j, y_j]) = [x_j, y_j]$ for $j = 1, \ldots, n$.

To study the analytic properties of T, it is convenient to introduce the weighted harmonic mean of the distances s_{pj} by defining

$$h([p]) = \frac{\prod_j s_{pj}}{\sum_k (w_k \prod_{j \neq k} s_{pj})} \tag{9.18}$$

Note that, since the points $[x_j, y_j]$ are assumed to be distinct, $h([p])$ is a continuous function and $h([p]) = 0$ if and only if $[p] = [x_j, y_j]$ for some j. With this definition, a straightforward calculation confirms that the algorithm may be regarded as the repeated application of the transformation

$$T: [p] \rightarrow [p'] = [p] + h([p])[R_p]. \tag{9.19}$$

Remark that T is a continuous mapping of the entire plane into the convex hull of the points $[m_j] = [x_j, y_j]$. The continuity is a simple consequence of the fact that $[R_p]$ is continuous except at the vertices (and is bounded in the neighborhood of those points), while $h([p])$ is continuous in the entire plane (and vanishes at the vertices).

Thus:

Theorem 4. If the point $[p]$ *minimizes* $\Phi([p])$, *then* $[p]$ *is a fixed point of T. If* $[p]$ *is a fixed point of T that is not a vertex, then* $[p]$ *minimizes* $\Phi([p])$.

Proof. If $[p]$ minimizes $\Phi([p])$, then $[R_p] = 0$ and $T([p]) = [p]$. If $[p]$ is a fixed point of T that is not a vertex, then $T([p]) = [p]$, and $[p] + h([p])[R_p]$, and $h([p]) \neq 0$. Hence, $[R_p] = 0$ and $[p]$ minimizes $\Phi([p])$.

The size of the step is determined by the magnitude of $h([p])[R_p]$. It can be shown[11] that the iterative procedure will converge to the minimum if $\Phi([p^{i+1}]) < \Phi([p^i])$ for all i. Thus, a sufficient condition on the step length is that this inequality be satisfied. We conjecture that it is satisfied for the step lengths given by the algorithm proposed in this paper, but have been unable to prove it.

Added in proof: In the period between the submission of this paper and its publication by the Journal, the results have been extended and further literature bearing on the problem has been discovered. We shall sketch the nature of the extensions and list only those references that bear directly on the material developed in the body of the paper. First of all, the algorithm of the paper has been considered, independently of the present account, at least three times. The first published version seems to be that of E. Weiszfeld in the *Tohoku Mathematical Journal*, 43 (1937), pp. 355–86. His convergence proof contains an error but may be revised to provide the following valid theorem:

Either (1) $T^i([p]) = [x_j, y_j]$ for some i and j or (2) $T^i([p])$ converges to the unique minimum point. Case 1 can occur for at most a denumerable number of points $[p]$ in the convex hull of the $[x_j, y_j]$.

Secondly, there is a problem dual to the Weber problem, with all of the desirable properties of linear programming duality. This dual can be stated:

Find n vectors (u_j, v_j) such that $\sum_j u_j = \sum_j v_j = 0$, $u_j^2 + v_j^2 \leqq w_j^2$ for all j, and minimizing $\sum_j(u_jx_j + v_jy_j)$. Although this duality was believed to be new, the special case of $n = 3$ and all $w_j = 1$ was published by Ed. Fasbender in *Journal für Mathematik*, 30 (1846), pp. 230–1. Indeed, the history of the original problem goes back farther than is commonly believed, having been posed by Fermat (as an exercise in maxima and minima) and solved geometrically by E. Torricelli before 1650! A complete account will be published elsewhere.

Notes

1. We shall solve a problem in Section III concerned with location of a population focal point for clusters of cities. In the recent literature, the problem has been phrased in terms of warehouse location (Eisemann [4]) and of the location of a switching center for a communication system (Palermo [11]).
2. That is, when all cost elements other than those which vary *regularly* with space are held constant. Labor costs are equal at all potential sites, no economies of agglomeration occur at presently industrialized sites, and so forth.
3. To avoid confusion, we shall denote all points in space as vectors (e.g. $[p]$ and $[m_j]$) to distinguish them from scalars and components of vectors.
4. See Courant and Robbins [1] for Steiner's proof. See also M. Klamken's note to Eisemann [4].
5. See also Isard [6], p. 225.
6. As, for example, in Polya [12] pp. 147–8 and Miehle [8].
7. This section may be omitted by the reader who is primarily interested in the technique for obtaining the numerical solution. It is recommended that he proceed immediately to Section III.
8. Courant and Robbins [1], p. 360. The proof is not given there, but we shall demonstrate the validity of the theorem in an example solved by our algorithm in Section III.
9. The population data are estimates for 1961 from [9].
10. Of course, despite our use of latitudes and longitudes, we are not using great circle distances, but rather are using two-dimensional straight-line distances.
11. See Curry [2].

References

[1] Courant, R. and H. Robbins, *What is Mathematics?* New York (1941).
[2] Curry, H. B., 'The Method of Steepest Descent for Non-Linear Minimization Problems', *Quarterly of Applied Mathematics*, 2 (1944), pp. 258–60.
[3] Dean, W. H., Jr., *The Theory of the Geographic Location of Economic Activities*, selections from the PhD thesis, University of Michigan, Ann Arbor (1938), p. 19.
[4] Eisemann, K., 'The Optimum Location of a Center', *SIAM Review*, 4 (1962), pp. 394–5.
[5] Hoover, E. M., *Location Theory and the Shoe and Leather Industries*, Harvard University Press, Cambridge, Mass. (1937), pp. 42–52.
[6] Isard, W., *Location and Space Economy*, New York (1956).
[7] Launhardt, W., *Mathematische Begründung der Volkswirtschaftslehre*. Leipzig (1885).
[8] Miehle, W., 'Link-Length Minimization in Networks', *Operations Research*, 6 (1958), pp. 232–43.
[9] *National Economy, U.S.S.R., 1960 Statistical Yearbook*, Moscow.
[10] Palander, T., *Beiträge zur Standortstheorie*, Uppsala (1935).
[11] Palermo, F. P., 'A Network Minimization Problem', *IBM Journal of Research and Development*, 5 (1961), pp. 335–7.
[12] Polya, G., *Induction and Analogy in Mathematics*, Princeton University Press (1954).
[13] Weber, A., *Über den Standort der Industrien*, Tübingen (1909). Translated as *Alfred Weber's Theory of the Location of Industries*, University of Chicago Press (1929), by C. J. Friedrich.

10 Exact and Approximate Solutions to the Multisource Weber Problem*

I THE WEBER PROBLEM SET

The concern of modern spatial economists with the optimal location of variable points (*sources*) in 2-space with respect to a set of fixed points (*sinks*), when the co-ordinates of the sources may vary continuously, dates from the publication of Alfred Weber's work in industrial location theory [15]. Weber's analysis was largely confined to the location of a single source, and, although he offered no method of solution, recent work has led to an efficient algorithm for exact solution of the problem (the single-source algorithm). The present paper treats the multisource Weber problem and presents (*1*) a branch-and-bound algorithm for the exact solution of the problem, which, to the best of our knowledge, is original (MULTIWEB), and (*2*) an approximate algorithm, to be used in support of MULTIWEB or in its place when appropriate, for which we claim no priority (CROSSCUT).

The Single-Source Weber Problem

The specifics of Weber's original formulation provide a useful introduction to the problem.

* Co-authored by Richard M. Soland and published originally in *Mathematical Programming*, 3 (1972), pp. 193–209, and reproduced with permission of Elsevier Science Publishers.For a more detailed discussion of the analyses contained in the present paper, the computational results, and for detailed program listings, the reader is referred to [8]. Tapes for the two programs discussed in the present paper are available at cost of reproduction from Institute for Defense Analyses, Alexandria, Virginia, USA.

Given:

(*i*) $n-1$ point locations (sinks) of the $n-1$ different raw materials used in the production of a product, with co-ordinates (a_j, b_j), $j = 1, 2, \ldots, n-1$;

(*ii*) sink n, at (a_n, b_n), a point market to which the product is shipped;

(*iii*) the weights of raw materials, w_j, $j = 1, 2, \ldots, n-1$, required per unit weight of product ($w_n \equiv 1$); and

(*iv*) shipping costs which are proportional to weight and Euclidean distance only.

Find: (x, y), the co-ordinates of a point source such that the total shipping cost for transferring the raw materials to the source and moving the product from the source to the market is a minimum.

The production point determined in this single-source problem was termed the *Weber point* by Kuhn and Kuenne [9], and its co-ordinates are obtained by minimizing

$$\Phi = \sum_j w_j[(x - a_j)^2 + (y - b_j)^2]^{1/2} \tag{10.1}$$

with respect to x and y.[1]

An algorithm for the x and y that minimize 10.1 was developed independently by Cooper [3] and Kuhn and Kuenne [9], and a somewhat similar iterative technique for a slightly different problem was evolved earlier by Miehle [11]. We shall term this algorithm the *single-source algorithm*. Its convergence to the global minimum has been proved (Kuhn and Kuenne [9]), and computational experience with it has been very good.

The Multisource Weber Problem

Consider, now, a generalization of the single-source Weber problem.

Given:

(*i*) n sinks in the plane at (a_j, b_j), $j = 1, 2, \ldots, n$;

(*ii*) the weights w_j which must be moved between the sinks and the sources; and

(*iii*) shipping costs proportional to weight and Euclidean distance only.

Find: (x_i, y_i), $i = 1, 2, \ldots, m$, the co-ordinates of m sources such that the sum of the weighted distances between sinks and sources – allowing for the possibility of fractional-weights movements – is a minimum.

Let $z_{ij} \in [0,1]$ be the proportion of w_j moving between sink j and source i. The problem may then be stated as

$$\text{minimize } \Phi = \sum_i \sum_j z_{ij} w_j [(x_i - a_j)^2 + (y_i - b_j)^2]^{1/2}, \tag{10.2}$$

subject to

$$\sum_i z_{ij} - 1 = 0, \quad j = 1, 2, \ldots, n. \tag{10.3}$$

By virtue of the linearity of transport costs when source locations are fixed and the lack of constraints on the volumes shipped between sinks and sources, the z-terms will in the solution be 0 or 1, and the problem may therefore be immediately interpreted as the determination of the optimal assignment of sinks to m clusters. Once the optimal clusters have been determined, the single-source algorithm could be used to solve for the optimal locations of the m sources within the m clusters. For the moment, however, it will be convenient to ignore this insight into the nature of the problem, and to treat the z_{ij} as continuous variables.

The relative simplicity of the single-source algorithm and the explanation of its speedy convergence spring from the strict convexity of the objective function 10.1 when the n sinks are not collinear (in which case the function is piecewise linear). The algorithm chooses an initial position in the convex hull of the sink locations (the center of gravity is recommended) and rolls speedily downhill to the global minimum.

Such mathematical nicety is not revealed, however, by the Φ function 10.2 of the multisource problem; as shown by Cooper [5], pp. 9–11, it is neither convex nor concave.[2] The multisource Weber problem, therefore, is a much more difficult one than its single-source progenitor, but two possible approaches to its solution are evident.

First, the insight that only values of 0 and 1 need be considered for the z_{ij} suggests a combinatorial programming approach, and the availability of the single-source algorithm to obtain optimal source locations for any allocation of sinks to clusters enhances the attractiveness of seeking to obtain optimal sink groupings. The branch-and-bound algorithm (MULTIWEB) presented here employs such an approach.

Second, a simple extension of the single-source algorithm has been suggested by several investigators, although it is recognized that only local minima are obtainable. Miehle seems to have been the first to suggest the approach, and it has been used by Cooper [5] and Eilon and Watson-Gandy [7]. We have programmed it as CROSSCUT and will discuss its use in Section III.

II PRELIMINARY PROBLEMS: THE CHOICE OF CO-ORDINATE SYSTEM AND METRIC

Before we turn to the approximate algorithm, however, we briefly consider the alternative co-ordinate systems and metrics employed in our methods and programs. At the present time the single-source algorithm cannot cope directly with great circle distances; however, by transformation of the latitude co-ordinate it is possible to approximate the great circle metric quite closely. Further, two other metric options are available to the users of the algorithms discussed in this paper.

When sink locations are given in Cartesian co-ordinates, Euclidean distances are used (the parameter METRIC in the programs is set equal to 1). When sink locations are stated in minutes of latitude and longitude (METRIC = 2), they are transformed into co-ordinates on a Mercator projection, distances are computed as *rhumb-line* map distances, and these distances are converted into statute or nautical miles. A rhumb line between two points on the earth's surface intersects intervening meridians at a constant angle, and straight-line segments on a Mercator projection are rhumb lines. We have found that for our purposes the simple rhumb-line distance between two points is a sufficiently good approximation to great circle distance, but piecewise linear rhumb-line routes between meridians spaced arbitrarily closely can be used to approximate such distances to any desired level of accuracy.

To obtain the latitude co-ordinate transformation in the Mercator mapping in minutes, we compute p, the colatitude of the spherical co-ordinate expressed in minutes, or 5400' less the given latitude. Then the Mercator co-ordinate is

$$X = 7915.704468 \log_{10} \cot (0.5\, p).$$

The longitude co-ordinate of the Mercator projection is identical to the spherical co-ordinate longitude. Map distances are then computed between Mercator co-ordinate points according to the Euclidean formula, and these map distances are corrected to nautical miles by multiplying the distances by the cosines of the mid-points of the spherical co-ordinate latitudes.[3]

Another metric of some importance in spatial work is the *metropolitan metric* between points on a Cartesian grid:

$$d_{ij} = |x_i - x_j| + |y_i - y_j| = [(x_i - x_j)^2]^{1/2} + [(y_i - y_j)^2]^{1/2}$$

Such distances may be relevant in analyses where transportation occurs on streets laid out as rectangular grids. Because the single-source algorithm is dependent upon the objective function being everywhere differentiable, the usual method of solution will not apply. However, we have programmed a method of solving the single-source algorithm when such distances are used, and it may be used in either MULTIWEB or CROSSCUT by setting METRIC = 3.

III THE CROSSCUT ALGORITHM FOR APPROXIMATE SOLUTIONS TO THE MULTISOURCE WEBER PROBLEM

Although the Φ function of the multisource problem is not well behaved, two of its characteristics recommend it for analysis by the extension of the single-source algorithm. The first is empirical and stems from the experience of Cooper with Weber problems. He has found the function to be quite flat in the region of the global minimum in many examples, so that the probability of arriving at a local minimum value reasonably close to the global minimum value by any good method may be quite high. Our own experience supports this conjecture.

The second characteristic is analytical. Although the Φ function is not convex in x, y, and z variables, *(i)* if the z terms are given at 0 or 1, the function is convex in x and y, and *(ii)* if the x, y variables are given, the function becomes linear in the z variables and therefore trivially convex. Let us view these two parameterization steps as taking crosscuts on the function in the *sink-allocation* and *source-location* directions. Once sink allocations are assumed, the Φ function decomposes into m single-source Weber problems, solvable by the single-source algorithm. And once the sources are located, the allocation question is merely that of choosing for each sink that source nearest it. The more complicated of these two crosscut operations is that involved in the iterations of the single-source algorithm for m Weber point problems, and computational experience with it leads us to be optimistic about the speed of convergence. We conjecture, therefore, that the obtaining of local minima will be rapid even for large problems. Indeed, for very large problems its usage may be the only practical method of analysis which can be used.[4]

An Illustrative Problem

To demonstrate the methods developed in this section we have chosen the following large problem. As sinks we take the hundred largest urban areas in the contiguous United States by 1960 United States census populations ($n = 100$), and as weights we use the relative populations of the areas [14]. The co-ordinates (a_j,b_j) are north latitudes and west longitudes respectively, computed to tenths of a degree, and expressed in minutes.[5] We seek the $m = 10$ source locations and associated sink clusters that minimize Φ in 10.2 for Cartesian co-ordinates, for rhumb line distances, and for the metropolitan metric. We shall place greatest emphasis upon the results for rhumb line distances (METRIC = 2).

An important input to the algorithm, both in terms of the computational time required for convergence and the solution value obtained, is the set of initial source locations. We chose these in quite arbitrary fashion in order to simulate use of the method under poor conditions. It will be understood, then, that had a more considered choice of initial source estimates been made the results would have been improved. The urban areas were ranked in descending order of their normalized weights and divided into ten sets, with set 1 containing the largest ten areas, set 2 the next largest ten areas, and so forth, to set 10 with the ten smallest urban areas by population. These ten sets were used as initial sources in ten separate runs of the algorithm for each of the three metric options. The Φ-values obtained are recorded in Table 10.1.

One striking characteristic of the local solution values for all three metrics is the relative tightness of the objective function values. The solutions for the first metric fell within 22 per cent of the best solution value obtained, and, for the second metric, within 24 per cent. In the case of the metropolitan metric, one extreme variation did occur, and widened this band to 43 per cent, but if this is excluded the solution values lie within 14 per cent of the best solution value obtained. These results support the conjecture that such objective functions are relatively flat-bottomed and that CROSSCUT will give good approximations to global minima.

Table 10.1 Objective function values for local solutions obtained from CROSSCUT

Initial source set	*Objective function value*		
METRIC =	1	2	3
1	114.955	118.703	198.674
2	129.412	127.959	181.941
3	115.662	114.723*	174.512*
4	125.618	124.357	189.425
5	110.986*	121.072	183.379
6	125.794	136.375	249.499
7	121.430	142.094	182.405
8	121.725	135.259	186.397
9	135.293	128.561	186.257
10	126.270	128.925	197.355

* Best solution obtained

This is further emphasized by consideration of the variety of sink clusters and source locations obtained in the solutions. For METRIC = 2, we defined a source to be located at a sink if its longitude and latitude each fell within 5 minutes of the sink, and to be located approximately at a sink if the co-ordinates fell within 10 minutes of the sink. On that basis, of the 100 sources located by the ten runs, 40 were located at sinks, 11 were approximately located at sinks, and 49 were located at intermediate points. The exact-sink locations included 12 distinct sinks: New York (10 locations); Los Angeles (6); Chicago (8); Philadelphia (1); Boston (3); Washington (4); Cleveland (1); Minneapolis (2); Cincinnati (1); Miami (1); New Orleans (2); and Spokane (1). In addition, 4 approximate locations occurred at Cleveland, 2 at Seattle, and 1 each at Washington, St. Louis, Denver, Ft Worth and Tacoma.

Another encouraging feature of the runs was their cheapness in terms of computer time; on an IBM 360/91, the ten runs for METRIC = 1, 2 and 3 required 15.32 seconds, 24.25 seconds and 17.30 seconds respectively. For practical problems of such dimension it is feasible to generate a large number of local solutions at low cost and with a rather high likelihood that the best solution value generated will be a good approximation to the global minimum.

The smallest Φ value occurred for the run with the third set of initial sources. It yielded 7 exact sink-locations (New York, Cleveland, Chicago, New Orleans, Dallas, Denver and Los Angeles), 1 approximate sink-location (Seattle) and 2 non-sink locations near Miami and Atlanta respectively.

In prior analyses with algorithms similar to CROSSCUT we made experimental runs with rather bizarre choices of initial sources, some of which included 10 identical sources at a patently non-optimal point (e.g. Death Valley, California or Newport, Vermont). Some of those runs converged to solutions which were prima facie nonoptimal in having fewer than 10 sources or containing sources with no assigned sinks. While we do not believe that CROSSCUT will exhibit this behavior for initial source selections that are distinct, it has been equipped to generate new solutions that have m distinct sources and to continue its search for a local minimum. Moreover, in order to speed convergence, if any source receives no sink assignments on an iteration, a new source location is generated for the next iteration.

An interesting result was that the best solutions for all three metrics in Table 10.1 were not those that included as initial sources the cities from which many of the final sources of the best and very good runs came. For example, among the METRIC = 2 runs, that one which used the largest ten urban areas as initial sources included only four sinks which appear as sources in the best solution found. Evidently, the bias against the south, west and northwest in set 1 was sufficient to pull its solution away from the best one. The case against set 1 or 2 for the other metrics is even clearer, and points up the lesson that it seems better to select initial source locations that permit all portions of the space to be represented than to choose them from the set of likely optimal selections.

The approximate algorithm CROSSCUT, therefore, may be used to obtain large numbers of local minima cheaply, to gain insights into the nature of the function and to provide guidance or an initial feasible solution for the branch-and-bound algorithm. The data sets in both programs have been made compatible in format, and we highly recommend that for problems of significant size CROSSCUT be used for these purposes.

IV A BRANCH-AND-BOUND ALGORITHM FOR THE MULTISOURCE WEBER PROBLEM

Branch-and-bound algorithms have been developed recently for many combinatorial and nonconvex optimization problems; two good expositions and surveys of branch and bound are those of Agin

[1] and Lawler and Wood [10]. The most rigorous and general treatment of the subject is that of Mitten [12].

The general method of branch-and-bound is as follows. The set of feasible solutions to a minimization problem is partitioned into subsets (usually disjoint), and for each subset a lower bound on the best solution value is computed. At each iteration a promising subset (usually one with the smallest lower bound) is further partitioned. During the course of the computations some particular feasible solutions are generated, and the one with smallest objective function value is retained. It is usually trivial to prove that after a finite number of iterations a condition will be reached in which the lower bound associated with every subset under consideration equals or exceeds the value of the best feasible solution generated. The process may then be terminated since it is impossible for there to be a feasible solution with smaller value.

In the following paragraphs we will characterize the subsets of solutions used by MULTIWEB, our branch-and-bound algorithm, explain how the lower bounds are computed and the feasible solutions are generated, and detail the rules for selecting and partitioning a subset at each iteration of the algorithm. Several computational considerations will also be discussed.

We partition the set of solutions to the multisource Weber problem on the basis of the assignment of sinks to sources. Each subset is characterized by its particular assignment of some of the sinks. For example, with two sources and five sinks a possible subset consists of those solutions in which sink 1 is assigned to source 1 and sinks 3 and 5 are assigned to source 2. In this subset sinks 2 and 4 are not assigned. Because of the symmetry inherent in the problem we may arbitrarily assign some sink to source 1; suppose this is sink 1. We may then take the entire set of solutions as the set that assigns sink 1 to source 1.

The subset A of solutions is partitioned into at most m disjoint subsets in the following way. A sink, called the branching sink, is selected that is unassigned in A. Suppose, first, that A is such that at least one sink is assigned to each of the m sources. The subset A is then partitioned into the m subsets $A_1, \ldots, A_m$, where A_i is obtained from A by assigning the branching sink to source i. If A is such that at least one sink is assigned to sources $1, 2, \ldots, k$, where $k < m$, then A is partitioned into $A_1, \ldots, A_{k+1}$, where A_i is as above (symmetric solutions are thus eliminated).

For each subset that is obtained from the partitioning of a larger subset we must compute a lower bound. If that lower bound is less than Φ^* (the value of the best solution found thus far) we must select a branching sink. A feasible solution is generated as a byproduct of the branching sink determination, and if this solution has a cost that is less than Φ^* its value replaces Φ^*.

The lower bound for subset A is the sum of $m + 1$ quantities. The i-th of these, $i = 1, 2, \ldots, m$, is obtained at the cost of optimally locating source i among the sinks assigned to source i in subset A. This quantity is positive unless the number of sinks assigned to source i in subset A is 0 or 1; it is easily computed by the single-source algorithm (subroutine WEBER in CROSSCUT and MULTIWEB). A tentative location for source i is determined at the same time (unless no sinks are assigned to it); this is utilized later in choosing a branching sink and generating a feasible solution. The final quantity needed for the lower bound is obtained by considering the set of unassigned sinks, adding to this the sinks (if any) that are the sole assignments to various sources, and then underestimating the cost of optimally locating m sources among the sinks. If r is the number of sources that have at least two sinks assigned to them in subset A, then we are, in effect, underestimating the cost of optimally locating $m + r$ sources among the n sinks, with the assignments of subset A for r of the sources. Such an underestimate is clearly a lower bound on the cost of optimally locating m sources among the n sinks. The following paragraphs detail our calculation of an underestimate of the cost of locating m sources among p sinks.

Consider the problem of locating one source among the first s sinks. Suppose the source is optimally located, and consider two of the s sinks, a and b. The costs due to sinks a and b are

$$w_a d_a + w_b d_b,$$

where d_a is the distance (according to the appropriate metric) between the source and sink a. Then we have

$$\begin{aligned} w_a d_a + w_b d_b &\geq (\min\{w_a, w_b\})(d_a + d_b) \\ &\geq (\min\{w_a, w_b\})(d_{a,b}) \end{aligned}$$

by the triangle inequality, where $d_{a,b}$ is the distance between sinks a and b. Now sum over all unique pairs a and b to get

$$\sum_{a<b} (w_a d_a + w_b d_b) \geq \sum_{a<b} d_{a,b} (\min\{w_a, w_b\}).$$

The left-hand side of the above equation is $(s - 1)$ times the cost because each quantity $w_a d_a$ appears $(s - 1)$ times. Thus

$$(s - 1)^{-1} \sum_{a<b} d_{a,b}(\min \{w_a, w_b\}) \tag{10.4}$$

underestimates the cost of optimally locating one source among the first s sinks (see [4], theorem 1).

Let us return to the problem of m sources and p sinks. If $p - m \leq 0$, the appropriate underestimate is clearly zero. MULTIWEB uses specific underestimating formulas for $p - m = 1, 2, 3, 4$ and 5, but we shall here derive only the result for $p - m = 3$; it is sufficiently illustrative. With three more sinks than sources three possibilities for the optimal allocation exist: (*i*) one source has four sinks assigned to it and all others have just one; (*ii*) one source has three sinks assigned, another has two, and all others have one; and (*iii*) three sources have two sinks assigned and all others have one. For case (*i*) the underestimate is obtained from (4) as $\frac{1}{3}(q_1 + \ldots + q_6)$, where q_i is defined to be the *i*-th smallest of the $\frac{1}{2}p(p - 1)$ quantities $d_{a,b}(\min \{w_a,w_b\})$. For case (*ii*) we apply 10.4 twice (once with $s = 2$ and once with $s = 3$) to get $q_1 + \frac{1}{2}(q_2 + q_3 + q_4)$. Note that we must choose the subscripts to make the expression as small as possible. For case (*iii*) we apply 10.4 three times to get $q_1 + q_2 + q_3$. We do not know which case actually occurs, so we must choose the minimum of the three values. The underestimate is therefore the minimum of[6] (a) $\frac{1}{3}(q_1 + \ldots + q_6)$, (b) $q_1 + \frac{1}{2}(q_2 + q_3 + q_4)$, and (c) $q_1 + q_2 + q_3$.

Similar underestimates are used for $p - m = 1, 2, 4$ and 5; the appropriate expressions to be evaluated and compared follow directly from an examination of the partitions of $p - m$ into at most m parts. For $p - m$ large, however, the number of expressions to be compared becomes too large (for $m = 6$ the number of expressions to be compared for $p - m = 5, 10, 15$ and 20 are 7, 35, 110 and 282, respectively), so a different approach is used for $p - m > 5$.

It may be shown that a lower bound to the cost of optimally locating m sources among p sinks is

$$G_1 = 0.5 \sum_{i=1}^{t} q_i, \tag{10.5}$$

where $t = p - m$ ($t = p + m + 1$) if $p \leq 2m$ ($p > 2m$), and q_i is, as before, the *i*-th smallest of the $\frac{1}{2}p(p - 1)$ *quasi-distances* $d_{a,b}$, $(\min \{w_a,w_b\})$. The proof follows closely that of [4] theorem 2, and is therefore omitted. The significant difference between 10.5 and Cooper's result is that Cooper's assumes the optimal solution assigns at least three sources to each sink. The same approach yields an additional lower bound:

$$G_2 = 0.25 \sum_{i=1}^{2p-3m} q_i. \tag{10.6}$$

If $p - m > 5$, therefore, MULTIWEB uses the greater of G_1 and G_2 as an underestimate of the cost of optimally locating m sources among p sinks.

We have now explained how the lower bound for a subset is computed; for each subset whose lower bound is less than Φ^* we select a branching sink and generate a feasible solution. Suppose that subset A assigns sinks to the first k sources; each of these sources has been given a tentative location during the computation of the lower bound. Our hypothesis is that these locations will not change significantly in subsets of A;[7] to generate a feasible solution we therefore fix the source locations at their tentative ones and assign each unassigned sink to the closest source.

If we must partition subset A into smaller subsets we would like the lower bounds for them to be as large as possible. Our heuristic is as follows: if the branching sink weight multiplied by its distance from the tentative location of a particular source is large, the lower bound of the subset of A that assigns the branching sink to the given source will exceed that of A by approximately the same large amount. We should choose the branching sink to be one that is *far away* from all tentatively located sources and has a large weight. We therefore compute, for each unassigned sink, its minimum and average distance from the tentatively located sources. The computer code allows two options in choosing the branching sink in MULTIWEB: under option 1 (IBRAN = 1) the sink is chosen whose weight times distance is a maximum, and under option 2 (IBRAN = 2) *minimum distance* is replaced by *average distance*.

When all subsets have lower bounds that equal or exceed Φ^* the algorithm terminates. As a computational device, comparisons may be made with $\Phi^*(1 + \varrho)^{-1}$, where $\varrho \geq 0$, instead of Φ^*. This may speed the algorithm considerably, and it guarantees a solution whose value is within 100ϱ per cent of the value of the optimal solution. As another computational device, the tentative locations of all sources used in a subset A are retained for use as starting points in subsequent uses of the single-source algorithm. Also, for each subset

Table 10.2 Computational experience with MULTIWEB

Problem number	*Size (n, m)*	S(*n*, *m*)	*Sink weights*	IBRAN	*Nodes created*	*Maximum active nodes*	*Time (sec)* (CDC 6400)
1–8	(7,2)	63	1.0	1	28.3*	6.1*	0.5*
9–16	(7,2)	63	random	1	26.8*	5.4*	0.5*
17A	(15,2)	16,383	1.0	1	451	85	54.2
17B	Same as 17A			2	425	88	
18A	(15,3)	2.4×10^6	1.0	1	661	116	
18B	Same as 18A			2	664	122	
19A	(15,4)	42×10^6	1.0	1	1038	132	
19B	Same as 19A			2	1306	147	
20A	(15,2)	16,383	random	1	511	90	16.5
20B	Same as 20A			2	435	83	
21A	(15,3)	2.4×10^6	random	1	796	112	82.7
21B	Same as 21A			2	649	119	
22A	(15,4)	42×10^6	random	1	2870	313	
22B	Same as 22A			2	2441	278	
23A	(20,2)	5.2×10^5	random	1	1267	209	52.1
23B	Same as 23A			2	1115	192	
24A	(15,2)	16,383	random	1	293	38	10.2
24B	Same as 24A			2	249	42	
25A	(15,3)	2.4×10^6	random	1	2272	264	53.8
25B	Same as 25A			2	1456	210	

* Averages

obtained from a partition of subset A, only one single source Weber problem need be solved because the others are identical to ones solved for subset A or its predecessors.

Finally, the program sets Φ^* to ∞ initially unless a feasible solution is entered as input, and it has an advanced start feature allowing the user to make an arbitrary number of preassignments of sinks to sources; only solutions containing these preassignments are then considered.

V COMPUTATIONAL EXPERIENCE WITH MULTIWEB

In debugging MULTIWEB a number of test problems have been solved, and this computational experience is summarized in Table 10.2. In it, S(n, m) is the Stirling number of the second kind and is equal to the number of unique allocations of n sinks to m sources. It is given as a reference point. The entries in the column labeled *nodes created* are indicative of the amount of computation since the creation of each node entails the solution of one single-source Weber problem and some bookkeeping to calculate the lower bound. The entries in the column labeled *maximum active nodes* are indicative of the storage requirements of MULTIWEB. The times shown are only approximate, and are totals for groups of problems. For all problems in the table METRIC = 1 and $\varrho = 0.005$; i.e., Cartesian co-ordinates were used and solution values were found to within 0.5 per cent of an optimal solution value.

Problems 1–8 were taken from Cooper [3]. Our solutions agree with his except in the case of problem 6; this is probably due to a misprint in the data in [3]. Problems 1–8 were also solved by CROSSCUT, with sources initially at the first two sinks. CROSSCUT found the optimal solution in four problems, found solutions within 5 per cent of optimal in two other problems, and found solutions in error by 17 and 26 per cent in the remaining two.

Problems 9–16 are identical to problems 1–8, respectively, except that the sink weights were generated randomly, and independently for each problem, according to a uniform distribution over [0, 10].

Problems 17–22 are based on the 15-sink problem analyzed in [4]. Problems 20–2 are identical to problems 17–19, respectively, except that problems 20–2 have sink weights that were generated randomly, once for all three problems, according to a uniform distribution over [0, 10]. In problems 17–25, for the one designated A the minimum distance option was used and for the one designated B the average distance option was used; the A and B versions are otherwise identical.

The sink co-ordinates in problem 23 were generated randomly according to a uniform distribution over [0, 50]. The sink weights were generated randomly according to an exponential distribution with mean 1.0. We attempted to solve problem 23 with 3 sources but found that the number of active nodes exceeded the maximum number allowable at that time (350). Problems 24 and 25 were then obtained by using the first 15 sinks from problem 23.

Upon comparing the results obtained with the two branching options in problems 20–5, we tentatively conclude that using average distance results in a somewhat more efficient solution procedure for problems with unequal sink weights.

The other significant problem attempted with MULTIWEB was the formidable problem with 100 sinks and 10 sources discussed in Section III. Its size makes the optimal solution unattainable at present, but we have made significant progress on the problem. Table 10.1 shows that the best CROSSCUT solution yielded $\Phi = 114.723$. The problem was run with MULTIWEB on an IBM 360/91 computer, and after 2017 nodes had been generated the number of active nodes exceeded the maximum allowable when core alone is available for storage (1800 nodes). The best solution obtained in this run yielded $\Phi = 109.048$, a significant improvement over the CROSSCUT solution, and required 8 minutes, 18 seconds of computer time. This solution was then used as an initial solution for CROSSCUT, and the algorithm converged in 4 seconds to a solution with $\Phi = 108.154$.

At this point the runs with both algorithms were studied with some care, and on the basis of the recurring patterns of sink-assignments, 77 of the 100 sink assignments were made to 8 sources. These preassignments were read into MULTIWEB as input. The remaining 23 sinks were permitted complete freedom in their assignments. Also, the CROSSCUT solution mentioned above was used as an initial feasible solution.

The algorithm generated a total of 2030 nodes before the number of active nodes exceeded 1800, and this time yielded the improved solution $\Phi = 107.532$ in 2 minutes, 42 seconds.

Next we fed the source locations of this MULTIWEB solution into CROSSCUT as input, and an improved solution with $\Phi = 106.425$ resulted. Finally, this improved solution was used as an initial feasible solution for an extended version of MULTIWEB which stores source assignments on disk, permitting 5598 active nodes. It ran for 9 minutes and did not improve the initial feasible solution after 6390 nodes had been generated.

In order to find better solutions and, ultimately, an optimal one, we are proceeding on three fronts. First, as noted, the maximum number of allowable active nodes in MULTIWEB has been increased by auxiliary storage on disk. Second, the lower bounds given by 10.5 and 10.6 have been improved slightly by considering separately some special cases of the partition of p sinks into m groups. And, lastly, consideration is currently being given to an alternative method of selecting nodes for branching that should reduce node storage.

Meanwhile, we have scaled down the problem by deleting some sinks to gain further experience. A 30-sink, 2-source problem was solved to within 5 per cent in 79 seconds and 2033 nodes. A 30-sink, 3-source problem, however, exhausted the node storage available at the time and was not completed in 80 seconds and 2665 nodes. A 50-sink, 2-source problem ran for 322 seconds and 3807 nodes without success. Finally, the 100-sink, 2-source problem ran for 360 seconds and 2000 nodes without reaching an optimal solution. In the 30-sink problem $\varrho = 0.05$ and in all others $\varrho = 0.001$.

Clearly, at least on the 360/91, time is not a deterrent factor in seeking answers to large problems; even the 100-sink, 2-source problem took only 6 minutes to exhaust available node storage. As m rises, with n fixed, time and node storage requirements rise to a maximum and then begin to fall; although we do not know much about the location of the maximum, some consolation may be taken from the fact that the requirement functions do not continue to rise as m is increased.

Beyond that, however, it is highly likely for any problem of practical size that a solution value close to the optimal one will be achieved within the bounds of current node storage, so that MULTIWEB and CROSSCUT used together yield very good answers, even to quite large problems. And, for many common problem sizes – in the neighborhood of 25 sinks and 1 to 5 sources – the existing programs will provide an exact solution.

Notes

1. The same mathematical problem is obtained if the number of raw material locations is r and the number of markets is $n - r$, where $0 < r < n$.
2. The objective function meets none of the usual criteria for well-behaved functions in an optimality context (e.g. see [13]).
3. For proofs and details see [6], pp. 112–20 and [8], pp. 11–14.
4. CROSSCUT and MULTIWEB have been coded in FORTRAN IV.
5. These data are listed in [8], pp. 20–3.
6. Case *(iii)* cannot occur if $m = 2$; in that event, only the first two expressions are compared.
7. The validity of this hypothesis should be high if A assigns most of the sinks to sources and low if it assigns only a few.

References

[1] Agin, N., 'Optimum Seeking with Branch-and-bound', *Management Science*, 13 (1966), pp. B176–B185.

[2] Bellman, R., 'An Application of Dynamic Programming to Location-

Allocation Problems,' *SIAM Review*, 7 (1965), pp. 126–8.

[3] Cooper, L., 'Location-Allocation Problems', *Operations Research*, 11 (1963), pp. 331–43.

[4] Cooper, L., 'Heuristic Methods for Location-Allocation Problems, *SIAM Review*, 6 (1964), pp. 37–52.

[5] Cooper, L., 'Solutions of Generalized Locational Equilibrium Models', *Journal of Regional Science*, 7 (1967), pp. 1–18.

[6] Deetz, C. H., and O. S. Adams, *Elements of map projection*, US Department of Commerce, Special Publication no. 68, Washington, DC (1945).

[7] Eilon, S., and C. D. T. Watson-Gandy, 'Models for Determining Depot Location', Imperial College of Science and Technology Report No. 69/4 (1969).

[8] Kuenne, R. E., and R. M. Soland, *The Multisource Weber Problem: Exact Solutions by Branch and Bound*, IDA Economic Series, Institute for Defense Analyses, Arlington, Va. (1971).

[9] Kuhn, H. W., and R. E. Kuenne, 'An Efficient Algorithm for the Numerical Solution of the Generalized Weber Problem in Spatial Economics', *Journal of Regional Science* 4, (1962), pp. 21–33. Reprinted as Chapter 9 of this volume.

[10] Lawler, E. L., and D. E. Wood, 'Branch and Bound Methods: a Survey', *Operations Research*, 14 (1966), pp. 699–719.

[11] Miehle, W., 'Link-Length Minimization in Networks', *Operations Research*, 6 (1958) pp. 232–43.

[12] Mitten, L. G., 'Branch and Bound Methods: General Formulation and Properties', *Operations Research*, 18 (1970), pp. 24–34.

[13] Ponstein, J., 'Seven Kinds of Convexity', *SIAM Review*, 9 (1967), pp. 115–19.

[14] *United States Census of Population*, 1960, US Summary, Number of Inhabitants, Washington, DC (1960).

[15] Weber, A., *Über den Standort der Industrien* Tübingen (1909). Translated as: 'Alfred Weber's *Theory of the Location of Industries*' (University of Chicago, 1929).

11 Approximate Solutions to a Dynamic Combinatorial Problem in Space*

I THE SEQUENTIAL COMBINATORIAL PROBLEM

A family of important problems in spatial economics may be particularized in the following terms. A set of vehicles, V, exists with elements $i = 1, \ldots, m$, each of which carries a given initial supply of product. We identify a sequence of 'legs', $t = 0, 1, \ldots, N$, during which we assume (fictitiously) that vehicle movement occurs. Finally, we specify a set of stations, S, with elements $j = 1, \ldots, n$, located at co-ordinates $[x_j, y_j]$ on the plane; for convenience, and without loss of generality, we assume that $j = 1, \ldots, f, f \leq m$, are the initial locations of the vehicles, or the stations at which they are located on Leg 0.

At the start of each Leg t (end of Leg $t - 1$) demand at each station is symbolized ${}_tD_j$, which is a supply when negative. The supply carried by $i \in V$ located at $j \in S$ is designated ${}_tS_{i,j}$, where t denotes the relevant leg. At the start of Leg t the demand for product at Station k is defined as

$$ {}_tD_k = {}_{t-1}D_k - \sum_i {}_{t-1}S_{i,j}\ {}_{t-1}Y^k_{i,j},\ t \geq 1 \tag{11.1}$$

where ${}_tY^k_{i,j} = 0$ if Vehicle i located at Station j at the start of Leg t does not move to Station k, or $= 1$ if it does so move. We define the set ${}_tI_k$

* Published originally in the *Journal of Regional Science*, 8 (1968), pp. 165–80, and reproduced with permission.

This work was initiated while the author was a research associate of the Regional Science Research Institute under National Science Foundation Grant GS-617. He is indebted to Professor Michel L. Balinski, who provided valuable advice at several points during the research.

as those vehicle indices that do move on Leg t to Station k:

$$ {}_tI_k = \left\{ i \mid {}_tY^k_{i,j} = 1 \right\}. \tag{11.2} $$

If ${}_tD_k < 0$, supplies exist at Station k on Leg t, and we arbitrarily impute this supply to a single vehicle presently located at k. By virtue of an efficiency constraint we shall impose below (see constraint 11.17) the entire supply will be carried aboard only one vehicle. We follow the convention of identifying the supply vehicle in this case with the vehicle of minimum index. We then define:

$$ {}_{t+1}S_{i,k} = \begin{Bmatrix} -{}_tD_k, \text{ if } {}_tD_k < 0 \text{ and } i = \min i \in {}_tI_k \\ 0, \text{ if } {}_tD_k \geq 0 \text{ and/or } i \neq \min i \in {}_tI_k \end{Bmatrix}. \tag{11.3} $$

Among the data of the problem are the initial supplies, ${}_0S_{i,j}$, $j \leq f$, and the initial demands, ${}_0D_j$, $j > f$. In our presentation of the problem we assume that $\sum_i {}_0S_{i,j} = \sum_{j>f} {}_0D_j$, or that the total supply of product at the start of the problem equals and just equals the total demand for the product at the start of the problem.

We require that every vehicle with positive supply at the start of a leg move during the relevant leg, while all vehicles with zero supplies remain stationary:

$$ \begin{cases} 1. \ {}_tS_{i,j} = 0 \Rightarrow {}_tY^j_{i,j} = 1, \\ 2. \ {}_tS_{i,j} > 0 \Rightarrow {}_tY^j_{i,j} = 0, \end{cases} \tag{11.4} $$

where

$$ {}_tY^k_{i,j} = 0 \text{ or } 1, \quad 0 < t \leqq N,\ k = 1, \ldots, n, \tag{11.5} $$

$$ \sum_k {}_tY^k_{i,j} = 1, \quad 0 < t \leqq N,\ k = 1, \ldots, n, \tag{11.6} $$

$$ {}_ND_j = 0, \quad j = 1, \ldots, n. \tag{11.7} $$

We define Leg N as the first leg on which all demands have been reduced to zero, and it is to be noted that we have not required empty vehicles to move to their Leg 0 station, although no bar to such a constraint exists in the nature of the problem or our methods of solution. Were we to require this type of Leg N movement our vehicular assignments would be complete traveling salesman circuits, but we have elected to retain the more general 'open circuit' assumption. In this form, the analysis is likely to be more relevant to problems in which *time* of travel (perfectly correlated with distance) is the important cost, but it must be stressed again that our method permits the closed circuit case to be attacked by requiring that ${}_{N}Y^{0}_{i,j} = 1$, where $o \leq f$ is the vehicle's Leg 0 Station.

Costs of movement between stations are given and proportional to Euclidean distance $d_{jk} = \sqrt{(x_j - x_k)^2 + (y_j - y_k)^2}$ between Stations j and k. The objective of the analysis is to assign vehicles to stations over the sequence of legs in such manner that we minimize total distance cost traveled by all vehicles, subject to 11.1 through 11.7:

$$\min C = \sum_{t=1}^{N} \sum_{i=1}^{m} \sum_{k=1}^{n} {}_{t}Y^{k}_{i,j}\, d_{jk}. \tag{11.8}$$

Some Family Resemblances

In general, it is not possible to assign an integral number of vehicles to every station in order to satisfy and only satisfy the initial demand at such stations. Were this possible we should have $N = 2$, which is the classic transportation problem of Hitchcock [4] in all its simplicity. In our statement of the problem, we shall term the case of only one nonstationary leg – through an abuse of terminology – a 'static' combinatorial problem in space, with the purpose of distinguishing it from our own multi-leg 'dynamic' problem. The static problem is the simplest near-relative of the dynamic case, and it serves the useful function of highlighting the peculiar complications of the latter.

The fundamental difficulty involved in seeking an optimal solution to the dynamic problem is the sequential dependence of the movement of vehicles compounded by the multiplicity of assignment possibilities if more than one vehicle is capable of assignment to a given station on a given leg. Whether or not a given vehicle at a given station should move on Leg p to Station q depends upon what that vehicle and all other vehicles have done on the previous $p - 1$ legs and upon what that vehicle and all other vehicles will do on Legs $p + 1$ through N. We are caught in a web of longitudinal (number of legs) and cross-sectional (number of vehicles) temporal interdependence, and the transportation problem simplifies by compressing the first of these dimensions.

If we simplify the dynamic combinatorial problem in the second dimension, which is to say that we permit only one vehicle to be used but permit more than one nonstationary leg to exist, a second simpler relative of our problem is obtained. It is the well-known 'traveling salesman' problem with its classic statement as that of moving a salesman from one of n points in a complete circuit going through the $n - 1$ other points and terminating at the initial point in such wise as to minimize the total distance traveled. We have already pointed out that were we to close the circuit of vehicular assignments, the dynamic problem resolves itself into a multiple traveling salesman model, and the objective may be stated as that of simultaneously partitioning n points into m (not necessarily disjoint) closed-circuit sets and obtaining the optimal sequencing of stations within each such circuit.

The great potential for complication that such work in this second dimension entails is well illustrated by the complexity of the simpler and limiting one-salesman problem. The formal statement of the traveling salesman problem as a linear programming problem betrays its relationship to our problem by virtue of its similarity to relations 11.1 through 11.8, at the same time that it belies the computational complexity involved in its solution. It seeks to minimize an objective function

$$\min C = \sum_{j} \sum_{k} Y_{j,k}\, d_{jk}, \tag{11.9}$$

where the existence of only one 'vehicle' permits us to ignore our generally useful device of the leg, and therefore where the sums are indexed over stations of origin (j) and destination (k). The search for a minimum is constrained by the following:

$$Y_{j,j} = 0 \tag{11.10}$$

$$Y_{j,k} = 0 \text{ or } 1, \tag{11.11}$$

$$\sum_{j} Y_{j,k} = \sum_{k} Y_{j,k} = 1, \tag{11.12}$$

for any subset $I = \{ j_1, j_2, \ldots, j_r\}$ of the integers 1 through n, (11.13)

$$Y_{j_1,j_2} + Y_{j_2,j_3} + \ldots + Y_{j_{r-1},j_1} \left\{ \begin{array}{l} < r, \text{ if } r < n, \\ \leqq n, \text{ if } r = n, \end{array} \right.$$

Requirement 11.13 states that if any given set of $Y_{j,k}$ contains a complete cycle with fewer than the n distinct stations, that set of $Y_{j,k}$ is not feasible. Moreover, if the subset under consideration is the whole set of stations, then the set of $Y_{j,k}$ cannot sum to more than n. If it sums to less than n, constraints 11.12 – that require every station of destination and origin to have one and only one nonzero term in the set of $Y_{j,k}$ – will be violated and render the set infeasible. Thus the inequality sign is included as a formal requirement of linear programming, and the equality sign will in fact hold for any feasible chain.

The fact that the problem can be stated so compactly in a linear programming formulation tempts to the view that the optimal solution will be readily obtainable, but in fact the number of the convexification constraints in 11.13 for problems with rather small n rapidly becomes astronomical. These constraints serve to define the convex hull of the set of feasible solutions of which the optimal solution is an extreme point, but the required number of them grows rapidly with the number of cities n. For example, Kuhn [6] found that the convex hull for a 5-city problem required the statement of 390 inequalities. Such experience leads to an abandonment of the integer programming approach to a solution and into searches for more specific algorithms possessing greater efficiency.

Gilmore and Gomory [3] have succeeded in solving a special case of the problem in n^2 steps; Lin [7] has devised an algorithm to solve the problem for $n \leq 13$; and Dantzig, Fulkerson and Johnson [2] have attacked successfully rather small-scale problems using inspectional methods not readily programmed. More success, however, has been obtained through the use of enumeration techniques such as the branch-and-bounds methods of Little *et al.* [8]. Although the time required expands at least exponentially with the number of cities, branch-and-bound has permitted solutions to 40-city problems at an average time of about 8.5 minutes. However, an added 10 cities at this level multiplies the time required by a factor of 10. We shall suggest an important, if subordinate, role for branch-and-bound in our method of attack upon the more complicated dynamic problem below.

A third set of problems with a similar structure to that of the dynamic combinatorial problem is a class of weighted set covering problems, which may contain both dimensions of our problem (Balinski [1]). One such closely related problem, for example, discussed by Balinski, simplifies by assuming that all vehicles depart from a single point, that a rather small set of potential routes exists, and that each client obtains his supply in one delivery. In terms of the present problem, such simplifications permit the reduction of the case to a one-leg variation of the classic transportation problem: that is, the fundamental simplification is that of the essential reduction of the longitudinal dimension to a one-nonstationary-leg size.

The Nonavailability of Exact Solution Algorithms for the Dynamic Problem

The complexities of the simpler dynamic close relatives of the generalized dynamic combinatorial problem and of the computational experience obtained in using various *ad hoc* or systematic algorithms prompt the assumption that no algorithm yielding exact solutions to a problem of reasonable scale and generalizable complexity in acceptable computation time may be expected in the foreseeable future. Therefore, we urge that in such situations it is wise to develop approximative methods whose results may or may not yield the optimum solution in a specific case, but whose use yields a high probability of finding a solution that is close to the optimum.

It is proper in such circumstances as these to abandon the perfectionist search for *the* optimum and to accept the need to develop workable if less intellectually satisfying methods that generate efficient if not provably optimal solutions. When these methods are developed heuristically, they sometimes yield the dividend of insights into the structure of the optimal solution and of the economic mechanism that underlies the optimum, even though a knowledge of the solution remains unattainable. Moreover, in all honesty it must be admitted that frequently obtaining the optimal solution is not that important in practical economic matters, particularly if we are assured that our own solutions lie within acceptable bounds of the optimum. Lastly, there is a certain excitement and richness of experience to be obtained in the exploration of inexactitude that the profession is only now awakening to in the development of such Gordian-knot methods as we suggest in this paper.

II AN INITIAL APPROACH

The author became interested initially in the dynamic combinatorial problem in his study of efficient fleet ballistic missile submarine deployment for *Polaris* missile usage (Kuenne [5]). The problem there presented itself in the following form: given 30 submarines each carrying 16 missiles, and given 480 optimal stations from each of which a given number of missiles is to be launched, route the submarines among the firing stations in such fashion as to minimize the total distance (time) of movement of the fleet.

The method adopted attempted to take into account (however imperfectly) the necessity of future legs of the vehicles by defining the variable

$$ {}_tF^k_{i,j} = \max\left[\,|\,{}_tD_k - {}_tS_{i,j}\,|,\, 1\right] \cdot d_{jk},\ {}_tD_k > 0, \tag{11.14}$$

which, for each leg, enters the objective function,

$$ C_t = \sum_i \sum_k {}_tF^k_{i,j}\, {}_tY^k_{i,j}, \quad 0 < t \leq N, \tag{11.15}$$

which is to be minimized subject to the constraints 11.4 through 11.7 and a further pair to be presented as 11.16 and 11.17 below.

The problem is thus made manageable by suboptimizing leg by leg, assigning vehicles such that the absolute value of 'product-miles' is minimized. The product term in 11.15 takes into account the over-supply of product (missiles) implied by a given movement of a vehicle (submarine) to a station with positive demand, or the amount of undersupply implied by such a vehicle movement which would have to be made good by the movement of other vehicles on other legs with implied distance costs. If a given submarine's supply just exhausted the demands of a station without overage, yielding a zero value for the absolute value term, then the F-term is equated to distance traveled.

This is admittedly a simple method of taking future legs into account with some attempt at minimizing their number and the implied distance movement that occurs within them, but it seemed to offer a better prospect of approaching the optimal solution than straightforward minimization of distance alone leg by leg, on the one hand, or, on the other, obtaining a feasible solution to the whole set of legs and attempting to shift vehicles to improve the total distance traveled.

To use this method of moving to a minimum on each leg it was necessary to find an initial feasible solution for that leg; however, one further set of constraints had to be placed upon the minimization, since their violation was a *prima facie* failure to effect a minimization of C_t. A solution that was feasible considering only constraints 11.4

through 11.7 could assign to a given station more product than necessary, in the sense that it would be possible to remove one or more vehicles from the station and still meet the demand requirements of the station in full. Since straight-line movement from station to station was always possible, such movements even if they get closer to stations with net positive demands could not be as distance-cheap as direct movement to such stations would be.

It is efficient, therefore, to restrict the feasible set of assignments further by defining the variable

$$ {}_t\lambda_k = {}_tD_k - \sum_i {}_tS_{i,j}\, {}_tY^k_{i,j} + \min\, [{}_tS_{i,j}\, {}_tY^k_{i,j} > 0], \tag{11.16} $$

which, more simply, subtracts from the demand at the station the total of supplies that move to it on the leg in order to get the net demand at the station. From this net demand is subtracted the supply brought by the vehicle with the smallest load. If, after the removal of the smallest supply component net demand is zero or negative, obviously that vehicle (and possibly others) is not necessary to meet the demand at that station at the start of the leg. We require, therefore, that to be λ-feasible, any solution meet the condition that

$$ {}_t\lambda_j > 0, \quad 0 < t \leqq N, j = f + 1, \ldots, n. \tag{11.17} $$

To obtain the minimum C_t the following procedure was followed. First, each vehicle was assigned to that station that minimized ${}_tF^k_{i,j}$, and, of course, to that station only. This assured that constraints 11.4 through 11.6 were met. The solution was then checked against 11.17, and if all ${}_t\lambda_j$ were positive, the allocation was the minimum C_t solution. If one or more of these terms were nonpositive, reallocation of the vehicles on the given leg was necessary, and was obtained as follows. For each vehicle the ${}_tF^k_{i,j}$ for movement to all other stations with positive demand were calculated, and the (non-negative) differences from the minimum-F term for that vehicle obtained. These differences were collected for all vehicles and were then ranked from lowest to highest.

A search was then begun for a feasible solution which guaranteed that the departure of the new solution from the distance-minimizing but infeasible solution would be minimal. The vehicle associated with the smallest ΔF is reassigned from its minimal-F assignment to the implied station, and the resulting solution is checked for λ-feasibility. If all ${}_t\lambda_j$ are thereby rendered positive (and certain checks made of 2-combinations to be defined below) the best feasible solution is at hand. If not, recourse is had to the second smallest ΔF term and the same procedure repeated. If no feasible solution is discovered after all such '1-combinations' (that is, movements of one vehicle only to a station other than its minimal-F station), a search is begun among '2-combinations' (movements of pairs of vehicles from their minimal-F stations) in such manner as to obtain a feasible solution. If one was obtained it is checked for assurance that no '3-combination' would yield smaller departures from minimum-F cost, and if this test is passed, the resulting feasible solution is selected as a minimal-F approximation to the optimum solution's Leg t assignments. The procedure then moves to Leg $t + 1$, repeats the performance, and continues such legs until the Nth leg when constraint 11.7 is satisfied.

In the work cited, the author used a variant of this algorithm that depended upon inspectional methods not readily programmed and which, while they did not yield such minimum-F solutions, did have certain other advantages of peculiar importance in problems of warfare. However, the general algorithm was programmed, and subsequent experience with it obtained on other problems. This usage revealed a strong form of the usual computational time problem with combinatorial problems: the computer time required for solution rose very rapidly in function of the number of stations in the problem. For example, although the solution to a 10-station problem was obtained in 1 minute, 6 seconds, a computation with 20 stations could not find a feasible solution to the first leg after an hour of time had been used up on the IBM 7090.

Another lesson from experience with the algorithm was that often the same algorithm acting to minimize F^* terms, these being defined as simple distances,

$$ {}_tF^{*k}_{i,j} = d_{jk}, \tag{11.18} $$

outperformed the usage of the F-terms. It must be emphasized that this occurred primarily on simple problems with the number of legs five or less; we conjecture that the F-term analysis would show significant relative advantages as the scale of the problem increased, both in terms of the number of vehicles and the number of legs.

Lastly, it was found that frequently some improvement in costs could be obtained by trying various sequences of the station-assignments of the vehicles. That is, if Vehicle 1 is assigned to Stations 4, 37, 89, and 9 in Legs 1 through 4 by the solution, it was wise to try the permutations 4, 9, 37, 89; 4, 89, 37, 9; 37, 4, 89, 9; and so forth, testing the total distances involved for each permutation, and adopting the permutation yielding the least distance cost for the vehicle's circuit.

III AN ALTERNATIVE APPROACH

The great cost in computation time for moderate-sized problems led the author to develop an alternative approach involving use of an initial feasible solution over all legs simultaneously, the generation of a set of alternative feasible solutions (perhaps as many as can be obtained during a given period of computer time), and the choice of that solution whose cost is least. This generation of a set of feasible solutions sacrifices leg-by-leg suboptimization, and is similar to an approach to the solution of the traveling salesman problem devised by Lin [7].

A Simple Illustration of the Method

Suppose the set of stations and demands displayed in Table 11.1 to exist, with one vehicle on Leg 0 at each of Stations 1, 2, and 3, and suppose further the distances between stations to be those recorded there.

We choose an initial feasible solution as follows:

> Vehicle 1 moves to Stations 4, 6 for a distance of 100 miles;
> Vehicle 2 moves to Stations 5, 6 for a distance of 70 miles;
> Vehicle 3 moves to Station 6 for a distance of 100 miles.

Its distance costs – the sum of the total of miles moved by each vehicle – are seen to be 270 miles. We shall now proceed to improve this solution, outlining the method in stepwise fashion.

Step 1

Compute the F-costs of movement of the vehicles on Leg 1, using 11.14, and the F^*-costs of such movement using 11.18. These are recorded in Table 11.2 with F-costs the first entry in the cell and F^*-costs the second entry.

Step 2

Choose Vehicle 1 and move it to its minimum-F (minimum-F^*)

Table 11.1 The distance matrix

Stations	*1*	*2*	*3*	*4*	*5*	*6*	$0^D{}_j$
1	0	20	40	80	20	30	−100
2	–	0	15	50	40	60	−100
3	–	–	0	5	25	100	−50
4	–	–	–	0	15	20	+50
5	–	–	–	–	0	30	+80
6	–	–	–	–	–	0	+120

Table 11.2 First revised feasible solution: *F*-costs and *F**-costs for Leg 1

Vehicle	*Stations*		
	4	*5*	*6*
1	4000;80	400;20	600;30
2	2500;50	800;40	1200;60
3	5;5	750;25	7000;100

station. If this movement results in a λ-infeasible solution, choose from the vehicles assigned to this station on Leg 1 of the initial feasible solution the vehicle whose index is the minimum of ${}_1I_k$ (defined in 11.2) and move that vehicle to its minimum-*F* (minimum-*F**) station if that station is not Station *k*, in which event we move the vehicle to its second-best *F*-cost (*F**-cost) station. If this new movement results in an infeasible solution at (say) Station *q*, we displace that vehicle whose index is the minimum of ${}_1I_q$ in the initial feasible solution, and repeat such performances until a new feasible solution has been obtained for Leg 1 in so far as these vehicles – those that have moved in consequence of Vehicle 1's movement – are concerned.

For the case at hand, we reassign Vehicle 1 from Station 4 to Station 5. But this results in a λ-infeasible solution at Station 5, and we reassign Vehicle 2 to its second-best *F*-cost station, or Station 6. This does not result in a λ-infeasible solution at Station 6, and thus we are finished with this step for the *F*-cost example.

Step 3

In the order of their indices, check the λ-feasibility of all remaining vehicles that retain their initial feasible assignments. Where the constraint is violated move them in ascending order of their *F*-costs (*F**-costs) until they receive a λ-feasible assignment, taking care never to disturb a prior assignment of Step 2 or of this step. Continue until all vehicles have λ-feasible assignments. Leg 1 of the first revised feasible solution has now been obtained. For our problem, we obtain:

First Revised Feasible Solution: Leg 1

F-Costs: Vehicle 1 moves to Station 5
Vehicle 2 moves to Station 6
Vehicle 3 moves to Station 6
*F**-Costs: Vehicle 1 moves to Station 5
Vehicle 2 moves to Station 4
Vehicle 3 moves to Station 6.

Step 4

Move to Leg 2 of the initial feasible solution and recompute the *F*-costs (*F**-costs) on the basis of the Leg 1 assignments just completed. Move Vehicle 1 to its minimum-*F* (minimum-*F**) station and follow the same procedures outlined in Steps 2 and 3. These yield the new Leg 2.

In our illustrative case, for *F*-costs, Vehicle 1's best assignment is the only assignment available to it: Station 4. This is also true for Vehicle 2. For the *F**-cost case, Vehicles 1 and 2 must move to Station 6. The results may then be written:

First Revised Feasible Solution: Leg 2

F-Costs: Vehicle 1 moves to Station 4
Vehicle 2 moves to Station 4
*F**-Costs: Vehicle 1 moves to Station 6
Vehicle 2 moves to Station 6.

Step 5

Repeat the procedures of Steps 3 and 4 until revised Legs 3, 4, . . ., *N* have been obtained. When these have been obtained the first revised feasible solution is complete. For our illustrative example, we record below the first revised feasible solutions using both *F*-costs and *F**-costs.

First Revised Feasible Solution:

F-Costs: Vehicle 1 moves to Stations 5, 4 for a distance of 35 miles;
Vehicle 2 moves to Stations 6, 4 for a distance of 80 miles;
Vehicle 3 moves to Station 6 for a distance of 100 miles.

*F**-Costs: Vehicle 1 moves to Stations 5, 6 for a distance of 50 miles;
Vehicle 2 moves to Stations 4, 6 for a distance of 70 miles;
Vehicle 3 moves to Station 6 for a distance of 100 miles.

The distance cost of the first revised feasible solution using *F*-costs is 215 miles and using *F**-costs is 220 miles.

Step 6

Permute the assignments of the first revised feasible solutions for *F*-costs (*F**-costs) by treating each vehicle's circuit as fixed but altering the sequence in which the stations are visited. Where it is impossible to perform such permutations by inspectional methods, it is recommended that branch-and-bound methods (Little *et al.* [8]) be employed. Where a saving in distance costs may be made by such permutations substitute the relevant sequence for that of the first revised feasible solution. This solution is called the 'first permuted feasible solution'.

For the case at hand, the first revised feasible solution using *F*-costs permits the permutation for Vehicle 1 of Station 1 to Station 4, Station 4 to Station 5, whose distance cost is 95 miles, which is greater than the existing sequence. Vehicle 2's circuit may be permuted to the sequence Station 2 to Station 4, Station 4 to Station 6, for a distance cost of 70 miles, which *is* cheaper than the 80 miles logged by the present sequence. Therefore, we substitute the permutation for the *F*-cost first feasible solution sequence for Vehicle 2. A check of the *F**-cost first revised feasible solution will reveal that no permutations will better the solution's cost. We record below the first permuted feasible solution using *F*-costs:
First Permuted Feasible Solution:

F-Costs: Vehicle 1 moves to Stations 5,4 for a distance of 35 miles;

Vehicle 2 moves to Stations 4, 6 for a distance of 70 miles;
Vehicle 3 moves to Station 6 for a distance of 100 miles;

The total distance cost of this solution is 205 miles.

An important point must be made at this juncture. The permuted solutions permit vehicles to move and make only partial contributions of their loads to the existing demands of the stations on given legs. This points up the essential fiction of the leg and its adoption as an analytical convenience.

Step 7

Compare the distance costs of the final first revised solutions for F-costs and F^*-costs of Step 6, and select the one that yields the minimum distance cost. If this cost is no smaller than the initial feasible solution, discard the revised solutions. If it is smaller and no permutations were made in it, adopt it as the new initial feasible solution. If it is smaller and permutations were made, adopt the unpermuted first revised feasible solution from which it was obtained as the new initial feasible solution.

In our example, the first permuted feasible solution yields cheaper distance cost than the initial feasible solution and the F^*-cost revised solution. Therefore, we adopt its unpermuted revised solution form as the new feasible solution.

Step 8

Repeat Steps 1 through 7 treating Vehicle 2 as the favored vehicle.

Step 9

Repeat Steps 1 through 7 treating Vehicles 3, 4, . . ., m in turn until all vehicles are adopted for preferred movement.

For the case in hand, the final and best solution obtained is the third revised feasible solution using both F-costs and F^*-costs:
Third Revised Feasible Solution

F-Costs and F^*-Costs:
Vehicle 1 moves to Station 6 for a distance of 30 miles;
Vehicle 2 moves to Stations 5, 6 for a distance of 70 miles;
Vehicle 3 moves to Station 4 for a distance of 5 miles.

Thus, the cheapest cost solution we obtain is this solution with a cost of 105 miles, which can be shown to be, by complete enumeration, the optimum feasible solution.

Step 10

(Optional) To improve the solution, if computer time is available, every feasible solution obtained in Steps 1 through 9 can be treated as an initial feasible solution, and steps 1 through 9 can be applied to it to generate more solutions. The best solution of all those obtained can be selected as the approximation to the optimum.

Some Comments on the Results

Our method has led us to what may be shown to be the optimum feasible solution, and the F-cost, involving the less readily interpretable criterion variable, has proved its worth by giving us at every step of the way no worse a result than F^*-costs, and a better solution in one instance. Further, the algorithm took us from a most uneconomic initial feasible solution to a markedly improved one in one step. Of course, much more computational experience will be necessary to judge the performance of the method, especially for larger scale problems. Nevertheless, in addition to its basic role in demonstrating the method in simplest terms, our example encourages us in its rapid improvement of results and in its choice of F-term solutions.

IV A LARGER-SCALE PROBLEM

Let us move to a larger-scale problem to gain more experience with the method. We shall assume the set V contains $m = 7$ elements located on Leg 0 at the following j:

1. New York, New York
2. New Haven, Connecticut
3. Worcester, Massachusetts
4. Boston, Massachusetts
5. Bangor, Maine
6. Burlington, Vermont
7. Albany, New York

The stations $j = 8, \ldots, 20$ are stations with net demands on Leg 0, and are identified as follows:

8. Newport, Rhode Island
9. Hartford, Connecticut
10. Providence, Rhode Island
11. Lowell, Massachusetts
12. Keene, New Hampshire
13. Augusta, Maine
14. Manchester, New Hampshire
15. Montreal, Canada
16. Quebec, Canada
17. Rutland, Vermont
18. Portsmouth, New Hampshire
19. Pittsfield, Massachusetts
20. Greenfield, Massachusetts

The distance matrix is reproduced in Table 11.3 with distances being approximate highway distances between the stations. Also, in the last column of the table we have listed the initial demands and supplies at each station. The spatial configuration of the stations may be seen in Figure 11.1, in which the nature of the problem is clear as

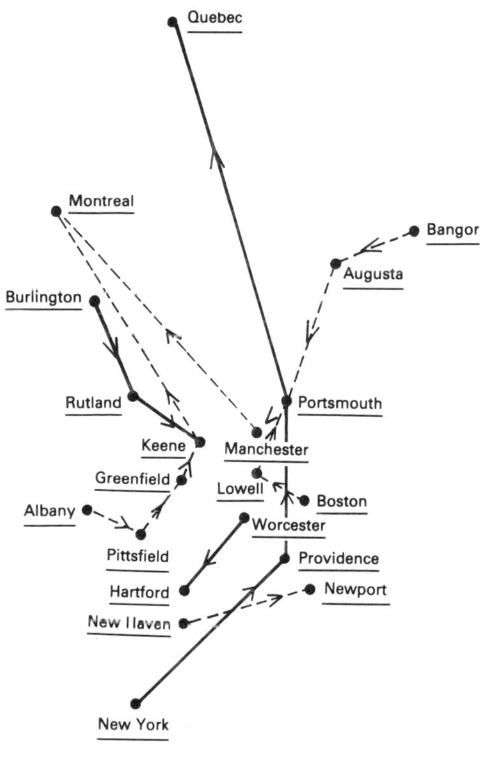

Figure 11.1 The best approximate solution

Table 11.3 The distance matrix

Stations	*8*	*9*	*10*	*11*	*12*	*13*	*14*	*15*	*16*	*17*	*18*	*19*	*20*	$_oD_j$
1	170	115	175	215	200	375	235	365	535	235	265	145	170	−1,000
2	95	35	100	140	125	300	160	345	455	180	190	90	95	− 300
3	65	60	40	45	55	205	60	295	395	130	95	95	60	− 350
4	65	100	45	25	85	165	55	310	370	155	55	130	60	− 800
5	295	325	275	225	260	65	215	295	230	290	175	350	295	− 100
6	260	225	230	200	140	220	165	95	235	65	185	155	160	− 200
7	185	105	165	155	100	295	155	225	380	95	200	35	80	− 400
8	0	85	30	92	130	235	121	370	445	205	127	160	130	300
9		0	75	105	90	265	120	320	425	155	155	70	60	350
10			0	60	90	210	90	330	415	165	100	130	95	600
11				0	63	159	32	298	345	138	51	134	78	250
12					0	202	54	240	375	75	99	91	35	150
13						0	154	267	205	231	108	293	237	80
14							0	266	315	111	46	144	88	175
15								0	230	165	291	254	260	500
16									0	300	315	390	460	380
17										0	146	94	95	100
18											0	185	129	75
19												0	56	100
20													0	90

Table 11.4 An initial feasible solution

Vehicle	*Station sequence*	*Distance (miles)*
1	9, 10, 15	520
2	20, 12, 15	370
3	10	40
4	11, 14, 15, 18, 16	929
5	13, 18	173
6	17, 16	365
7	19, 8	195
	Total	2592

that of supplying the New England and Canadian region with product from the first seven stations.

By the application of Steps 1 through 9 we obtained seven distinct revised solutions from the use of *F*-costs and two distinct revised solutions using *F**-costs. These are identified by their distance costs as follows: (*1*) *F*-costs – 2662; 2091; 2502; 2617; 2415; 2494; and 2732; (*2*) *F**-costs – 2277; 2295. The minimum-cost solution obtained yielded a distance cost of 2091, and resulted from an *F*-cost analysis. It is substantially lower than the initial feasible solution in distance cost.

When we applied our optional Step 10, using each of these 9 distinct solutions as an initial feasible solution and repeating Steps 1 through 9 for each, we obtained 34 more distinct solutions, of which only two yielded smaller distance costs than our previously best solution whose costs are 2091 miles. The best of the 44 distinct solutions obtained was the result of a permuted *F*-cost analysis, and the second best was the result of a permuted *F**-cost solution, the first yielding costs of 1812 miles and the second 1951 miles. The former is adopted as the final estimate of the optimal assignments, and is detailed in Table 11.5 and charted on Figure 11.1.

It should be pointed out that in the derivation of these 44 solutions we did not study each to see if permutations could be obtained to lessen costs, but did so only for those solutions which were very close to the best solution of Table 11.5. With this understanding that the results could be improved in all likelihood by the application of systematic branch-and-bound techniques to each vehicle's assignments in all solutions, we present in Table 11.6 the frequency distribution of the 44 distinct solutions.

Table 11.5 The approximate least-cost solution

Vehicle	*Station sequence*	*Distance*
1	10, 18, 16	590
2	8	95
3	9	60
4	11, 18, 14, 15	388
5	13, 18	173
6	17, 12	140
7	19, 20, 12, 15	366
	Total	1,812

Table 11.6 Frequency distribution of feasible solutions

Classes	*Frequency*
1,800 to 1,899	1
1,900 to 1,999	1
2,000 to 2,099	1
2,100 to 2,199	1
2,200 to 2,299	6
2,300 to 2,399	5
2,400 to 2,499	10
2,500 to 2,599	6
2,600 to 2,699	6
2,700 to 2,799	5
2,800 to 2,899	1
2,900 to 2,999	1
	44

The efficient solution yielded by Steps 1 through 9 of Section 3 was within 15 per cent of the very best of the 44 solutions, which would seem to be an acceptable deviation in many realistic situations. Of course, we have no hint of whether a better solution than we have exists, and if so, how much better it is. It is also interesting that both this efficient solution and the very best solution came from *F*-cost analyses, while the second-best solution came from an *F**-cost analysis. This is further justification for the *F*-cost calculation as well as the simultaneous use of the *F**-cost index, for it would seem that either may out-perform the other in specific cases.

V CONCLUSION

The availability of the computer has made possible a method of generating feasible solutions to combinatorial problems in quantity such that, if an efficient algorithm for obtaining feasible solutions is employed, the probability of obtaining a good approximation to the optimum solution in an acceptable amount of computer time is high. Mathematical purism frequently leads economists to seek the very best where very good suffices for all practical purposes, and to reject approximative methods even when exact ones cannot be derived. We have developed an approximative method which rejects this perfectionism.

It is the disadvantage of such methods, of course, that deductive generalizations are not easily obtained, so that their usefulness can be judged only after a large amount of computational experience has been accumulated. However, as each investigator attacks his specific problems, it will be possible to alter the method to obtain better results in detail. And, hopefully, from such experience will come insights into the structure of the problems which will permit the optimum to be attained efficiently one day.

References

[1] Balinski, M. L., 'Integer Programming: Methods, Uses, Computation', *Management Science*, 12 (1965), pp. 253–313.

[2] Dantzig, G., D. R. Fulkerson and S. M. Johnson, 'On a Linear Programming Combinatorial Approach to the Travelling Salesman Problem', *Operations Research*, 7 (1959), pp. 58–66.

[3] Gilmore, P. C. and R. E. Gomory, 'A Solvable Case of the Travelling Salesman Problem', *Proceedings of the National Academy of Sciences*, 51 (1964), pp. 178–181.

[4] Hitchcock, F. L. 'The Distribution of a Product from Several Sources to Numerous Localities', *Journal of Mathematics and Physics*, 20 (1941), pp. 224–30.

[5] Kuenne, R. E., *The Polaris Missile Strike: A General Economic Systems Analysis*, Ohio State University Press, Columbus, Ohio (1966).

[6] Kuhn, H. W., 'On Certain Convex Polyhedra', *Bulletin of the American Mathematical Society*, 61 (1955), pp. 557–8.

[7] Lin, S., 'Computer Solutions of the Traveling Salesman Problem', *The Bell System Technical Journal*, 44 (1965), pp. 2245–69.

[8] Little, J. D., K. G. Murty, D. W. Sweeney and C. Karel, 'An Algorithm for the Traveling Salesman Problem', *Operations Research*, 11 (1963), pp. 972–89.

12 Exact and Approximate Solution of Constrained Dynamic Combinatorial Problems in Space*

I THE SEQUENTIAL COMBINATORIAL PROBLEM REVISITED

In a recent issue of this Journal the author confronted an important set of spatial combinatorial problems of a sequential nature and suggested a method to obtain approximate solutions rather efficiently (Kuenne [4]). In that article the problem is defined formally; its structural resemblances to the transportation, travelling salesman, and weighted set covering problems were discussed, and several examples were solved heuristically. Although existing combinatorial approaches such as branch and bound were mentioned briefly, and a role assigned them in the approximative algorithm, the view was expressed that such paths to exact solution were infeasible computationally.

That view was overly pessimistic, at least for problems of moderate dimension, and, further, eliminates an even more efficient method of generating approximate solutions where exact solutions may be beyond reach. The present article describes a branch-and-bound algorithm, DYNCOM, which yields exact solutions to moderate-sized problems within realistic computer time and memory limitations and which gives good approximate solutions with rather limited time expenditures for larger problems.[1] A key contribution to this correction of pessimism is that the imposition of *simple* constraints upon combinatorial problems contributes to the ease of solution by speeding the elimination of potential solutions. In contradistinction to such

* Published originally in the *Journal of Regional Science*, 12 (1972), pp. 1–22, and reproduced with permission.

areas as linear and nonlinear programming, where the multiplication of constraints increases the difficulty of solution in general, constraints in combinatorial programming may have the opposite effect.

The formal definition of the problem will not be repeated but restated informally in Section II for present purposes. In the belief that branch-and-bound techniques will become an increasingly important tool for the economist and regional scientist, Section III is devoted to an exposition and illustration of them. Section IV contains an exposition of DYNCOM without, however, being a user's manual, and demonstrates that the algorithm constitutes an explicit or implicit complete enumeration of all feasible solutions. Next, Section V is devoted to discussion of such computational experience as has been acquired in applying DYNCOM to date. In that section we address explicitly some of the examples discussed in the earlier article. Finally, we conclude with a brief summary in Section VI.

II THE DYNAMIC SPATIAL COMBINATORIAL PROGRAMMING PROBLEM

The recent exact solution of the multisource generalized Weber problem by branch and bound (Kuenne and Soland, [5, 6]) opens some rather exciting prospects for regional scientists. Hopefully, it will lead to intensive efforts to extend the algorithm to permit solutions when constraints are present, and thereby gain power over classic locational point sink–point source analyses.

It also suggests the extension of such techniques into dynamic least-cost path problems of the following type. Assume (*1*) a set of vehicles located initially at m sources i, with specified supplies S_i, and (*2*) a set of n sinks j with designated demands D_j. The sum of supplies over sources must be at least as great as the total demand at all sinks. The problem is to route the trucks among the sinks in such manner as to satisfy demands within the individual source supply constraints and in a manner that minimizes the sum of distances travelled by the vehicles.

In the original article the following simple example was employed to demonstrate the heuristic techniques, and variants of it will be discussed in Section V to demonstrate DYNCOM. There are 3 sources, with fixed supplies $S_i = 100$, 100, and 50 respectively, as well as sink demands $D_j = 50$, 80, and 120 respectively. The distances of sinks to sources and between pairs of sinks are given in Table 12.1.

Table 12.1 The distance matrix

	Sink 1	*Sink 2*	*Sink 3*	*Quantity*
Source 1	80	20	30	S_1 = 100
Source 2	50	40	60	S_2 = 100
Source 3	5	25	100	S_3 = 50
Sink 1	0	15	20	D_1 = 50
Sink 2	15	0	30	D_2 = 80
Sink 3	20	30	0	D_3 = 120

A Matter of Co-ordinates and Metrics

We shall enrich the problem in three directions in obtaining exact solutions by DYNCOM. First, we discussed in the earlier article the possibility of requiring that the trucks return to their source locations after completing deliveries, making the problem a 'closed-circuit' constrained multiple travelling salesman problem. This requirement was not imposed in the original presentation, but we include it as an option in DYNCOM and will solve it as well as the original case in Section V.

Secondly, we shall permit certain sinks to be preassigned sources as constraints on the problem, so that only those solutions with the specified preassignments are feasible. In building this option into the algorithm a step is taken toward inclusion of constraints other than demands and supplies.

Finally, we shall permit the sources and sinks to be identified by spatial co-ordinates and compute the distances between them by a choice of co-ordinate transformations and metrics. The options are indicated briefly below, and are discussed more fully in Kuenne and Soland [5].

(1) *Cartesian Co-ordinates*. A first option is simply that the $[X, Y]$ co-ordinate pairs define location on the Cartesian grid, with the Euclidean metric $D_{a,b} = [(X_a - X_b)^2 + (Y_a - Y_b)^2]^{1/2}$ measuring cost of movement.

(2) *Spherical Co-ordinates*. When co-ordinates are stated in *minutes* of latitude (north positive, south negative) and longitude (west positive, east negative), DYNCOM transforms these co-ordinates into points that can be manipulated by the Euclidean metric and uses another transformation to convert the resulting distance measurement to approximate statute miles. Because the spatial economist,

geographer and regional scientist frequently find their data in spherical co-ordinates and the computation of great circle distances by the law of cosines clumsy or inappropriate, we shall present the technique briefly because of its independent interest.

The Mercator projection of the earth's surface is characterized by the fact that a line drawn on it connecting two points with different longitudes will intersect intervening longitudinal meridians at the same angle. A quick check on a globe will indicate that the great circle route connecting the two points will not do so (except on the equator). Such lines are called 'rhumb lines', and we shall term the Euclidean distances along such lines between Mercator co-ordinates 'rhumb line map distances'.

In practical navigation rhumb lines have a substantial role to play, given the centuries-old aversion of sailors to nonlinear courses. To travel between two points on the earth's surface it is optimal to follow the great circle route between them. In practice this is accomplished by plotting the great circle route on a special projection, noting where the intersections of the route with a set of meridians occur (say, meridians spaced 5° apart), plotting these intersection points on a Mercator map, and connecting them with straight line segments. By sailing these latter, a piecewise linear approximation to the great circle route is obtained.

If we imagine a huge Mercator map whose width is 21 600 nautical miles at the equator, we may convert a given latitude, λ, to Mercator map coordinate X by the following transformation:

$$X = 7915.704468 \log_{10} \cot .5p,$$

where p is the colatitude of the point (in minutes):

$$p = 5400 - \lambda.$$

The map co-ordinate corresponding to latitude θ is unaffected by the transformation to Mercator co-ordinates:

$$Y = \theta.$$

The map distance along the rhumb line between two such points, $[X_a, Y_a]$ and $[X_b, Y_b]$, is then $D_m = [(X_a - X_b)^2 + (Y_a - Y_b)^2]^{1/2}$.

and this may be corrected for the distortions of the Mercator projection beyond the neighborhood of the equator by multiplying it by the cosine of midlatitude:

$$D \text{ (in nautical miles)} = D_m \left[\cos .5\ (\lambda_a + \lambda_b)\right].$$

To obtain statute miles this must be multiplied by 1.1515.

In prior work with the MULTIWEB algorithm for the solution of multisource Weber point problems, this approximation was sufficiently accurate. That accuracy may be increased to any desired degree of approximation to great circle distances by reducing the interval between meridians when drawing the rhumb-line approximation to the great circle. The reader is referred to Kuenne and Soland [5] for more detailed discussion of the technique and to Deetz and Adams [2] for proof of the validity of the transformations.

(3) *The Metropolitan Metric.* In solving some problems – for example, those framed in terms of travel along the rectangular grid of city streets – the 'metropolitan metric' is more appropriate. It is defined as:

$$D_{a,b} = [(X_a - X_b)^2]^{1/2} + [(Y_a - Y_b)^2]^{1/2},$$

where the $[X, Y]$ pairs are assumed to be co-ordinates in a Cartesian grid. This option is also provided by DYNCOM.

Lastly, the algorithm permits reading in the sink-to-source and intersink distances as inputs.

III COMBINATORIAL PROGRAMMING AND BRANCH AND BOUND

The Conceptualization of the Method and Feasibility Considerations

This spatial problem may be seen immediately to be one in constrained combinatorial programming: the core of the solution is the m not necessarily disjoint sets of sink sequences which are the optimal source assignments. The difficulties of finding those optimal routes for problems of moderate dimensionality are dealt with in the earlier article in some detail. We may confine ourselves here to the comment that although for the simple example of the previous section there are only 27 possible two- and three-sink routes for the sources before taking into account demand-supply constraints on movements, for the 7-source, 13-sink problem with which we shall deal in Section V there are 43.589×10^9 routes that are potential candidates.

Branch and bound is designed to find the optimal source routes by explicitly or implicitly eliminating all non-optimal combinations. It is an approach to problems rather than a methodology, for ordinarily its specific rules of procedure must be tailored with the peculiarities of the problem at hand fully measured. At the conceptual level, therefore, it is necessary to perform three problem-conditioned tasks:

1. The problem must be framed combinatorially so that the objects can be grouped into subsets that are meaningful in terms of the solution sought;
2. A means must be devised to obtain a lower bound (for a minimization problem) for each such subset of solutions; and
3. A set of rules must be chosen for moving from subset to subset in such manner that every feasible combination of objects is explicitly or implicitly tested for optimality.

The nature of branch-and-bound operations may be understood from these conceptual requirements. Combinations are formed on the basis of the rules for grouping objects, successive combinations (*nodes*) becoming increasingly well specified as the solution continues. For example, we move from the subset of solutions assigning sink 3 to source 5, to the subset which assigns sink 3 to source 5 *and* sink 1 to source 3. Thus, each node generally gives rise to a branch of nodes as one continues, the whole hierarchy of nodes resembling an inverted tree in graph theory. At each node the bounding rules are employed to calculate an underestimate for all solutions characterized by it, available for checking against the best feasible solution value for the objective function so far attained. If the lower bound is greater than that feasible value, the branches of the tree originating from the node need not be followed, for the nodes thereon cannot yield the optimal solution. As the algorithm continues down the tree, feasible solutions with lower values are substituted for those that yield higher values, and at every point the best available feasible solution is retained to filter the lower bounds of nodes.

The movement from one node to the next, according to the third specification above, must guarantee that all nodes will have come under consideration, if only implicitly by the elimination of the more general subsets of solutions in which they are embedded. Conceptually, the solution procedure guarantees that the optimal solution can be found in a finite number of steps: in effect, the lower bounds of nodes farther and farther down the tree rise as the search proceeds, and the feasible solution values, which are upper bounds, descend. When the smallest lower bound coincides with or exceeds the lowest upper bound, the optimal solution has been reached.

But the troublesome questions of computational feasibility confront us: while the search may be guaranteed conceptually to end at the optimum in finite time, the amount of computer time expended or the number of nodes requiring storage at any particular time may exceed feasible limits. Practically speaking, therefore, the algorithm must be computationally efficient in both dimensions if the optimum is to be attained.

However, in conformance with the spirit of the original article, which argued for the general acceptability of approximate solutions, DYNCOM has two attractive features. First, it prints out each feasible solution as it improves the existing feasible solution, so that even if an optimum is not reached the user has obtained a good approximate solution. And second, the user has an ability to control the degree of deviation of the approximate solution from the optimal; that is, he may specify a degree of tolerance he is willing to accept, and the algorithm will attempt to find a feasible solution which is guaranteed to be in the specified area of deviation from the optimum. Given the nature of combinatorial programming and the badly-behaved objective functions of spatial economics, both features are valuable characteristics of branch and bound.

The branching and bounding procedures adopted are most germane in determining the efficiency of the algorithm. The lower bounds should be as 'tight' as possible, every bit of ingenuity being employed to obtain lower bounds that are as high as possible, even at the expense of time-consuming calculations. This is very important in the early computations, for the higher the lower bound, everything else being equal, the earlier in the search branches can be broken off for exceeding the best feasible solution value, and the fewer will be the demands on time and computer memory. Generally speaking, with third-generation computers, tight lower bounds are worth very large numbers of calculations in tradeoff, so that the intricacy of computation involved in obtaining the lower bound becomes a countervailing consideration only in very large problems.

The second contributor to the feasibility of calculation is the set of rules for moving 'down the tree', or the manner in which the algorithm chooses a next node from its present position. Several general methods are available for adoption (see Agin [1]), two of which warrant discussion for relevance to the spatial problem. Perhaps the manner most plausible, prima facie, is that of choosing that active node (that is, a node that has not been branched from) whose lower bound is the minimal lower bound in storage. The rationale behind such choice, of course, is that such solutions should have a high probability of containing the optimum or at least good feasible solutions.

One difficulty with such a procedure, however, especially in large problems where node storage is huge and computer memory limitational, is that it may entail storing a large number of active nodes at every point in the search. Time of solution may be minimized by the rule where node storage is not constrained, but for larger problems solution may not be possible by virtue of memory limitations. Therefore, a second manner of procedure may be recommended: *backtracking*.

In this manner of movement the algorithm chooses a node and branches from it to increasingly specific solutions further down the tree until it obtains a feasible solution, then backtracks to a higher branch and follows the same single-minded search for another feasible solution. Frequently, as in the dynamic combinatorial problem with which we are concerned, a primary node (two-combination of a source and a sink) can be generated and all of its branches investigated before the next primary node is generated. When movement is made to successive primary nodes, all nodes from the previous search may be eliminated, so that computer storage is freed. Moreover, the rapid generation of feasible solutions gives high probability that branches can be broken off high in the tree, offering some compensation for penalties in time suffered in searching exhaustively every primary node's branches.

A Demonstration of the Method

To demonstrate branch and bound and to illustrate the considerations discussed, we shall solve a simple problem following various rules. The problem selected is one designed by Harvey Wagner to illustrate dynamic programming, but we shall use it as an object for solution by branch and bound (Hillier and Lieberman [3], pp. 239–40). A salesman whose territory was the Old West was faced

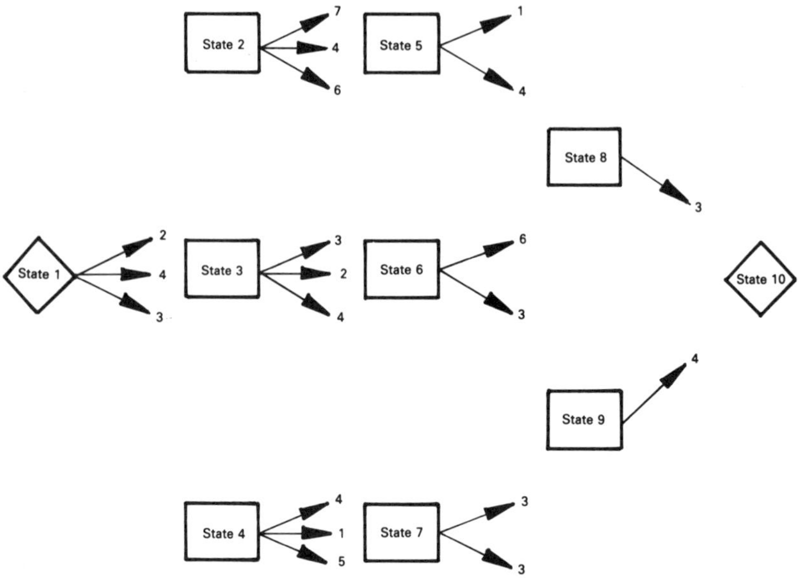

Figure 12.1 The stagecoach problem

with the need to go from State 1 to State 10 in Figure 12.1, by stagecoach. To do so, however, he had to change coaches three times, and at State 1 and the first two changes he faced a choice of intermediate destination. State 1 could have led to States 2, 3, or 4, depending upon his choice of stagecoach; his second stage could have been boarded for States 5, 6, or 7; his third stage was either State 8 or 9; and his last stage went to State 10.

His choice was dictated by the desire to minimize the relative risk of attack by Indians on each of the steps of the routes. It is assumed that life insurance companies set the premiums listed in Table 12.2 on the basis of those risks. By clever terminological design, Wagner's problem is to find an optimal 'policy' involving different 'states' of the system at a sequence of 'stages', the optimum being, of course, that route whose insurance cost is a minimum.

We shall solve the system first by following the rule of always branching from the active node with smallest lower bound. Further, the initial method of determining lower bounds will be purposely loose: for each stage we shall determine the cheapest possible outcome of the stages yet to be chosen, whether or not the combination of stages is a feasible solution. We have circled these outcomes in

Table 12.2 The cost matrix

From \ To	State 2	3	4	5	6	7	8	9	10
State 1	(2)	4	3	–	–	–	–	–	–
2	–	–	–	7	4	6	–	–	–
3	–	–	–	3	2	4	–	–	–
4	–	–	–	4	(1)	5	–	–	–
5	–	–	–	–	–	–	(1)	4	–
6	–	–	–	–	–	–	6	3	–
7	–	–	–	–	–	–	3	3	–
8	–	–	–	–	–	–	–	–	(3)
9	–	–	–	–	–	–	–	–	4

Table 12.2: the cheapest possible cost of Stage 1 is 2, of Stage 2 is 1, of Stage 3 is 1 and of Stage 4 is 3. These costs are not those of a feasible combination of stagecoaches, but we shall neglect this fact.

Figure 12.2 depicts the tree actually followed in the solution under these branching and bounding procedures. Each bubble on the diagram records the pertinent information about a node, or subset of solutions. The first number is its position in the sequence as it was generated; the second numeral represents the destination stage of the stagecoach; the third numeral is the lower bound of the node; and the last numeral is the value of the objective function for the best feasible solution ruling when the node was generated. The numbers on the lines or *activities* of the graph are the costs of the insurance policies for the relevant routes, as shown in the figures by the arrows of Figure 12.1 and in Table 12.2.

Node 1, therefore, is the start of the tree, and represents all solutions to the problem. The salesman is located at State 1, and if we sum the cheapest possible insurance costs for each stage to follow we obtain a lower bound of 7. That is, the cheapest cost of the first stage is 2 (to State 2), of the second stage is 1 (to State 6), of the third stage is 1 (to State 8), and of the fourth stage is 3 (to State 10). The fact that this is not a feasible solution does not concern us at the moment – it is enough that it does constitute a lower bound, for no solution can be cheaper than this figure. Finally, because we assume we have no feasible solution initially, the value of the solution (upper bound) is ∞ to start.

From State 1 three subsets of solutions are created: (*1*) those that go through State 2, (*2*) those that go through State 3, and (*3*) those

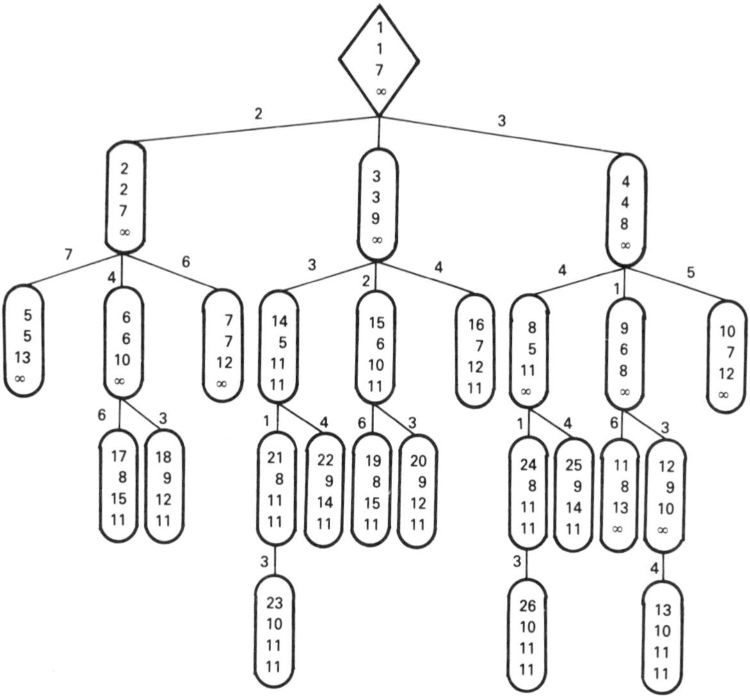

Figure 12.2 Initial solution tree for the stagecoach problem

that go through State 4. Each subset is represented in Figure 12.2 by a (primary) node. The lower bounds of the nodes are composed of two parts. First, the actual costs of the move from State 1 to the relevant state, and, second, the sum of the cheapest costs of the stages that lie ahead. The sum of the two components yields a lower limit to the solution subset costs. For example, in node 2, the actual cost of moving from State 1 to State 2 is 2, and the cheapest possible costs of Stages 2, 3, and 4 are 1, 1, and 3 respectively. Consequently, the lower bound for node 2 is 7. In like manner, the lower bounds for nodes 3 and 4 are calculated.

There are now three active nodes on the graph, or three nodes from which no branching has occurred. Which sink should be selected for branching? In following the hard and fast rule of selecting that node to branch on that has the lowest lower bound, we branch on node 2, and create nodes 5, 6 and 7, with lower bounds 13, 10 and 12. There are then 5 active nodes, and because node 4 has the cheapest lower bound, we branch from it to nodes 8, 9, and 10.

We next choose node 9 for branching, and then branch successively to nodes 12 and 13, the last of which is a *final node* (feasible solution). It yields a value for the objective function of 11, and this constitutes the value of the best feasible solution to this point. The lower bounds of all subsequent nodes will now be compared with this value, until a better feasible solution is discovered (if discovered). Moreover, we may compare existing active nodes with this value of 11: any active node whose lower bound is above 11 may be eliminated as a candidate for further branching, since it cannot generate more specific subsets that contain the optimal solution. We simply break off the branch of the tree at this point. In the case at hand, nodes 5, 7 and 11 are so terminated.

The lowest underestimate in the graph is now 9 for node 3, and we form nodes 14, 15 and 16 by branching from it, the last of which may be ignored as being above the best feasible solution so far obtained. We continue in this manner to branch and bound: when all active nodes are greater than or equal to the best feasible solution obtained so far we may stop, for we have determined the optimal solution(s). In the present case we see that there are three optimal routes whose objective function cost is 11: States 1, 3, 5, 8, 10; States 1, 4, 5, 8, 10; and States 1, 4, 6, 9, 10.

The solution required the generation of 25 nodes (the first need not be produced). However, once a node has been branched from, it may be omitted from further consideration; in practical terms, it no longer need be stored in the computer's memory, and its space may be filled by an active node. For large problems the maximum number of active nodes stored simultaneously, as mentioned above, is an important index of the efficiency of the rules for branching adopted. In our solution that number was 8. Would it be possible to conduct the search in a manner that reduces storage needs?

Improvement by Backtracking and More Efficient Bounding

By resorting to the backtracking technique for extracting all feasible solutions from a primary branch of the tree, leaving it only when no active nodes remain for further branching, we may frequently reduce storage space (and not always at the expense of time or money when peripheral storage devices are used in conjunction with a computer's central processing unit). We simply follow the branching rule of choosing that node with smallest lower bound in the same primary branch and at the same hierarchical level, tracking down the tree until we reach a final node and a feasible solution. We then 'backtrack' up the tree to the lowest level of active nodes, and search for final nodes unless we may eliminate them. This is continued until all active nodes have been eliminated from the primary node branches, so that only the best feasible solution(s) to that point need be stored, in addition to those primary nodes which are still active.

Had we followed this procedure in the solution to the stagecoach problem, we should have emerged from the first primary node (node 2) with 4 nodes in storage (two feasible solutions and the other primary nodes) and a maximum number at any one time of only 6. We should have abandoned the search in the second primary node (node 4) with only 3 nodes for storage and a maximum of 6. And the analysis of the subsets of solutions associated with the third primary node (node 3) ends with three feasible solutions stored and at no time requires more than 4. Thus, by lowering maximum storage requirements to 6, we have reduced the storage provision by 8 as compared with the previous method which did not use backtracking.

Let us turn now to a second method of increasing the efficiency of search, especially in so far as minimization of storage is concerned: tightening the lower bounds. We purposely used a very loose bounding technique which can be improved easily. For example, an alternative for deriving the second component of the bound at each stage is to consider in detail the values of the solution if the salesman moved to each of the alternative states of the next stage, and thereafter benefited from the cheapest possible cost of successive stages whether or not feasible.

To illustrate from Figure 12.2 let us bound node 2 in this improved manner. The first component of the bound – actual costs undergone on the movement from State 1 to State 2 – is unchanged at 2. Now, however, let us make three estimates of the lower bound for this node by assuming successively that the salesman moves to States 5, 6, and 7. If he moves to State 5 and then on the cheapest route in the following stages he will have costs of 7 for movement from State 2 to State 5, plus a minimum cost of 1 from Stage 3 and of 3 for Stage 4, or a second component total of 11. If he moves to State 6, this second component total will be 10, and if he moves to State 7 it will total 12. Then, the potential lower bounds, found by adding to these alternatives the first component value of 2, are 13, 12, and 14, and we choose the minimum of these to be the lower bound for node 2. This bound is raised from 7 in Figure 12.2 to 12, with the promise of enabling us

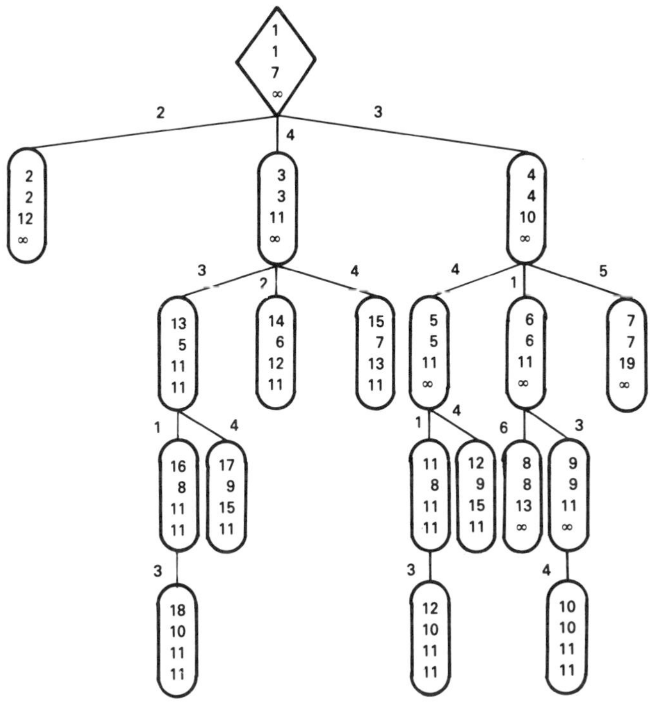

Figure 12.3 Solution tree with tighter bounding and backtracking.

to break off branches somewhat earlier in the solution.

In Figure 12.3 we illustrate the tree obtained when the new method of bounding and backtracking is employed. Note that the tighter bounding procedure led us to branch initially from node 4 (instead of node 2) and therefore gave a feasible solution more quickly. The three optimal solutions are obtained in only 18 nodes instead of the 25 of Figure 12.2, and the largest number of active nodes requiring storage at any time was 5. The extra time necessary to compute the lower bounds, therefore, has paid handsomely in node storage savings and the elimination of branches.

A third source of improved efficiency may result from the provision of a feasible solution at the start of the calculations with objective function value replacing ∞. This can result in a pruning of branches early in the game; however, in the present example the determination of a feasible solution other than the optimum would not help in the solution.

Branch and Bound in Summary

Branch and bound has been shown to be a method of procedure in combinatorial programming that moves from subset to subset of solutions in such manner as to guarantee that all solutions will be considered explicitly or implicitly, that bounds each such subset from below, and that eliminates subsets as these lower bounds exceed successively lower objective function values for feasible solutions. Time and node storage considerations are crucial in determining whether the technique is feasible, and either or both of these may usually be lessened by adopting efficient manners of bounding and branching from node to node. Further, a good initial feasible solution to the problem is a boost to computational efficiency. With this basic introduction to the method, we may turn now to the description of DYNCOM, particularly with respect to these elements.

IV A DESCRIPTION OF DYNCOM

DYNCOM is a branch-and-bound algorithm using backtracking to locate successively better feasible solutions to the spatial combinatorial problem, and, if the dimension of the problem lies within the storage capacity of the computer and the time available for solution, to locate the optimal solution(s). It currently exists in two versions. The first version relies wholly upon the computer's core for all memory, and, on an IBM 360/91, with 1200K bytes of core, it can store about 7000 active nodes. The second version stores most of the variables associated with nodes on a single disk file, and thereby essentially eliminates the storage problem. For example, when it is programmed to store 36 000 nodes it takes up 600K bytes of core and it stores 135 000 nodes when all of core is used. In all of our experiments with this largest program, time set limits to our usage and storage was not a factor.

Let us illustrate some of its features by returning to the illustrative problem involving the allocation of three sinks to three sources. DYNCOM begins by forming all possible initial source-to-sink allocations, ranking them in descending order in terms of the distances between them, and chooses as the first primary node that source-to-sink assignment that minimizes distance. This is designated the first active node and the only node stored at this point.

The procedure then becomes one of branching within this primary node following the rule of the smallest bound until all active nodes have been rendered inactive and all final nodes on this branch have been explicitly or implicitly examined. Departure from the primary node is then made with only the best feasible solution(s) stored, and a second primary node is obtained by taking that source-sink combination whose distance is the second lowest. The procedure is repeated until all $m \times n$ primary nodes have been searched and feasible solutions obtained. The optimal solution is that final node whose objective function value is minimal.

Let us demonstrate that this procedure will in fact permit all feasible solutions to be considered. Our problem has three sources and three sinks, so that the primary nodes (two-combinations) are nine in number. For example, consider the assignment of source 1 to sink 3, or 1–3; this means that source 1 moves to sink 3 *as a first assignment*, and is located at sink 3 upon completion of this initial delivery. Let us ignore demand-supply constraints and assume that all possible sequences are feasible at each step of the way, also assuming, however, that no source visits any sink more than once. Then, stage 2 may result in states representing movements to sinks 1 or 2, and the three-combinations are 1–3–1 and 1–3–2. Finally, 1–3–1 can give rise to one four-combination, 1–3–1–2, and 1–3–2 can lead to 1–3–2–1, exhausting the possible four-combinations. Therefore, given a two-combination, branching at every step to all nonrepetitive sinks leads us to examine every feasible combination springing from the two-combination. By repeating the process for every two-combination (primary node) we will be exhaustive in our search. By placing supply-demand restrictions upon our deliveries, many of these combinations will prove infeasible, but we have shown that they will be considered at least implicitly by our method of branching.

We have said that primary nodes are chosen in the sequence determined by ascending order of distance from sink to source. This procedure should lead to good feasible solutions early in the search and therefore minimize node formation. To experiment with the increased efficiency this manner of selecting primary nodes yields, we have, for the 3 × 3 problem of Table 12.1, solved the example by choosing primary nodes in descending order of source-to-sink distance. The resulting number of nodes formed will be compared with those formed in straightforward use of DYNCOM in Section V.

Once a primary node has been chosen for branching, the algorithm follows the rule of branching on that node with the smallest lower bound. A path is charted among the nodes to reach a feasible solution, and once reached all nodes with lower bounds above this solution value are eliminated as candidates for branching and their storage space freed for future nodes. The program then backtracks to the node with the smallest remaining lower bound and tracks to another feasible solution, repeating the process until all nodes within the branch of the tree originating with the primary node have been explicitly or implicitly investigated. It exits the primary node branch with all nodal storage space freed, retaining only the best feasible solution(s). The primary node with the smallest remaining source-to-sink distance is then chosen for complete branching, and the search is repeated.

Lower bounds are determined straightforwardly. The first component consists of the distances actually traversed to reach the node. The second component is arrived at in the following manner. If the source is located at its original site, the smallest distance of that source to a sink is taken. This is a lower bound to the cost of a source assignment whether or not the supply of the source is exhausted at the sink. For example, suppose the source is assigned to sink 1, and, with unexhausted supplies, let us suppose that it will in fact move to source 2. Then, by the triangle inequality relation in a metric's definition, $D_{i,2} \leqslant D_{i,1} + D_{1,2}$, so that considering only the $D_{i,j}$ distances and minimizing yields lower bounds for the second components of all nodes, even with potential multisink deliveries in future stages. If the problem involves the return of the source to its original location ('cycling') and the supply of the source is reduced to zero by the original movement, this distance value is doubled. If the source is located at a sink, the lowest distance from that sink to another is taken as the lower bound second component distance, and, if cycling is required, the distance from recipient sink to the source is added when source supply becomes zero. The sum of the first and second components constitutes the lower bound of the node.

This method of bounding does not give rise to a notably efficient lower bound, but to date all attempts to tighten it have proved unworkable. Although it is fanciful in the large to assume in essence that every source at each stage will send its supplies to the nearest demand point and go no farther in future stages, it cannot be ignored that a particular source may in the optimal solution do exactly this, and therefore the possibility must be retained for every source at every node. In practice, however, our limited computational experience to date indicates that DYNCOM, despite this loose bounding, is workably efficient within the context of combinatorial methods, as we shall discuss to the limited extent we can in Section V.

Lastly, DYNCOM affords the user the ability of specifying a tolerance factor, RHO, so that the best solutions obtained by the algorithm can be guaranteed to be within [(1 − RHO) × 100] percent of the true optimum. In combinatorial programming problems one frequently deals with objective functions that are quite flat-bottomed, so that the algorithm converges rather rapidly upon a feasible solution that is very close to the optimum, then spends huge amounts of time trying to improve the solution by very small amounts. If the user is willing to settle for a solution within 5 per cent of the true optimum (RHO = .05), he may save large amounts of computation time. The algorithm exits the search, then, when all active nodes have lower bounds which are greater than [1/(1 + RHO)] times the best feasible solution found. The user is thereby given a manner of controlling the 'goodness' of his non-optimal feasible solution, tailoring it to the importance of accuracy in the situation and the available money for its solution.

V COMPUTATIONAL EXPERIENCE

A limited amount of experience has been accumulated in the use of DYNCOM at this writing, leading us to suspect that its ability to obtain good feasible solutions by exact methods will be more important than its potential ability to locate optima. We shall detail this conclusion below.

Problem 1: The 3 × 3 Example

The simple problem whose data are presented in Table 12.1 was solved in the original article, and, because of its simple structure, it was used to test the various options of DYNCOM most extensively. The solution to the problem when distances were read into the program, no cycling was required, no feasible solution was provided initially, and no preassignments were made is shown in Table 12.3.

On Princeton University's IBM 360/91 this solution was obtained in about 3 seconds, with a total of 94 nodes generated. It was rerun with a change in the manner of selecting the primary node for investigation, the method being the choice of the primary node whose source-to-sink distance was *greatest* (rather than least). This hiked the time to about 4 seconds and increased the number of nodes produced to 129.

Table 12.3 The 3 × 3 solution

Source	*Sink sequences*	*Deliveries*
1	3	100
2	2	80
	3	20
3	1	50
	Optimal cost: 105 miles	

Finally, the manner of bounding the nodes was changed to a method that was in general illegitimate, but which did not inhibit the solution to this simple problem. These tighter bounds permitted the solution to be reached in 45 nodes and about 2 seconds. These experiments give us some idea of the importance of good branching and tight bounding to the efficiency of solution.

The various options of DYNCOM were next employed in the solution of the problem. The inclusion of a feasible solution with a cost of 310 had no effect upon node storage or time of solution, as expected. By requiring that the vehicles return to their original sources after their supplies were exhausted (cycling), the solution is altered only by the reversal of the sink sequence for source 2: sink 3 is visited first and sink 2 last. The cost of the solution rises to 200 miles. However, 149 nodes were generated in the problem, leading us to suspect that for larger problems the requirement that circuits be closed will lead to a substantial increase in the number of nodes generated and stored. Time requirements were about 3 seconds.

We next introduced some preassignment requirements, arbitrarily directing that in the optimal solution sink 1 be assigned to source 2 and sink 2 be assigned to source 3. The optimal solution is reproduced in Table 12.4. This solution required 7 seconds to attain – a notable increase in time – and generated 244 nodes. We conjecture, therefore, that with preassignments both efficiency indexes fall rather drastically. No doubt the increased node storage was the result of the longer time it took to generate a feasible solution, and would fall if such a solution were among the inputs.

Finally, the various metric options were tested. We placed arbitrary Cartesian co-ordinates on the sources and sinks, and solved the problem in a little more than 4 seconds and with 159 nodes. When these same points were used and the metropolitan metric employed, 153 nodes were required and 5 seconds. Generating distances rather

Table 12.4 The 3 × 3 solution with preassignments

Source	*Sink sequences*	*Deliveries*
1	3	100
2	2	30
	1	50
	3	20
3	2	50
	Optimal cost: 130 miles	

than having them in core seems to have accounted for the slight increase in time. To experiment with latitudes, longitudes and our rhumb-line distances, we assigned the co-ordinates of New York, Boston and Washington, DC, to sources 1, 2 and 3 respectively, and of Buffalo, Cincinnati and St Louis to sinks 1, 2 and 3 respectively. The optimal solution, reached in a surprising 6 seconds with only 122 nodes produced, is given in Table 12.5. The small number of nodes produced and the slight increase in time is probably accounted for by the change in relative placements of sinks and sources. In any event, the Mercator transformations do not seem to increase time expenditure greatly – a result that reinforces our experience in using MULTIWEB (Kuenne and Soland [5, 6]).

In brief, then, our experience with the simple set of 3 × 3 problems is that the optimum solutions can be solved with DYNCOM in a matter of 2 to 7 seconds with fewer than 200 nodes being generated. The use of co-ordinates rather than distances does not increase time and node requirements greatly, and seemingly the three metrics do not differ greatly among themselves in either dimension. The use of preassignments will probably increase the need for storage and time,

Table 12.5 The 3 × 3 solution with latitudes and longitudes

Source	*Sink sequences*	*Deliveries*
New York	Cincinnati	80
	St. Louis	20
Boston	St. Louis	100
Washington	Buffalo	50
	Optimal cost: 2228.1 statute miles	

Table 12.6 The 2 × 7 problem data

Source no.	*Name*	*Supply*	*Sink no.*	*Name*	*Demand*
1	New York	1,000	1	Newport	300
2	Burlington	905	2	Hartford	350
			3	Providence	600
			4	Lowell	250
			5	Keene	150
			6	Augusta	80
			7	Manchester	175

but inclusion of a feasible solution should reduce the increments substantially.

Problem 2: The 2 × 7 Example

In the previous article we presented a 7-source, 13-sink problem for attack by our heuristic methods. The sources and sinks of that problem are presented in Problem 5 below, although we shall not detail the distance matrix again, referring the reader to the earlier presentation. From that larger problem we abstracted a 2 × 7 problem with the data shown in Table 12.6. The reader is referred to the distance matrix of Kuenne [4] for exact distances, but the relative placements of the cities may be seen in Figure 12.4, which depicts the optimal solution to the problem. The solution required 6636 nodes to be generated and 68 seconds of computer time, with no feasible solution entered. The solution is presented numerically in Table 12.7. Both speed and node storage requirements could probably have been improved with a reasonably good feasible solution.

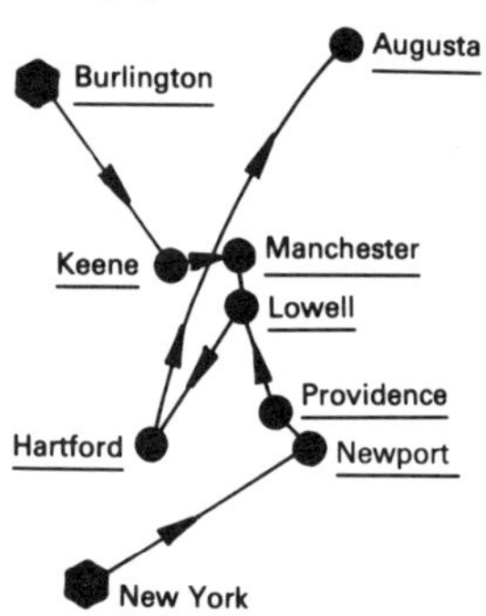

Figure 12.4 Optimal solution to the 2 × 7 problem

Table 12.7 The 2 × 7 optimal solution

Source	*Sink sequences*	*Deliveries*
New York	Newport	300
	Providence	600
	Lowell	100
Burlington	Keene	150
	Manchester	175
	Lowell	150
	Hartford	350
	Augusta	80
	Optimal cost: 696 miles	

Table 12.8 The 2 × 13 problem data

Source no.	*Name*	*Supply*	*Sink no.*	*Name*	*Demand*
1	New York	1,650	1	Newport	300
2	Burlington	1,500	2	Hartford	350
			3	Providence	600
			4	Lowell	250
			5	Keene	150
			6	Augusta	80
			7	Manchester	175
			8	Montreal	500
			9	Quebec	380
			10	Rutland	100
			11	Portsmouth	75
			12	Pittsfield	100
			13	Greenfield	90

Problem 3: The 2 × 13 Example

The next problem essayed was a 2-source, 13-sink problem with data depicted in Table 12.8.

To solve the problem DYNCOM was first adapted for use on the Time Sharing System of the IBM 360/67. It was possible to dimension the algorithm for 10 000 nodes storage on this TSS, but this proved to be insufficient for obtaining the optimal solution to the problem. With the best solution obtained from this analysis as an initial solution, the problem was put on the IBM 360/91 in a program with

Table 12.9 The 2 × 13 best feasible solution found

Source	*Sink sequence*	*Deliveries*
New York	Hartfold	350
	Newport	265
	Portsmouth	75
	Augusta	80
	Quebec	380
	Montreal	500
Burlinton	Rutland	100
	Pittsfield	100
	Greenfield	90
	Keene	150
	Manchester	175
	Lowell	250
	Providence	600
	Newport	35
	Cost of solution: 1296 miles	

135 000 active node storage capacity, and given 60 minutes of budgeted time. The best feasible solution obtained in the 109 615 nodes employed during that time period is listed in Table 12.9 and graphed on Figure 12.5.

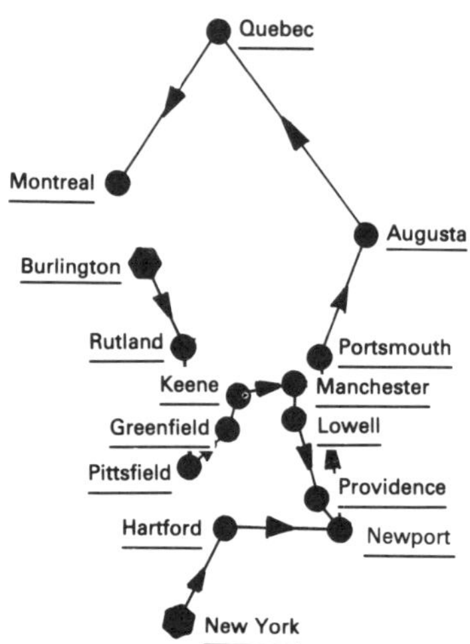

Figure 12.5 Best feasible solution to the 2 × 13 problem

Table 12.10 The 5 × 10 problem data

Source no.	*Name*	*Supply*	*Sink no.*	*Name*	*Demand*
1	New York	1 000	1	Newport	300
2	Albany	400	2	Hartford	300
3	Burlington	200	3	Providence	360
4	Bangor	100	4	Lowell	240
5	Boston	800	5	Keene	150
			6	Rutland	10
			7	Montreal	400
			8	Quebec	300
			9	Manchester	250
			10	Pittsfield	100

Problem 4: The 5 × 10 Example

In the desire to determine the sensitivity of DYNCOM to the number of sources, the next problem attempted was one in which the relative number of sources compared with sinks was substantially increased. The sources and sinks selected, together with demands and supplies, are listed in Table 12.10.

The problem was first run on the TSS console and, after 10 000 nodes of storage were exhausted, the best feasible solution obtained was used as an initial feasible solution for the 360/91 program. By virtue of budget limitations, the execution time was limited to 3.5 minutes, and, after this usage, with an estimated 16 170 nodes of storage space employed, the program exited without attaining the optimum explicitly. The initial feasible solution was not improved by the additional 6000 nodes of search. That solution is given in Table 12.11 and Figure 12.6.

Problem 5: The 7 × 13 Example

Finally, we approached the large problem originally devised for the earlier article. In that article, using our heuristic approach, we obtained a best feasible solution costing 1812 miles. After using that solution as an initial feasible solution, and permitting DYNCOM to search for 60 minutes on the IBM 360/91, with an estimated storage of 100 000 nodes, that solution was not improved upon. We conjecture that it is in fact optimal. The data of the problem are given in Table 12.12 and that best feasible solution in Table 12.13 and Figure 12.7.

Table 12.11 The 5 × 10 best feasible solution found

Source	*Sink sequences*	*Deliveries*
New York	Newport	300
	Providence	50
	Keene	150
	Rutland	10
	Montreal	200
	Quebec	200
Albany	Pittsfield	100
	Hartford	300
Burlington	Montreal	200
Bangor	Quebec	100
Boston	Lowell	240
	Manchester	250
	Providence	310
	Cost of solution: 1337 miles	

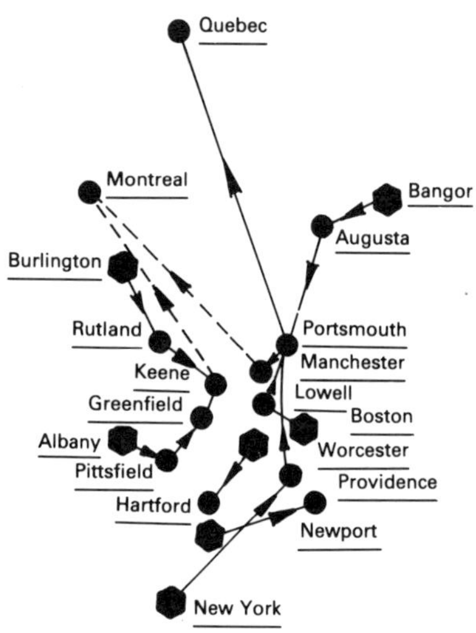

Figure 12.6 Best feasible solution to the 5 × 10 problem

VI A SUMMARY OF COMPUTATIONAL EXPERIENCE AND SOME CONJECTURES

On the basis of our limited experience with DYNCOM to date, it is felt that the algorithm will permit the solution of dynamic combina-

Table 12.12 The 7 × 13 problem data

Source no.	*Name*	*Supply*	*Sink no.*	*Name*	*Demand*
1	New York	1 000	1	Newport	300
2	New Haven	300	2	Hartford	350
3	Worcester	350	3	Providence	600
4	Boston	800	4	Lowell	250
5	Bangor	100	5	Keene	150
6	Burlington	200	6	Augusta	80
7	Albany	400	7	Manchester	175
			8	Montreal	500
			9	Quebec	380
			10	Rutland	100
			11	Portsmouth	75
			12	Pittsfield	100
			13	Greenfield	90

Table 12.13 The 7 × 13 best feasible solution found

Source	*Sink sequence*	*Deliveries*
New York	Providence	600
	Portsmouth	20
	Quebec	380
New Haven	Newport	300
Worcester	Hartford	350
Boston	Lowell	250
	Portsmouth	35
	Manchester	175
	Montreal	340
Bangor	Augusta	80
	Portsmouth	20
Burlington	Rutland	100
	Keene	100
Albany	Pittsfield	100
	Greenfield	90
	Keene	50
	Montreal	160
	Cost of solution: 1812 miles	

torial problems in space within feasible expenditures of time (say 90 minutes) and money on third generation computers in the neighborhood of 3 sources and 10 sinks. The experience acquired reinforces the conjecture made in Kuenne and Soland [5, 6] that the objective

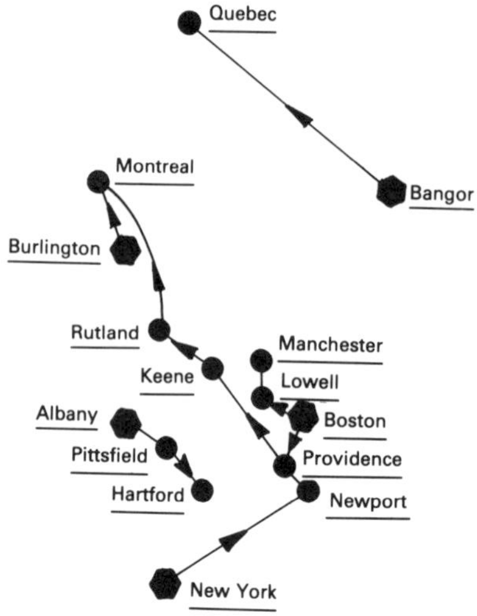

Figure 12.7 Best feasible solution to the 7 × 13 problem

functions in spatial combinatorial problems are quite flat-bottomed, and that the specification of a nonzero tolerance factor will permit close approximations to the optimum with substantial savings of time and money. Even if one opts for 100 per cent accuracy, DYNCOM has the distinct attraction of generating a series of successively improving feasible solutions, so that good solutions can be obtained even if the optimal solution proves to be out of reach.

It was the urging of the original article that good solutions are often nearly as satisfactory as the best solution, given the approximative nature even of the optimal solution to the realistic problem encountered. In that spirit, DYNCOM can be applied to much larger problems in the expectation of generating very good solutions in relatively short computational time.

Note

1. DYNCOM is programmed in FORTRAN IV and is available on tape at cost from the author.

References

[1] Agin, N. 'Optimum Seeking with Branch-and-Bound', *Management Science*, 13 (1966), pp. B176–B185.

[2] Deetz, C. H. and O. S. Adams, *Elements of Map Projection*. US Department of Commerce, Special Publication no. 68, Washington, DC, 5th edn (1945).
[3] Hillier, F. and G. J. Lieberman. *Introduction to Operations Research*, Holden–Day, San Francisco (1967).
[4] Kuenne, R. E. 'Approximate Solutions to a Dynamic Combinatorial Problem in Space', *Journal of Regional Science*, 8 (1968), pp. 165–80. Reprinted as Chapter 11 of this volume.
[5] Kuenne, and R. M. Soland, *The Multisource Weber Problem: Exact Solutions by Branch and Bound*, Institute for Defense Analyses, Economic Series, Alexandria, Va. (1971).
[6] Kuenne, R. E. and R. M. Soland, 'Exact and Approximate Solutions to the Multisource Weber Problem', *Mathematical Programming*, 3 (1972), pp. 193–209. Reprinted as Chapter 10 in this volume.

13 Economic Decision-Making in a Poisson Demand Space*

I INTRODUCTION

The attention of economists has been so fixated upon the notion of demand as a continuous function of price, with quantities purchased rather fully within the discretion of the buyer, that they have ignored other important demand regimes. One such type may be characterized as 'rare-event' contingency demand, in which sudden, unexpected interruptions in the individual's routine occur, initiating demands for goods or services to correct the situation with little discretion for denial or delay. Such events range from the truly catastrophic – the occurrence of death, sudden illness or accident, fire or crime – to the annoyance of a breakdown in important durable appliances in modern living.

Such demands have obvious relevance for both the private and public sectors of the economy. The location of hospitals, ambulance facilities, police and fire stations, as well as the determination of the optimal distributions of patrol activities by police must face these types of demand. In this paper, however, we shall limit our attention to firms in the private economy. Our discussions and examples envision the activities of small firms, with perhaps a plumber or physician as convenient prototypes.

For the individual or household, the likelihood of occurrence of such 'catastrophes' during a time period (say, a day) is very small, but rapid access to remediation may be most important, with the price of such services, while having some demand effect, definitely secondary.

* Published originally in Åke Andersson, W. Isard, and Tönu Puu, (eds), *Regional and Industrial Development Theories, Models and Empirical Evidence*, North-Holland, Amsterdam, 1984, pp. 331–46, and reproduced with permission of Elsevier Science Publishers. I am indebted to Professor Alan Karr, the Johns Hopkins University, for early guidance and a review that led to important improvements.

For some services such as police protection from victimization by criminals, the likelihood of occurence may be responsive to the expected time of response to need. Hence, the spatial structure of service may be most important.

Another aspect of the provision of such services over space that has been neglected in the spatial structure literature is the coincidence of the market areas served by competing firms. Typically, plumbers or physicians in an area will concentrate their locations in a population center and will respond to calls within the same region. Löschian hexagonal market area analysis is not appropriate to the understanding of spatial competition in such markets. Rather, an interpretation of individual firms' demand response areas during any one time period is characteristic of a Poisson demand space.

This paper applies a Poisson point spatial process framework to the analysis of such spatial demand structures. It is meant to be a tentative exploration of the potential for such stochastic models to yield insights into firms' spatial behaviour as it interrelates simultaneously with price-output decisions, and into the normative aspects of public services. A Poisson spatial process is described formally in Section II. In Section III the model of the monopolistic firm's pricing decision in a fixed service area is considered, and the firm's decision-making is enlarged to include the choice of a service radius. In Section IV analysis of competition within the same area is conducted. Lastly, Section V contains a brief indication of future paths of research.

II THE POISSON SPATIAL PROCESS

Given a planar space, S, and a family of subsets, A in S, we imagine a random scattering of demand points over S. Let λ be the average density of points in S per unit area, and let $N(A)$, $A \varepsilon \mathrm{A}$, be the integer number of points contained in subset A with area $|A|$.

Then $\{N(A), A \varepsilon \mathrm{A}\}$ is a *homogeneous Poisson point process* of intensity $\lambda > 0$ if (1) for each $A \varepsilon \mathrm{A}$, $N(A)$ has a Poisson distribution with parameter $\lambda |A|$, and (2) the variables $N(A_1), \ldots, N(A_n)$ are independent for any finite collection $\{A_1, \ldots, A_n\}$ in S, where $A_1, \ldots, A_n$ are disjoint (pairwise).

The Poisson process is a plausible description of spatial occurrences when certain postulates can be accepted as descriptive of underlying forces:[1]

Postulate 1 $N(A)$ can take only non-negative integer values and, when $|A|$ is nonzero and finite, the probability that $N(A) = 0$ is in the interval $]0,1[$. That is, no matter how large A is taken to be as a finite area, the probability of nonoccurence of the relevant event remains positive.

Postulate 2 $N(A)$ depends only on the area $|A|$, not its shape or location in S, with $Pr(N(A) \geqslant 1) = \lambda |A|$ in the limit as $|A|$ approaches zero.

Postulate 3 Given $A_1, A_2, \ldots, A_m$, disjoint regions in S, $m = 2,3, \ldots$, $N(A_1), N(A_2), \ldots N(A_m)$ are independent random variables with $N(A_1 \cup A_2 \cup \ldots \cup A_m) = N(A_1) + N(A_2) + \ldots + N(A_m)$. Hence, outcomes in one region cannot affect outcomes in any other disjoint region.

Postulate 4 Two occurrences cannot happen at the same point in space. Formally,

$$\lim_{|A| \to 0} \frac{Pr(N(A) \geqslant 1)}{Pr(N(A) = 1)} = 1.$$

If we imagine S to be a densely populated urban area, certain types of demand for services approximate the conditions of the postulates. By Postulate 1, such events must be 'rare', in the sense that for a given time period no matter how large the area within S we choose to take, the probability that no such demand will emerge remains positive. Certain catastrophic lapses from an accustomed tenor of living give rise to such demands, as listed above. Postulates 2 and 3 – area relevance only and independence of events – can be accepted as accurate descriptions of these demand contexts. Finally, the occurrence of two such events of the same type in a given time period at a given point is rare enough to permit us to accept Postulate 4.

For our spatial analysis we shall deal with a circular space S with area πr^2, where r is the radius of the service area. With probability λ of demand occurrence per unit area in a given time period (say, the day), the expected number of occurrences in S is $\lambda\pi r^2$, or, defining $y = \pi r^2$, λy. The probability of an integer number, k, of demand occurrences is the Poisson distribution

$$Pr\,\{N(S) = k\} = \frac{(\lambda y)^k}{k!}\, e^{-\lambda y},\ k = 0, 1, \ldots \tag{13.1}$$

For simplicity, we consider a firm with perfect information about its service calls at the start of the day. For example, we assume that only calls made before 9 a.m. are serviced that day, subsequent calls being backlogged for the following day. The firm then determines a sequence of service calls that moves successively to the closest point remaining to be serviced.[2]

That is, the service facility moves from one point to a neighboring point in space S. Let r_1 be the distance to that nearest point. The cumulative density function for r_1 is

$$Pr\ \{r_1 \leq x\} = 1 - e^{-\lambda\pi x^2}.$$

Hence,

$$f(r_1) = 2\lambda\pi r_1 e^{-\lambda\pi r^2} 1,$$

and the expected value of r_1 is

$$E(r_1) = 2\lambda\pi \int_0^{\infty} x^2 e^{-\lambda\pi x^2}$$

$$= \frac{\gamma\ (1.5)}{\lambda^{.5}\pi^{.5}}.$$

Since $\pi^{.5} - 2\gamma\ (1.5)$, we may simplify this expression to

$$E(r_1) = \frac{.5}{\sqrt{\lambda}}. \tag{13.2}$$

Therefore, the expected distance between neighboring service calls falls as the square root of demand density λ. A quadrupling of this density halves the expected distance to the next demand point to be serviced. Competition, to the extent it reduces density for each firm, raises the distance between calls and with it the number of vehicles needed, the amount of labor employed, and the amount of labor wasted between jobs. In this sense it is 'inefficient', but much less than linearly with the reduction in density.

We have defined the notion of a Poisson process in space and attempted to establish the plausibility of its postulates in terms of the underlying demand forces for 'catastrophic' services. Let us now turn to a monopolistic firm's decision-making in this spatial demand structure.

III MONOPOLY IN A FIXED SERVICE AREA

Let us assume that a firm specializing in the provision of such services is a monopolist in S, where $|S| = \pi r^2 = y$. The amount of demand in the area for its services each day follows the Poisson distribution:

$$D(d, S) = \frac{(\lambda y)^d}{d!} e^{-\lambda y}, \; d = 0, 1, \ldots \tag{13.3}$$

with

$$E(d) = \lambda y$$

Average cost consists of three components:

1. A constant variable cost, c, per unit of demand;
2. A quadratic variable cost term, a, that is multipled by the square of the amount of services demanded to capture the rising average production costs incurred as the number of points serviced rises. This includes the need to expand employees and vehicles as demand rises in a spatial context.
3. Distance cost, t, reflecting the cost of movement from customer to customer, including cost of labor time.

Total expected costs, therefore are[3]

$$E(C) = c \cdot E(d)^2 + a \cdot E(d)^2 + t \cdot E(d) \cdot E(r) \tag{13.4}$$

where $E(r)$ is defined in 13.2. Substituting, we obtain

$$E(C) = c\lambda y + a\lambda^2 y^2 + .5t\lambda^{.5}y. \tag{13.5}$$

Finally, we assume that demand has some sensitivity to price. If customers consider price too high they may depend upon their own skills for service or may decide to purchase a new unit of equipment rather than repair the old. Specifically, we assume that λ, the demand density, is a linear function of delivered price, p:

$$\lambda = k - bp. \tag{13.6}$$

The firm's objective is taken to be the maximization of profit, ϕ, through its price policy only, as we assume initially that it views its service area $|S|$ as fixed. We use the Langrangean form

$$Z = p\lambda y - c\lambda y - a\lambda^2 y^2 - .5t\lambda^{.5} y - \theta(\lambda - k + bp), \qquad (13.7)$$

and obtain the first order conditions:

$$\frac{\delta Z}{\delta p} = \lambda y - \theta b = 0 \qquad (13.8.1)$$

$$\frac{\delta Z}{\delta \lambda} = py - cy - 2a\lambda y^2 - .25t\lambda^{-.5} y - \theta = 0 \qquad (13.8.2)$$

$$\frac{\delta Z}{\delta \theta} = \lambda - k + bp = 0 \qquad (13.8.3)$$

The second-order sufficiency condition for a constrained maximum, derived from the familiar conditions on the bordered principal minors of the bordered Hessian of the system, will be dealt with in more detail in Section IV. There it will be seen that its holding depends upon the expression

$$\lambda^{1.5}(1 + aby) > .0625tb,$$

which, with b small relative to λ and y large we expect to hold true in realistic conditions. Hence, the first-order conditions yield a global profit maximum.

From 13.8.1, θ is marginal profit with respect to λ, demand density. Therefore, 13.8.2 assures that marginal profit with respect to price (λy) must equal marginal profit with respect to λ (θ) times marginal change in λ with respect to price (b).

By manipulation of the first-order conditions we obtain

$$(k - bc)\lambda^{.5} - 2\lambda^{1.5}(1 + aby) - .25bt = 0.$$

Because b is the impact of a unit change in price upon λ, a small value, and because t, the distance charge, may be assumed to be relatively small, let us assume $.25bt$ can be ignored. Then,

$$\lambda^0 \cong \frac{k - bc}{2(1 + aby)} \qquad (13.9)$$

from 13.8.3,

$$p^0 \cong \frac{k(1 + 2aby) + bc}{2b(1 + aby)} \tag{13.10}$$

where p^0 and λ^0 are optimal values.

Then

$$\frac{dp^0}{dy} \cong \frac{a(k - bc)}{2(1 + aby)^2} = \frac{a\lambda^0}{(1 + aby)} > 0 \text{ when } k > bc.$$

The condition $k > bc$ is a viability condition in the absence of which $\lambda^0 < 0$.

Thus, the monopolist is led to increase his price as his service area increases, but at diminishing rates. For a given density λ^0, p^0 rises at the rate $a/(1 + aby)$, which is concave in y. We would expect the price of a monopolist in a larger service area to be greater than that of a monopolist in a smaller area, but by amounts that are far less than proportional to the increase in area.

Consider, for example, a firm with the parameters listed in Table 13.1.

Table 13.1 Base case parameter set for examples

1. $k = .40$	4. $a = .10$
2. $b = .02$	4. $t = .05$
3. $c = 10.00$	4. $y = 113$ $(r = 6)$

We shall treat these parameters as a base case to illustrate the theorems derived throughout the paper. For the present case, the solution is:

1. $\lambda^0 = .082$
2. $p^0 = \$15.92$
3. $E(d) = 9.27$
4. $E(r_1) = 1.7$
5. $E(\phi) = \$46.41$

However, if r is doubled to 12, so that $y = 452$, then the firm's solution becomes:

1. $\lambda^0 = .053$
2. $p^0 = \$17.37$
3. $E(d) = 23.96$
4. $E(r_1) = 2.2$
5. $E(\phi) = \$116.63$

A doubling of service radius with a consequent increase in service area by a factor of 4.6 results in only a 9.1 per cent increase in price. Nonetheless, our general theorem holds: a firm that finds it desirable to increase the size of its market area will find it profitable to reduce the demand density of that area by raising price. This holds true as an approximation in the usual case when b is very small.

IV MONOPOLY IN A VARIABLE-SIZED AREA

Assume that the monopolizing firm can vary not only price but the radius r of its service area. To the first-order conditions of 13.8 we then add

$$\frac{\delta Z}{\delta y} = p\lambda - c\lambda - 2a\lambda^2 y - .5t\lambda^{.5} = 0. \tag{13.8.4}$$

The second-order sufficient condition for a maximum cannot be established unambiguously in this case. The relevant bordered Hessian is

$$[H] = \begin{bmatrix} 0 & 1 & b & 0 \\ 1 & A & y & B \\ b & y & 0 & \lambda \\ 0 & B & \lambda & -2a\lambda^2 \end{bmatrix},$$

where $A = (.125t\lambda^{-1.5}y - 2ay^2)$ and $B = (p - c - 4a\lambda y - .25t\lambda^{-0.5}y)$.

The sufficiency condition for $d^2Z < 0$ is that the bordered principal minors have the indicated signs:

$$|H_1| = \begin{vmatrix} 0 & 1 & b \\ 1 & A & y \\ b & y & 0 \end{vmatrix} > 0, \; |H| < 0.$$

In Section III we have indicated that $|H_1|$ can be expected to be positive. The sign of $|H|$, on the other hand, depends upon the sign of

$$-2B\lambda b - 4a\lambda^2 yb + 2a\lambda^2 Ab^2 + \lambda^2 + b^2B^2.$$

For practical values of the parameters, A will be negative. By subtracting 13.8.4 from B it is seen that

$$B = -2a\lambda y - .25t\lambda^{-.5}y + .50t\lambda^{-.5},$$

which rather clearly will be negative given practical sizes of y. Hence, $|H| < 0$ if

$$\frac{4ay}{b} - 2aA > \frac{1}{b^2} + \frac{B^2}{\lambda^2} - \frac{2B}{\lambda b},$$

which our studies indicate would be likely only for relatively large λ. It does not hold for the base case parameters of Table 13.1.

If we solve 13.8.1–13.8.4 we obtain:

$$\begin{aligned}
\lambda^0 &= \sqrt{(.25bt)^2} \\
p^0 &= (k - \lambda^0)/b \\
y^0 &= \frac{(p^0 - c - .5\lambda^{0-.5})}{2a\lambda^0} \\
E(\phi) &= E(\mathrm{d})(p^0 - c) - a(E(d))^2 - .5t\lambda^{0.5}y^0 \\
E(d) &= \lambda^0 y^0
\end{aligned} \tag{13.11}$$

Most interesting is the 'two-thirds power' law of demand density. It will rise when t rises, which is not surprising because t is the distance cost factor. When it becomes more costly to service a given area, the firm will reduce its price to increase λ in order to make the expected distance between customers smaller. But λ^0 will also rise when b rises. This implies that when b rises p^0 must always fall, so that $dp^0/db < 0$, and p^0 must fall by enough so that bp^0 falls. A firm which suddenly finds its demand much more sensitive to price than formerly will lower its price to increase the demand density of its service area.

For the base case parameters of Table 13.1, if we eliminate y as a parameter, 13.11 yields:

$$\begin{aligned}
&1.\ \lambda^0 &&= .004\\
&2.\ p^0 &&= \$19.80\\
&3.\ y^0 &&= 11{,}776\ (r^0 = 61.2)\\
&4.\ E(d) &&= 47.10\\
&5.\ E(\phi) &&= \$221.12\\
&6.\ E(r_1) &&= 1.7
\end{aligned} \tag{13.12}$$

When the monopolistic firm of Section II is free to determine its service area, it finds it profitable to expand it significantly. The radius rises by a factor of 8, and, as we expect from the analysis of Section II, the firm finds it advantageous to raise price and lessen the density of its demand. Profits rise 4.8-fold and sales 5-fold, but price increases only by a modest 24 per cent.

But suppose it were to become much more expensive to service the area, as might be expected in periods of rising energy costs. For example, let $t = .25$ rather than .05. Then the solution becomes:

$$\begin{aligned}
&\lambda^0 &&= .012\\
&p^0 &&= \$19.40\\
&y^0 &&= 2{,}015\ (r^0 = 25.3)\\
&E(d) &&= 24.20\\
&E(\phi) &&= \$141.32\\
&E(r_1) &&= 4.6
\end{aligned} \tag{13.13}$$

Even though a quintupling of distance cost triples demand density and more than halves the service radius, price responds only by a modest drop. The generality of this price behavior in our example is evident in Table 13.2, where we have listed the response of variables to changes in t. Note that as t rises, $E(r_1)$, the expected distance between calls, falls only as the square root of λ. Hence, although it pays the firm to constrict its service radius, it is also desirable that the demand density be increased by price reductions in order to reduce the intercall distance.

We are led to hypothesize that with a rise in the price of service travel, as has happened since 1973, servicers of this type of demand would tend to restrict their service radii and reduce the real price of their services, *ceteris paribus*. Price effects, however, would be expected to be less pronounced than service area reduction. It would be

Table 13.2 Variable sensitivities of change in t

t	λ^0	r^0	p^0	$E(d)$	$E(r_1)$
.05	.004	61.2	$19.80	47.10	7.9
.25	.012	25.3	19.40	24.18	4.6
.50	.018	21.8	19.10	26.87	3.7
1.00	.029	17.6	18.55	28.22	2.9
2.00	.046	13.6	17.70	26.73	2.3
5.00	.085	8.7	15.75	20.18	1.7
9.00	.127	5.3	13.65	11.23	1.4

interesting to perform empirical analysis to see if in fact these tendencies exist.

V THE FIRM IN COMPETITIVE CONTEXTS

We will now introduce competition into the Poisson space. We assume that the services of n firms are nondifferentiated and are perfect substitutes for one another, and that each firm is also identical in its cost structure. The firms compete pricewise, but do not engage in oligopolistic price cutting.

If λ is the demand density of the space for the industry, each firm will experience the density λ/n where $\lambda = k - bp$. Each firm will then act to maximize its profit, and the first-order conditions yield:

$$\begin{aligned}
&1.\ \lambda^0 = \sqrt{(.25bt)^2 n} \\
&2.\ p^0 = (k - \lambda^0)/b \\
&3.\ y^0 = \frac{(p^0 - c - .5t(\lambda/n)^{-.5}}{2a\lambda^0} \\
&4.\ E(\phi) = \lambda^0 y^0/n \\
&5.\ E(\phi) = E(d)(p^0 - c) - a\cdot(E(d))^2 - .5t(\lambda/n)^{.5}y^0 \\
&6.\ E(r_1) = \frac{.5}{(\lambda/n)^{.5}}
\end{aligned} \tag{13.14}$$

Note that demand density rises only as the cube root of the number of firms, but that it does *rise*. This is a result, of course, of the fall in price that each firm is led to adopt as its *initial* demand density decreases with entry of other firms. This will occur with or without a change in the size of the market area, for it is independent of y. This dependence of λ^0 solely on its price sensibility, distance cost, and the number of firms is one of the interesting insights obtained from the model.

Another interesting insight is the distance inefficiency that is brought about by competition. The average intercall distance, r_1, rises above its value for the same value of λ and $n = 1$. This inefficiency rises as the square root of n. It is offset somewhat, however, by the fact that a rise in n increases λ^0, and hence tends to reduce r_1 somewhat below what it would have been with $n = 1$.

The rise in λ^0 should tend to reduce the market area served. However, if the firms' average production cost function is rising, distributing a given number of service calls among n firms will reduce the average and marginal costs of the calls, and will lead the firms to expand sales and market area. Given λ^0, p^0 is set, and with these values production cost parameters c and a, distance cost t and n determine the service area size. The n in the numerator of the expression for λ^0 in 13.14.3 results from the term in parentheses being divided by a/n, where a is the coefficient of the quadratic term in production cost. Hence, y would rise directly with n were it not for the increase in distance between calls.

For the base case parameters of Table 13.1 (except, of course, for y) and when $n = 2$, the expressions of 13.14 yield:

1. $\lambda^0 = .005$
2. $p^0 = \$19.75$
3. $y^0 = 18{,}500$ ($r^0 = 76.7$)
4. $E(d) = 46.25$
5. $E(\phi) = \$213.91$
6. $E(r_1) = 10$

These results should be compared with those in 13.12, which are those of a monopolist ($n = 1$) in the same parametric environment. In the competitive case, sales have doubled, with a slight drop in price and large increase in market area. Each of the two firms makes almost as much in profit as the monopolist did by himself. However, had the monopolist serviced 92.5 customers, as the two firms do, profits would have been \$13.55 because of the rapid rise in average production costs.

In Table 13.3 we summarize the solutions as n takes a range of

Table 13.3 Solutions for the Poisson space model, base case parameters, for various numbers of competitors, t = .05.

n	λ^0	p^0	r^0	$E(d)$	$E(r_1)$
1	.004	$19.80	61.2	47.10	7.9
2	.005	19.75	76.7	46.25	10.0
3	.006	19.71	85.3	45.72	11.2
5	.007	19.66	101.1	44.96	13.4
10	.009	19.57	124.3	43.70	16.7
20	.011	19.46	155.9	41.98	21.3

values, given the relevant parameters of Table 13.1. Clearly, in our example, the very sharp rise in production costs with output relative to distance costs dominates the solution. Firms find it most profitable to reduce price and expand service radii in order to maintain output at about 45 units.

Suppose, however, that we increase transport costs to $t = 1$, holding all other parameters constant. The results are presented in Table 13.4. In these cases the market area rises for 2 and 3 firms, but then, as the expense of average distance travelled between calls begins to outweigh the savings in average production cost as output per firm falls, the service area begins to contract. These tendencies are made clearer in Table 13.5, where we list the production costs for the data of Table 13.4. Average total costs begin to rise for $n = 3$, at which point the decline in average production cost is overtaken by the rise in average distance, and the service radius to fall.

Table 13.4 Solutions for the Poisson space model, base case parameters, for various numbers of competitors, t = 1.00.

n	λ^0	p^0	r^0	$E(d)$	$E(r_1)$
1	.029	$18.54	17.50	28.00	2.9
2	.037	18.16	19.64	22.41	3.7
3	.042	17.89	27.15	20.41	4.2
5	.050	17.50	19.95	12.50	5.0
10	.063	16.85	11.80	2.75	6.3

Profit per firm almost disappears with 10 firms. If entry were to continue until a purely competitive equilibrium were achieved with price equal to average total cost, $n \cong 12$.

Table 13.5 Cost and profit data for various numbers of competitors, t = 1.00

n	*Total production costs*	*Total distance costs*	*Average production cost*	*Average distance cost*	*Average total cost*	*Profit*
1	$358.40	$81.20	$12.80	$2.90	$15.70	$79.50
2	274.32	82.92	12.24	3.70	15.94	49.73
3	245.76	85.72	12.04	4.20	16.24	33.66
5	140.63	62.50	11.25	5.00	16.25	15.63
10	28.26	17.33	10.28	6.30	16.58	.76

VI CONCLUSION

The case of Poisson demand for the services of firms or social agencies, especially in urban agglomerations with dense populations, servicing units from the same location, appears to be a relatively common spatial economic phenomenon. We have considered the decision-making of the monopolist with fixed and variable sized service areas, as well as firms in competitive contexts, when perfect information about the locations of a period's customers exists.

In these cases a Poisson space is relatively easy to analyze. For the case of a fixed service area, the demand density parameter is a function of all parameters, although only weakly affected by distance cost t. But when service area y becomes variable, λ is a function only of b – the price 'sensitivity coefficient' of λ – and t, as well as n, the number of firms. As firms enter the area, price is depressed to increase demand density and service area is increased until distance cost increases outweigh production cost decreases. Service area tends to be much more responsive to entry than prices, however.

A next step in the research is to introduce the complications of next-call demand responses rather than neighboring-call moves. Beyond that is the treatment of the oligopolistic firm in a crippled-optimization regime.[4] Several interesting questions concerning the nature of oligopolistic interdependence arise in a Poisson space, and merit further research.

Notes

1. See Michael F. Dacey [1], pp. 41–4, for an alternative formulation.
2. In future work this assumption will be dropped in favor of the more

realistic condition that calls are made in the order of their occurrence during the day. The distribution of distances to the *next* call is less amenable to analysis than the distribution of distances to the *neighboring* call, and therefore less appropriate to the exploratory nature of the present work.

3. It is arguable that the second term should be written $a \cdot E(d^2)$, i.e. with the expected value of the square of demand. We assume, however, that the firm has an expected demand which it uses as an argument in all three terms in the cost equation when making its decisions, and that it does not independently have expectations of the square of that demand.
4. See Robert E. Kuenne, [2] – [5].

References

[1] Dacey, Michael F., 'Two-dimensional random point patterns: A review and an interpretation', *Regional Science Association Papers*, 13 (1964).

[2] Kuenne, Robert E., 'Towards an Operational General Equilibrium Theory with Oligopoly: Some Experimental Results and Conjectures', *Kyklos*, 27 (1974), pp. 792–820.

[3] Kuenne, Robert E., 'Toward a Usable General Theory of Oligopoly', *De Economist*, 122 (1974), pp. 471–502.

[4] Kuenne, Robert E., 1979, 'Rivalrous consonsance and the power structure of OPEC', *Kyklos*, 32, (1979), pp. 695–717.

[5] Kuenne, Robert E., 1980, 'Dupoly reaction functions under crippled optimization regime', *Oxford Economic Papers*, 32 (1980), pp. 224–40.

Part III

Of Time: Austrian Capital Theory

Introduction to Part III

One of the least satisfactory performances by Walras, as noted in the introduction to Part I, was the theory of savings and investment, which is not to dispute some very important contributions. His use of a perpetual annuity unit to homogenize all capital goods and money was a most important step forward. But the motivation behind saving and investment was not examined in depth, and the genesis of the market functions for investment was unclear. I feel that economic theory has an important obligation to ground economic motivation explicitly in preferences and technology – extra-economic phenomena – in order to provide a complete and symmetrical general equilibrium theory. The reader will note that in this Part and in Part IV I have spent a great deal of effort in the attempt to integrate capital and money in the primitives of economic behavior.

Austrian capital theory has always seemed to me to come closest to explaining these foundations of the existence of interest, saving and capital, even though it failed to build a satisfactory general equilibrium model employing them. It is, in a sense, poetry that has never been satisfactorily translated into prose, and it retains the insights of the former without sustaining the transformation to the utilitarian purposes of the latter. In chapter 4 of my *Theory of General Economic Equilibrium* I present a variety of capital theories and try to indicate the shortcomings of each. And, in 1971, I was invited by the Columbia University Press to write a monograph on Böhm-Bawerk's theories. It is presented as Chapter 14.

Böhm-Bawerk attempted to expand upon Menger's vision of distribution theory as an extension of utility theory, and to weave a theory of interest and capital into the body of economic thought. To do so was to integrate that theory into a model of general equilibrium, for utility theory was viewed by Menger and Walras to be the motivating core of such grand constructions. To Böhm-Bawerk, interest was not a payment for a separate factor service that enters capital as a cost, nor was it a return for the psychological pain of 'abstinence', since no such disutility exists. It arose because time distorts subjectively the objective marginal utility of future goods and hence prevents capital goods from being produced to the point that their 'true' marginal utility of product should dictate. Interest and capital theory are manifested in markets where present goods are exchanged for future goods at a consequent premium.

But demands for current goods by capitalists could not be wholly divorced from technology. Capital goods, as inchoate forms of future consumer goods, were constructed because it was an empirical fact that they allowed man greater leverage over the forces of nature. They permitted more product to be produced from a given number of today's resources than could be obtained from the application of those resources to production directly. But interest is determined at the margin, and only the psychological myopia of consumers prevents the construction of capital goods to the point where the marginal value of such constituent factor services in congealed form equals the marginal value when used directly. Were that to occur the marginal advantage of capital goods would disappear and the interest rate become zero. Even, however, were no capital goods to exist, a positive interest rate would arise in the consumer loan market because of such short-sightedness. Hence, it is ultimately in consumer preferences that the phenomenon of interest is grounded.

The longitudinal view of capital through time, as an intertemporal set of consumer goods at different stages of completion, is a distinctive feature of Austrian capital theory. It differs from the cross-sectional view of capital as a stock of equipment and inventories existing at a moment of time, yielding independent factor services over a period of time. All such theories implicitly accept time extents as an important part of their explanation of capital and interest, although some (e.g. Knight's) seek to abstract from it in assuming the capital stock, once created, is eternalized through reinvestment, frequently by assuming the stationary state.

It is a distinctive feature of Austrian theory that time is a vital element in the creativity of capital and the ultimate explanation of interest. Note that money as a good which is never consumed over time but yields a flow of services eternally does not fit this framework well, and hence is not extensively treated by Böhm-Bawerk. In this sense, Walras's cross-sectional view of capital as a stock of factors yielding saleable factor services period by period accommodates money (and land) and unifies the treatment of factors for greater ease of incorporation in a general equilibrium system.

Böhm-Bawerk strove hard to keep productivity out of the theory of capital and interest, but failed in the end to do so. His Third Ground for the existence of interest is simply the marginal productivity of investment that time activates. One of the greatest faults in his theory of interest and capital is this attempt to present it as wholly the result of consumer preferences. He ends, as Wicksell and Fisher recognize, with both positive time preference and the productivity of capital as co-ordinate forces in the determination of interest and the explanation of the motivation for investment. I shall return to this criticism of the Third Ground in Chapters 15 and 16 to follow.

The theory propounded by Böhm-Bawerk sought to integrate capital and interest into the whole of value and distribution theory in a manner that was Walrasian in spirit. It was in this sense a general equilibrium theory, but faulty because of an inability to conceptualize the role of time in such measures as the period of production except under the simplest of conditions and because of his lack of technique. Examples of the latter which hampered his theory were his failure to understand the notion of simultaneous, interdependent causation and to adopt production functions which permitted continuous substitution among factors. It was left to Wicksell and Fisher to start from his model and to improve it in these regards and in the acceptance of productivity to accomplish the integration satisfactorily.

In this chapter I attempt to synthesize Böhm-Bawerk's concepts into the general equilibrium model at which he aimed. This must be done within the context of the stationary state and with the incorporation of other extreme departures from reality (point-input – point-output production functions, for example). In the end, the requirements necessary to formalize it and to render unambiguous his concept of the period of production are too unrealistic to accept, and modern capital theory has never been tempted by it.

Nonetheless, his concept of positive time preference, and the insights into the role of time in economic decision-making with the distinctive features it bestows upon capital goods, is an enduring legacy. The core vision remains, and after major adjustments and corrections, continues to inform economists' thoughts about capital and interest. Indeed, in the transmuted forms present in the work of Wicksell and Fisher, the imputed insights are invaluable contributions.

Selections 15 and 16 develop more fully the challenge to Böhm-Bawerk's 'third ground' for the existence of a positive interest rate: the so-called 'technological superiority of present goods over future'. He believed that if positive time preference did not characterize consumers' choices between present and future goods, the former would still be preferred to the latter because of the productivity of roundabout or indirect processes that employed time. The two papers show the conditions under which this would be true and reveal the unreality of those conditions. Even though the productivity of roundabout processes characterized production functions, provided they reached a maximum in function of time it would not bring about the third ground. That is, the technological superiority of present goods is not identical with the productivity of roundabout processes. For the former to occur, positive time preference would have to rule, which would destroy Böhm-Bawerk's notion of its independence from consumer attitudes toward the future.

Chapter 15 was written as a brief note commenting on a valuable article by Professor R. Dorfman on Austrian capital theory. Chapter 16 develops more fully arguments that were foreclosed by space limitations on Chapter 15, and seeks to isolate the assumptions that must be made for Böhm-Bawerk and his critics to be correct. It was an attempt to provide some definitive answers to a debate which frequently proceeded from different sets of implicit assumptions. The paper defines three different definitions of time preference and considers four cases in which the productivity of roundabout processes does or does not approach an upper bound and in which terminal consumption occurs or reinvestment is possible. In terms of realism, the invalidity of the third ground must be accepted.

Böhm-Bawerk's rather stubborn insistence on the third ground reflected his continuing desire to adopt a straightforward productivity of capital goods, but in the end merely reinforced his need to recognize it as of dual importance to time preference in his theory of interest. In the end he does so, and indeed with a vengeance: as indicated in Chapter 14, at times he seems to view productivity as the *dominant* factor and to arrive at a productivity theory of interest.

Austrian theory in its purest form, in seeking to measure the productivity of capital by the length of a vague 'period of production' which eliminated the value of goods from its calculation, is not a fruitful manner of incorporating capital and interest into a general equilibrium system, or, indeed, of dealing with partial analyses. Treatment of capital goods as separate factors of production providing flows of productive services is much more useful in these respects. Moreover, the desire to establish interest as the reflection wholly of consumer preferences was a vain and needless one, as Böhm-Bawerk effectively came to realize.

But the notions of positive time preference and of the productivity of roundabout or capital-based processes which accrued only at the cost of postponed consumption were valuable contributions to the understanding of the foundations of economic theory. And the practical outcome of that theory's early formulations were the improved theories of Wicksell and Fisher which illuminated the origins of demand and supply functions for savings and the determination of the rate of interest.

14 Eugen von Böhm-Bawerk*

I BÖHM-BAWERK IN RELATION TO CLASSICAL AND NEOCLASSICAL THEORY

For a century following publication of Adam Smith's *Wealth of Nations* in 1776, economic theory contained a paradoxical dissonance in its major themes. On the one hand, the classical thought of Smith, David Ricardo, James Mill and John Stuart Mill gave ideological support to economic liberalism and the free market economy. But, on the other hand, it succeeded neither in distinguishing the function of the key figure in that economy – the capitalist – from that of the entrepreneur, nor in providing a convincing rationale for the existence of the economy's distinctive factor share – interest.

It is not difficult to find passages in Adam Smith that view interest payments as exploitative subtractions from a national product created wholly by labor. Although Ricardo was insightful in discerning the price differences that production with fixed capital of different longevities or circulating capital with varying turnover periods introduced, he nonetheless fell back upon a simple labor-quantity theory of value that was embarrassed by such phenomena and ignored them in explaining prices. James Mill and McCulloch, Ricardo's most loyal disciples, attempted to complete their master's model with a theory of interest that treated capital as the source of a peculiar type of labor service, payment for which was somehow the proper due of the capitalist; Ricardo himself rejected their efforts. And thus it was that Karl Marx could propound his theory of capitalist exploitation within the mainstream of classical economic theory, drawing heavily upon it to 'prove' his propositions. It was against the set of technological and demographic assumptions in this theory – the law of diminishing returns to agriculture and the Malthusian population response to

* Published originally as Robert E. Kuenne, *Eugen von Böhm-Bawerk*, Columbia University Press, New York, 1971, and reproduced with permission.

income change – that Marx rebelled, not its economic substance.

The difficulty lay in the labor theory of value employed by Smith, Ricardo and James Mill, and its later adoption in thin disguise as a cost-of-production theory by John Stuart Mill. In these systems goods receive their values from the labor or primary factor costs that enter them, tempting theorists to resolve capital goods into their labor or primary factor content, and to merge the capitalist with an entrepreneur whose contribution could be explained as the labor of superintendence. Interest became confused with the profits of the firm, and capital was subsumed as a form of stored labor or subsistence for labor.

But beneath these dominant themes of the major theorists there was a sustained if confused counterpoint featuring the deeper implications of capital and interest, and seeking to achieve harmony with value theory. Adam Smith does see the capitalist as a 'saver' abstaining from immediate consumption. We have already noted Ricardo's perception of the role of time in valuation, and at one point he refers to interest as a 'just compensation' for abstinence. John Stuart Mill, in moving away from the quantity-of-labor theory of value, and in developing the apparatus of supply and demand, began to push theory in the directions that permitted an incorporation of the psychological aspects of capital formation, and his cost-of-production theory of price permitted abstinence to enter gracefully as an element of exchange value.

Moreover, outside the main currents of economic theory by the 1830s, economists who were less committed to the labor theory of value strove for an integration of the technological and psychological factors that underlie the phenomena of capital and interest and for a theory of value that could incorporate them. In 1834, Mountifort Longfield explained the existence of capital in terms of the productivity of roundabout processes. John Rae in the same year viewed capital goods as instruments that ripened into consumer goods, recognized the psychological aspects of saving but believed them to be non-effective because changing technology kept the interest rate above the subjective rate of discount, and even defined a 'center of gravity' for the period between which capital goods were built and transformed into consumption goods. And, most significantly, Nassau Senior in 1836 worked explicitly and consistently with abstinence and the productivity of roundabout processes. Indeed, in the area of capital and interest theory, John Stuart Mill's performance, though based upon Senior's work, was a regression from Senior's achievement.

The obstacle to understanding the role of capital and the place of interest in the pricing of the free market economy, therefore, lay not in the inability of the profession to isolate the causal forces, but rather in its reluctance to abandon a familiar if disproved value theory in the absence of an alternative. John Stuart Mill's clear-cut definition of demand and supply schedules was a step in the direction of building consumer preferences into value theory, and thereby of incorporating abstinence in the theory of capital and interest; but his determination to stay loyal to Ricardo's labor theory of value would not permit him to effect the needed revolution. Further, while the technological implications of time were perceived by Ricardo, the cost-of-production theory was not far enough away from the labor theory to permit a correction of the overemphasis upon the technological that inhered in the latter. Economic theory entered the third quarter of the century searching for a framework that would permit utility to be an active rather than an enabling partner in the determination of price, and that would absorb time as an integral part of the valuation process rather than a disturbing element in its explanation.

The Marginalist Revolution

That framework emerged from the marginalist revolution in economic theory which began in the 1870s with the independent enunciations of the marginal utility doctrine by W. Stanley Jevons in Great Britain, Carl Menger in Vienna, and Léon Walras in Lausanne. The extension of the new techniques in an effective way to production theory was completed by 1900, having begun with Jevons' and (most importantly) Menger's attempts to bring capital goods under the regime of marginal utility reasoning by treating them as future consumer goods. In this way both the primary factors of production and their creature capital goods could be viewed as possessing marginal 'social utility' products which ultimately determined the prices of their services. At the hands of Menger's Austrian School, this transitional phase of the marginalist theory evolved the theories of imputation of social utility to the factors of production and of opportunity cost.

Time and space conspired to place Eugen von Böhm-Bawerk well for the finding and accomplishment of his intellectual lifework. Born in 1851, he was in his early twenties when the revision of value theory began in earnest. At the University of Vienna he encountered the same environment that was inspiring Menger to rewrite the core of economic analysis in a manner that reasserted the relevance of theoretical reasoning in opposition to a strong historical school. In implementing his unifying vision of value and distribution as different forms of pricing grounded in marginal utility, Menger had placed a temporal dimension at the very core of utility theory with his distinction between goods of lower and higher order; yet obligingly he had left a yawning gap in his own structure by failing to construct a theory of capital. Thus, Böhm-Bawerk moved in an intellectual milieu that possessed the appeal of the revolutionary cosmology to the very young: a cosmology that found its gravitational principles in simple concepts making no great demands for a technical background, and that led, within its global design, to a Grand Challenge of large but manageable dimensions.

Nor was this all. Böhm-Bawerk was a born polemicist trained in the law, and the challenge to construct the missing theory of capital and interest was heightened by the publication in 1867 of the first volume of Marx's *Das Kapital*. Vienna, German in language and culture, was distant enough from the stultifying influences of historicism to permit this island of theoretical analysis to develop, and at the same time gave access to the stimulating if antagonistic work of one of the best Marxist theory groups in Europe. We shall note the ambivalent relation of Böhm-Bawerk to Marx's work that led Schumpeter to term him the 'bourgeois Marx'.

Böhm-Bawerk's academic career was a discontinuous one. After applying himself to the study of law at the University of Vienna, he studied political science at Heidelberg, Leipzig and Jena, entering the Austrian government's finance department in 1872. He left in 1881 to accept a professorship at the University of Innsbruck, but returned to the finance department in 1889. By this time – at the age of thirty-eight – he had written the two monumental works of his career: the *History and Critique of Interest Theories* [3] and the *Positive Theory of Capital* [5]. He served as finance minister in 1895, in 1897, and from 1900 to 1904, when he resigned to devote the remaining decade of his life to teaching and research at the University of Vienna. Sadly, however, his creative years were behind him, so he never had the opportunity to apply his full powers to the perfecting and polishing of his theories – effort that the primitivity of his technique required to bring them up to more demanding standards. His work is, therefore, hurried, prolix, polemic, and deficient technically, offering many openings for sallies by opponents; it bears silent witness to the life of a harried civil servant unable to devote the full energies of his creative potential to the tedious but valuable academic dialectic.

It was the dream of Böhm-Bawerk to weave into Menger's marginal utility fabric a theory of capital and interest that would meld seamlessly with it. Menger, in both the preface and the body of his major work, presented the unifying insight of the theory of distribution as an extension of commodity pricing to the factors of production, and his use of marginal social utility imputation pointed the manner of the integration. Böhm-Bawerk, following this lead, desired to derive the rate of interest and the capital structure of an economy simultaneously with all other prices and factor returns in the system, the whole set of such values reflecting the underlying marginal utilities of the society as final causes.

In his desire to achieve that ambition, a temptation existed to find in capital goods a productive factor co-ordinate with land and labor, the 'primary factors', and to view interest as a payment for the services of this factor in the manner that Menger's imputation theory ascribed marginal social utility products to the primary factors land and labor. The temptation to appeal to Menger's primitive marginal productivity theory must have been doubly acute because Karl Marx, the arch-antagonist, denied an independent productivity to capital, asserting that it merely enhanced the productivity of living labor. But Böhm-Bawerk followed Marx in rejecting capital as a third factor, seeking instead to ground the existence of interest in what appeared in the revolution's dawn light as the most fundamental economic substance: utility.

The Fundamentals of Böhm-Bawerk's Approach

Böhm-Bawerk's historic attempt to unify capital and interest theory with price theory proceeded from the view that the value of any good is dictated by its marginal utility, and that scarcity plays a supporting role of deciding where the margin will be. Because the cost of present consumer goods consists wholly of the marginal utility value of the services of the primary factors that enter them, so must it be with future consumer goods. With Karl Marx, Böhm-Bawerk objected to the attempts by Malthus, James Mill, McCulloch, Senior, and John Stuart Mill, to find another cost that enters future goods to be recouped when they are sold as present goods. It is 'clear', said Böhm-Bawerk, that interest is not a payment for real cost incurred –

it is an excess over the historical land and labor cost of present goods. To explain the existence of interest, therefore, is not to explain the presence of some factor service or psychological disutility that enters into future goods in manner proportionate to those goods' absorptions of primary factor services and the time required for their production. It is rather to explain the persistence of a premium, or agio, that the free market economy places above costs in these proportions on present goods.

The excess of the value of present goods over their costs arises in the most fundamental sense because of psychologically and technologically dictated attitudes of men to present and future consumption goods. Let us abstract from the technological basis of goods valuation for the moment. The values of present and future goods are then reflections of the perceived marginal utilities of the goods, and Böhm-Bawerk believed that for both rational and irrational reasons men consistently under perceive the true marginal utilities of future goods as they anticipate their receipt in the future. Time acts upon the psyche as a distorting lens.

The psychological distortions of a positive time preference prevent the production of future goods to be undertaken to the point that their 'true' marginal utility would dictate, but lead to their production short of such equality with present goods' marginal utility. When they ripen as present goods, future goods sell at a premium above their incurred costs, because they sell at prices that reflect their true marginal utilities. Since future goods are 'produced' in the active present by permitting the inchoate goods presently in capital good form to continue their passage through the present into the future with the addition of current resources where necessary, as well as by an increase in the number of such inchoate future goods, the operational manner in which the myopic effects of time preference operate is by restricting the capital goods formation that would otherwise take the marginal utilities of present and future goods to equality and eliminate the excess over cost received on the former.

At this point we may introduce the technological basis for present goods' value to exceed cost. We shall challenge the contention below, but Böhm-Bawerk believed that the greater physical productivity of capitalistic methods implied the value superiority of present goods over future goods. That is to say, technology permitted $1 in today's goods to be invested for $1.10 in tomorrow's goods – the excess in value being an agio on present goods. This led capitalists to seek to buy present goods for purposes of investing them in future goods to reap this premium.

The essence of both capital and interest, therefore, is the exchange of present for future consumer goods. This exchange occurs continuously in the markets where present consumer goods exchange against future consumer goods in the form of (*1*) current primary factor services and (*2*) loans, or the promise of future goods. Capitalists sell spot (give present goods) to the owners of primary services and buy future (obtain future goods) from them, or sell spot to borrowers for the promise of future goods. That is to say, capitalists with current goods purchase current labor and land services for their ability to contribute to future goods by adding their flows of current services to the stock of capital goods; and the owners of primary services sell them for a smaller number of present goods than those services will produce in future goods. Similarly, borrowers who wish to consume present goods in exchange for future goods obtain them at a premium in the loan market.

Karl Marx interpreted capitalism in the same market terms as the Austrian theorist, although the only primary factor he dealt with extensively was labor. In the socialist's system this discounting of labor's product was an exploitative process inherent in the property relations of capitalism's mode of production and would vanish with it. To Böhm-Bawerk it manifested the subject–object relations of free men to present and future goods, rooted in psychology and technology, and therefore it was fated to emerge in any rational economic society.

If current services of the primary factors are used to produce present goods – that is, if *direct* methods of production are used – the owners of such factors receive undiscounted returns. Why, then, do they not organize such production processes and receive the 'whole' value of the product? Since the owners of the land and labor services have the option in a free market society to use them in this way, it must follow that the undiscounted factor returns of direct methods are generally smaller than the discounted factor returns under *roundabout* methods of production. The *investment* of current primary factor services over a period of time and the reaping of their consumer goods production in the future tends to yield a larger value of such goods at fruition; the capitalist's return is not a deduction from the value of factor services contained in cost. Capitalists and the owners of primary factors benefit from the use of roundabout methods. Their existence implies the accumulation of current services in the form of capital goods stocks, and it was Böhm-Bawerk's hope that the enhanced productivity of such methods could be indexed by the amount of time the average unit of current service spent in the limbo of capital good form; further, he hoped to show that this time period was in turn rigorously correlated with the size of the capital stock.

In Böhm-Bawerk's system the enhanced productivity of primary factors induced by roundabout methods provides the surplus value from which interest is paid, but it is merely a necessary condition for such payment. It is theoretically possible that the costs of the primary factors will expand to absorb this surplus. A sufficient condition for the existence of interest is achieved if the psychological discount of future goods typifies human psychology. Together, these conditions are necessary and sufficient for the existence of interest in a production economy, and they guarantee that the employment of capitalistic or roundabout methods will be bounded. However, even if no roundabout processes existed, psychological discount would be sufficient for a rate of interest to exist, and in this sense the psychological are more fundamental causes of the premium on present goods than the technological . Realistically, the limit to which an economy can be made capitalistic is set by psychological discounting of future goods: the higher the rate of discount, the less attractive will be any given quantity of future goods relative to present goods, and the fewer will be the future goods produced.

The Task Ahead

In this framework Böhm-Bawerk strove to explain the construction of capital goods, the existence of a positive interest rate, the amount of capital goods in existence at a point of time, and the structure of that stock of capital goods in terms of a larger or smaller proportion ripening in the near or distant future. The features that the framework possesses in common with marginal utility pricing and allocation in nontemporal contexts are indicative of his desire to derive the values of these variables as part of the whole solution to the system. These are the broadest outlines of the theory I shall present in more detail and criticize in the remaining sections of this essay.

II OF TIME, CAPITAL AND INTEREST

Economists have defined capital in many ways, the variety explained by the differing roles it plays in their analytical systems. For a Ricardo, interested in the dynamic forces shaping the distribution of national income among laborers, landowners and capitalists, it was sufficient to abstract capital as a circulating fund of subsistence goods for the sustenance of labor in the current production period. Karl Marx, with other aims in mind, converted the concept to 'means of production' without which direct labor could not be productive, dividing the total into fixed capital goods and raw materials on the one hand, and a money value of subsistence goods on the other. The first form of capital became at once an indispensable complement to, as well as a substitute for, labor: an 'unemployer' of labor as well as its employer.

The Austrian school of economists – of which Böhm-Bawerk was a founder and prominent member – cannot be said to have had a common view of capital; but, in general, it viewed capital longitudinally through time rather than cross-sectionally, as a succession of intermediate products moving through the production process, changing form at every stage as they absorbed primary factor services, and ultimately destroying themselves in a harvest of consumption goods. For the productivity theorists capital became a permanent fund of produced factors, every element of which was mortal but which, for one reason or another depending upon the theorist, would yield an eternal flow of services co-ordinate with those of land and labor. For Keynes's shorter-run purposes, capital was identical with fixed capital goods, of interest only as the size of its stock affected the attractiveness of changes in its size in the current period to provide offsets to the community's savings. To modern mathematically oriented economists, constructing linear general equilibrium models, it may be defined simply as the whole set of produced goods that survive the period of analysis.

Specific interpretations of interest as a factor payment, of course, directly reflected the particular view of capital adopted. Ricardo was content to obtain a marginal product of labor-plus-capital, then determine interest as a residual left after the payment of subsistence to labor and rent to land. It had strong elements of what we would term 'normal profits' today, although we have seen that he tried to justify its receipt as a return to abstinence. To Marx, interest was an exploitative return to the ownership – not the productivity – of the means of production, payable because in capitalism such ownership permitted the capitalist to absorb all of labor's product above its subsistence needs. In the Austrian scheme of things, interest became a payment for the productive use of time that capital permitted, not for the productivity of capital itself, whereas to the productivity theorists interest was simply the marginal product of a factor of production co-ordinate with land and labor. Keynes, following Walras and Wicksell, treated interest as the net returns on the asset money (for the sacrifice of liquidity) and on fixed capital (for its productive services) which tend toward equality. Finally, in modern linear models interest becomes little more than a discount factor permitting the normally productive economy to distribute a larger product at the end of the period than it possessed at the beginning, with such dividends being earned proportionately with costs of production in the absence of capital stocks, and thus taking on the aspect of profit; or with the value of capital stocks employed over the period, in which case it is an interest payment.

This sample of the conceptual frameworks capital and interest have generated is by no means exhaustive, but a sufficient number has been reproduced to convince the reader of the existing diversity. Yet one common element does unify the definition: *time*. Each definition of capital and interest makes them, explicitly or implicitly, *dynamic* categories by implying that these phenomena are inextricably linked with the simple passage of time. In the various definitions the function of capital is to provide the means of support for the society (or more narrowly, labor) during the period of production; to permit a synchronization between the application of present primary factor services and the receipt of product from them; or to permit a shifting forward of the income stream, by storing productive services in superproductive form. Correspondingly, interest is the factor return rewarding men for sacrificing their desire to consume currently in order to permit future goods to be produced; permitting goods to be consumed currently before the product of present resources is ready; or merely allowing the uncertainties of the future to be hedged in the interests of greater individual security.

If these concepts are so related to time, is there a unique manner to depict 'the' essence of capital's relation to production? Is it 'best' to treat capital goods as durable assets that typically last over many periods of production, yielding services over those periods at the cost of a gross return including the means of their replacement, and whose physical embodiment is a stock of fixed capital assets existing in a cross-sectional slice of time? Or is 'the' substance of capital short-lived goods of no immediate interest to the consumer, entering each period of production merely to yield themselves up phoenix-like, being thereby reborn in more valuable form or emerging as finished consumer goods? Does interest find its source as a net payment for the productive services of fixed capital assets? Or does it arise from the power over natural forces which these transient forms confer upon land and labor as they pass from one phase to another? Or is it ultimately rooted in the perverse psychological myopia of human beings confronting the future?

It is necessary to avoid two extremes in seeking answers to these questions. The first of these is to assert that each of the existing explanations is equally good or bad because the questions themselves are meaningless or unimportant. I assert the essential philosophical propriety of the search for the bases of the existence of capital and the payment of interest. The inability to obtain unique or satisfactory answers to the questions may reflect upon the complexity of the matter and the imperfection of our methods, but not upon the meaningfulness of the questions. I do believe that it is possible, for example, to demonstrate the logical deficiencies of the Marxist answers.

But, on the other hand, too much time has been wasted in economic theory debating 'the' nature of capital and interest. There is no answer to such questions in the sense of searching for a unique 'essence' of capital and interest, and in this Grail-like quest Böhm-Bawerk was misled. It is possible to accept as co-ordinate theories interpreting capital as a body of long-lived productive assets, short-lived circulating productive means, or temporary means of subsistence for the population. Similarly, interest may be seen as a payment for the services of provisioning, or for the lapse of time arising from the needs of production, or for the productive services of a multiperiod factor. Time is the keynote in all such explanations, and each is fitting for particular circumstances; the term 'best', where it is possible of definition, can be construed only as 'most frequently germane' rather than 'uniquely appropriate'.

In this interpretation it is not a fruitful exercise to seek to demonstrate the universal validity of the productivity-of-capital, or wages fund, or Böhm-Bawerkian, viewpoints. Such explanations are complements, in that capital is best accepted as occurring in different forms, fulfilling different functions, and displaying different life-histories. Differing theories of capital and interest emphasize one or the other of these forms, and even if they all contain time as an essential, little may be gained in clarity by striving to use it analytically as a unifying theme. To study the similarities and differences of viewpoint of such theories and to examine their strengths and weaknesses in particular instances is not a meaningless or sterile task; but it should be undertaken with the idea that the complications of time need not have a unique substance, so that a search for a single unifying explanation need yield no satisfying result.

Why Does Capital Exist?

To illustrate our meaning we shall consider a most fundamental question: what are the reasons for the existence of capital? Why do men find it useful or requisite to accumulate stocks of produced goods? We approach the question by choosing a reasonably long period of time called a *time span* over which to view the continuous flows of economic activities; further, we shall find it convenient to divide the time span into a sequence of equally long *analytical time periods*, t, where $t = 0, 1, 2, \ldots, h$. Let us arbitrarily break into this sequence at the start of $t = f$, and classify the stocks of produced goods of concern to our capital theory.

First, we demarcate stocks of finished raw materials and consumer goods in existence at $t = f$ which do not survive to $t = f + 1$. They are, therefore, part of the *initial* stocks of $t = f$ but do not enter into its *terminal* stocks. We shall term the stocks the *synchronization* capital of the economy, for its function is to sustain the continuously ongoing productive processes of the economy and the population before the flow of finished consumption goods and raw materials begins during the period to support its processes. Such goods bridge the time gap between need and provision.

Second, we denote as capital a set of produced goods either in existence at the start of $t = f$ or begun during the period, but unfinished at $t = f + 1$. These are the unfinished goods, or *goods in process*, which again reflect the fact that time is an element in production, and that our arbitrary choice of length for t will be a fraction of the periods of construction of many goods. These goods, therefore, reflect the mere continuity of the production processes and the essential artificiality of any analytical period we may adopt for convenience.

Third, we identify those finished goods which are neither consumer goods nor raw materials, and which (*1*) exist at $t = f$ but not at $t = f + 1$; (*2*) exist at $t = f$ and also at $t = f + 1$; and (*3*) do not exist at $t = f$

but do exist at $t = f + 1$. These are *fixed capital* stocks whose purpose is to give the production activities protection from or control over natural forces in durable forms that yield services over more than one analytical period. Their ability to give man leverage over the forces of nature for more than one period distinguishes them from many intermediate goods with essentially the same function.

In this schema no sharp break occurs between the intermediate products of land and labor at one extreme and fixed capital goods at the other. The intermediate product is a good which comes into existence during $t = f$ and disappears before $t = f + 1$. As the duration of the analytical period is permitted to shrink, intermediate products tend to become capital goods of the first, second, or third types. As that duration grows to equal the length of the whole time span, on the other hand, capital goods of the first, second and third types tend to become intermediate products. The possibility of a theorist emphasizing one type of capital over another by choosing the length of the analytical period indicates the importance of that period's definition.

Most generally, however, capital serves the functions of synchronizing needs and provisions; sequencing productive processes through the time dimension; and permitting a multiperiod control over the forces of nature. Different theorists of capital have emphasized one or the other of these functions as more 'important' in some sense than the others, and that emphasis gives their theories distinctive flavors. But it is important to see, at this point of our presentation, that any such theory of the functions of capital must incorporate time in its framework or explicitly contrive mechanisms to permit abstraction from it. For if we choose the analytical period's length on the useful basis that it is the longest period for which we can neglect realistically the passage of time, the distinction between produced inputs that are capital and produced inputs that are not is that the former cross time boundaries and thereby force that dimension into the analysis while the latter do not. To the extent capital exists, therefore, time intrudes into the framework of analysis.

Böhm-Bawerk's Concept of Capital

Böhm-Bawerk's distinctive treatment of capital can be fitted neatly into this typology. At its most simple and fundamental, that treatment assumes: synchronization capital at $t = 0$ contains only raw materials, not subsistence goods, and is rather unimportant in the whole scheme of things; fixed capital goods are nonexistent at $t = 0$ and to the extent needed are created during the time span; and the time span of production for the individual product is chosen to be of that length that sees all fixed capital stocks created during the whole process destroyed by its end at $t = h$. In this way, effectively, except for some rather unimportant stocks of raw materials at the start of the time span, all capital can be treated from the viewpoint of the time span as capital of the second type – goods in process. Capital is a multiperiod intermediate good. This type, therefore, is the distinctive capital form in Böhm-Bawerk's purest vision of capitalistic process.

Capital is the sequence of unfinished goods from $t = 0$ to $t = h$ that leads from the initial investment of the primary factor services of land and labor in $t = 0$ through their subsequent investments in $t = 1$ through $t = h - 1$ to their exhaustive fulfillment in consumption goods at $t = h$. Capital is not means of production in the sense of Marx, but the evidence that time-consuming or roundabout methods of production are employed. As a longitudinal concept it embraces all intermediate stages of the production of consumer goods over the time span; the unfinished goods and fixed capital that pass from $t = f$ to $t = f + 1$ are capital in Böhm-Bawerk's definition. The time span for a single process is in the ideal wholly self-contained, so that all fixed capital goods can be viewed as variants of the second form of capital – goods in process – and, even more specifically, *consumer* goods in process.

One peculiar violation of this viewpoint occurs in his inclusion at one point of the stock of money as a part of social capital, on the grounds that it permits the use of roundabout processes. But, like the primary factors of Böhm-Bawerk's schema, money is unconsumed during the time span, yielding up only its services of synchronization to the individual. By incorporating money in the social capital stock, therefore, Böhm-Bawerk adopts a productivity or use theory of capital in its regard. We should expect, therefore, that money plays little further part in his formal system, and this turns out to be correct. For in that system, produced goods are distinguished as capital or non-capital on the basis of *not currently consumable* versus *currently consumable*: the *never consumable* produced good must be grouped with primary factors.

What appears to be a violation of the logic of his definitional criteria in the other direction – excluding a stock of goods from capital that fulfills the requirements – occurs with Böhm-Bawerk's rejection of the stocks of finished consumer goods at $t = 0$ needed to sustain the society (or its laborers) until the flow of consumer goods begins from the roundabout processes. It is not entirely clear whether he meant to exclude such goods if they were in the stocks of producers or merchants; but to have included an initial stock of produced goods in capital would have put such accumulated stocks on a par with primary factors in $t = 0$, and therefore would have led to a case of capital containing the services of capital that originated outside the process. The desire of Böhm-Bawerk to resolve all capital into land and labor services we shall see to be misplaced purism. Moreover, he does attempt to deal with initial capital stocks in the less-idealized parts of his analysis by resolution into land and labor services, ignoring such services' contributions in the distant past.

The cross-sectional view at t that Böhm-Bawerk's theory takes of capital is indeed that of a stock of subsistence goods in various degrees of readiness. Its function is to provide a phased supply of such means of support over the future and thus to permit primary factor services in t to be devoted to roundabout processes. This statement sounds a bit like the Puritan's urging that one work hard while it is day because soon the night cometh and one will not be able to work any more; we seem, that is, to be justifying the existence of capital goods on the grounds that if they exist more capital goods can be produced. But, in fact, the cross-sectional function of capital stocks is to permit primary factors at t to produce more capital, and, to enhance the seeming circularity, the size of this stock dictates the maximum size of the increase. The interpretation of the existing capital as consumers goods (especially for the near future) makes this outlook clearer, because the larger the stocks of consumer goods ascribable to past primary factor services ripening in the near future, the less dependent is the society currently on primary factor services for its subsistence, the longer it can postpone receipt of the consumer goods resulting from their use, and the greater the future flows of consumer goods will be.

In summary, Böhn-Bawerk's view of capital in his ideal type is what might be termed an *organic* as opposed to a *mechanistic* one. The consumer good exists initially as a seed, or cluster of primary services, in $t = 0$; it is planted and absorbs further services of the primary factors over the h periods of its development; then it ripens and yields a once-for-all harvest of consumption goods at $t = h$. Each irreversible stage of the plant's development is destroyed by passage into the next stage with the absorption of more primary factor services and/or time, and finally the process reaches a once-for-all stage of self-destroying culmination.

A mechanistic outlook would see the same production process in cross-sectional slices. In *t* the initial stocks of capital (synchronization, goods in process and fixed), of land, and of labor are viewed co-ordinately as contributing services to the production of the joint product. The services may include any or all of the good itself (depreciation may range from 0 to 100 per cent), but usually by one assumption or another the nature of the production process is to repair these incursions on the resource stocks. The intermediate products of the intraperiod production process are abstracted from as uninteresting, so that inputs of the factors of production at the start of *t* yield a production in $t + 1$, which product includes current consumption goods, replacements and maintenance for all factors to the extent required, and stocks to be passed along as unfinished or finished goods to the next period.

Why Does Interest Exist?

Because we have seen that capital is inseparable from the notion of time, interest as a payment to capital in some sense must then be a payment involving time. It might be thought that the most direct way to explain and determine it is to adopt the mechanistic view of capital with its emphasis on capital services, for then capital is a good whose services enhance the total output of the economy by permitting more goods to be produced, as in the case of any other factor. Interest must then arise by virtue of the marginal contribution. Indeed, economists today frequently argue as if the enhancement of the product that both Böhm-Bawerk and productivity theorists admit occurs when capitalistic methods are used is sufficient to establish a positive rate of interest. But two objections challenge the sufficiency of this condition.

First, the market economy prices factor services in relation to their marginal contribution to output, not their total contribution in some sense or other. Many cubic feet of air enter into the production of every ton of pig iron, so that air is infinitely productive in the sense that in its absence no pig iron could be produced. Yet air is a free good because its *marginal* product is zero. To establish, therefore, that the presence of a factor results in a larger physical output than would occur in its absence – even if output were zero in that nonevent – does not provide a sufficient basis for a payment to the owner of that factor from that surplus. A necessary condition it is, of course: the factor must be requisite in that its service contributes to the total product; but it may nonetheless be a free good.

It is conceivable, therefore, that capital in all senses defined above could exist, that it would be productive in the total sense, yet the net return to it would be zero. That is, all types of capital that yield themselves up totally during the period would return to their owners only the value of the land and labor services that went into their construction; types of capital that suffer only a fractional destruction would yield only that fraction of their land and labor value; and infinitely durable capital goods would receive no payment at all. To explain an interest rate, then, must be to complement the productivity analysis with an explanation of why the *marginal* product of capital remains above zero, so the total physical productivity is not given to the other factors exactly as Marx believed it should and would be in a socialist economy.

Our second objection is that interest is a *value* return, not a *physical* return, in that it involves a comparison between a complex of inputs and a complex of outputs that can be compared usefully only when their units have been reduced to a homogeneous value form. It is conceivable that capitalistic methods might produce larger physical products than would be produced in their absence, yet by virtue of the elasticity of demand for such products that these goods would sell for a total value that would not recompense society for the value of capital absorbed. In the limit, the extra product may reduce price to zero, if it satiates demand.

The implication of this second point is that no adequate theory of interest can be wholly phrased in physical terms. Since the phenomenon is grounded in value, and value has a psychological dimension, the attitudes of the economy to future consumption must find their place in the explanation of interest. As an extreme example, it is possible in a world in which the interest rate is zero to have goods which expand at some positive rate of growth with no absorption of economic resources. If, for example, cattle roamed over free land and expanded at a rate of 50 per cent per year, two animals in $t = 0$ must give rise to three in $t = 1$; but with a zero interest rate three animals in $t = 1$ are equivalent in value to two animals in $t = 0$; and thus we are led to the conclusion that the value of two animals in $t = 0$

is equal to the value of three animals in $t = 0$. They are, therefore, free goods, and the limiting case where society is satiated in cattle has been attained.

One criticism of Böhm-Bawerk's theory of interest is that, despite his waspish disparagement of other theorists for not establishing the productivity of capital in a value sense, he himself did not prove this value superiority. Only in a stationary state, with constant prices over time, will the marginal physical product of capital result in a provable positive value increment. But Böhm-Bawerk did not use the stationary state in his theory and consequently did not establish the value productivity of roundabout processes.

Böhm-Bawerk's View of Interest

We have seen that the ultimate source of interest in Böhm-Bawerk's theory – the generator of the excess of the value of present goods over their historical land and labor costs of production – is rooted in the psychology of man. Given a fixed amount of primary factors, all value must reflect the marginal utilities of consumer goods produced, the factor services that produce them receiving their values from these goods' marginal utilities. In the grand theory of value with which Böhm-Bawerk tried to integrate the theory of capital, the services of time are no exception to the rule, nor are the goods into whose production they enter. However, at the most fundamental level, time restricts current supplies.

Even in the absence of capital, which is to say, even if roundabout processes of production were not productive, interest must be expected to occur as a pure exchange phenomenon. But roundabout processes are productive, and thus to explain interest in the real world is to explain not the interest of mere exchange but of productive activity, or what Böhm-Bawerk called 'originary' interest. Nevertheless, the ultimate reason the surplus value exists must lie in the psychological constraints that inhibit capital from expanding to the point where the marginal value products of the primary inputs used directly and indirectly are equal.

The investment of time implied by the use of roundabout processes is also the explanation of the dependence of labor upon capital, for if the leverage over nature could be obtained from intermediate goods, the labor could appropriate these advantages for itself. Because they are not thus appropriable, the laborer sells his present services for their discounted future value, since present goods are more valuable than future. Nor is labor cheated in such trades by the judgment of the market system: for if the valuation basis of all goods is the psychological preferences of consumers (given the stocks of primary factors) in the unifying theory of value, and if the society accepts these valuations in general as legitimate expressions of social preference, the same legitimation must be projected onto future goods in terms of present goods. To dispute Marx's contention that labor is exploited, therefore, Böhm-Bawerk contends that the laborer exchanges the future goods resulting from his service for smaller physical amounts of present goods, but the values are equivalent in the sense of the market's judgment.

Value, Capital and Interest

We now have before us Böhm-Bawerk's underlying vision of economic process. The function of the market mechanism is to allocate fixed amounts of primary resources – land and labor – among current and future uses, and that mechanism employs one grand and unifying method to accomplish both purposes. The theory of value is a theory of marginal utility, and the theory of distribution a projection by imputation of marginal utility. Capital goods are the evidence of the use of time in productive processes to enhance the productivity of primary factors in physical output. This physical productivity is converted to a value productivity through the peculiarities of intertemporal marginal utilities (and here Böhm-Bawerk is on questionable ground). The owners of capital receive a positive value imputation not as a reward for the enhanced physical productivity of capital, but by virtue of the psychological valuations of present goods (consumer goods) over future goods (primary factor services) in the market place. Time, therefore, restricts the current goods supply.

III THE THREE GROUNDS AND THE INTEGRATION WITH VALUE THEORY

We have seen that in Böhm-Bawerk's theory the necessity for a positive interest rate inheres in the superiority of present goods over future goods. The sources of this superiority Böhm-Bawerk locates in three extra-economic grounds; the fact of it permits him to bring capital and interest theory under the regime of marginal utility, with hopes for the grand integration of the analysis with value theory. The first pair of grounds are psychological and subjective in nature, the third purportedly technological and objective. We shall see that the third, as propounded by Böhm-Bawerk, is not independent of the former two, and so we shall view his theoretical efforts as strongly biased toward the psychological grounds as an explanation of the existence of interest.

The First Ground: Provision for Present and Future

The First Ground for the existence of interest is the great frequency of desire for present goods to meet the current needs of people who will be more adequately provisioned with goods in the future. Typically, Böhm-Bawerk argues, many persons have a 'rational' psychological preference for present goods because their marginal utility of income in the present is higher than their truly foreseen marginal utility of income in the future. The marginal utility of today's income is greater than that of tomorrow's *before* any perceptual distortions associated with time are considered, because tomorrow's income will be larger.

But are there not many consumers in the opposite position? The worker approaching retirement, for example, should find future goods more attractive than present: why will he not lend present goods to younger members of society at a zero rate of interest, or even accept negative returns if necessary to obtain future goods? Böhm-Bawerk steps into a money economy for the first time at this juncture to bound the interest rate away from negative values; he answers that the person who finds future goods more attractive than present can accumulate money and hold it at zero interest, thereby controlling the preferred future goods with no need to pay for the privilege with negative rates of interest. Thus, the asymmetry between the situations of the two types of consumers provides a floor for the interest rate. However, this does permit the interest rate to fall to zero, so that in the exchange economy (with fixed stocks of goods for trading and no production), if only this factor operated, a preponderance of those who prefer present to future goods must be assumed in order to obtain a positive interest rate.

Let us call this First Ground for the existence of interest 'λ time preference', since the symbol *lambda* is frequently used to denote the marginal utility of income, and the existence of this type of time preference rests upon the desire to equalize over time the perfectly foreseen marginal utilities of income each period. It would be zero in a stationary state: if consumers are receiving a constant income, period after period, and if they are permitted to accumulate a stock of capital for their heirs, they will build that stock up to a desired level, after which a stationary state is attained, and an extra dollar of income today would be worth no more than one tomorrow because of λ time preference. Böhm-Bawerk's inclusion of the First Ground, therefore, is an indication that he is not wedded to the stationary state as an analytical concept in his theory.

But suppose we are in an expanding economy in which everyone looks forward until his death to a rising income, period after period. Then each consumer will desire to borrow present goods in order to bring about the equality of marginal utility of income each period, and there will be a strong force operating for a positive interest rate to keep those desires within the constraints of available present goods.

The Second Ground: Time Preference Proper

The more fundamental impact of time's passage upon the consumer's preferences between present and future income in Böhm-Bawerk's theory is of a different nature. The Second Ground for the existence of interest springs from the distorted image the typical person is assumed to possess of the marginal utility of future income. Most people, Böhm-Bawerk believed, discount future pleasures and pains significantly, and their actions are based upon comforting distortions of what is to be. Future goods are prized less than present, even if incomes over the future are prospectively constant, simply because man lives psychologically in the present.

The trait is not completely irrational: the view of future needs and events can be at best unclear, so failure to foresee accurately the intensity of those needs may be excused on plausible grounds. Moreover, the brevity of life and its uncertain time of ending may also provide some rational basis for the foreshortened view. But it is the subjective phenomenon existing after such objective risks and uncertainties are eliminated that truly interested Böhm-Bawerk: he was determined to build a theory of interest that was free of dependence upon such incertitude. Although it might be intensified by such realities, therefore, the Second Ground was not dependent upon the inability to see the future clearly, but rather upon a deep-seated psychological propensity of man to distort subjectively the objective implications of the clearly foreseen. (Do you smoke cigarettes?)

It is not that the *cost* of future goods is greater but that their *utility* seems less that explains interest. If we view primary factors as consisting wholly of labor for the moment, the true cost of such goods is either the marginal utility of the present sacrifice (if alternatives exist for the invested labor) or the marginal disutility of the expended labor (if such alternatives do not exist). There is no additional cost to be added to this magnitude explained by the pain of 'waiting' or 'abstaining'; rather, time exerts an indirect influence on value by distorting the anticipation of the future satisfactions such efforts will bring. In this sense, Böhm-Bawerk says, Senior and Marshall were wrong to view interest as the wages of abstinence or waiting, for in a causal sense time is not a cost but a consistent distorter of benefit. The ultimate explanation of the value of a good lies in its future, not its past.

Consider, for example, spending one day to plant an apple orchard which will begin to bear ten years after such setting. The true cost of the orchard is the opportunity cost or disutility of the single day of labor invested, not that day plus ten years of waiting as Senior would have it. For suppose the orchard were destroyed or sold the day after planting: is the cost not simply the day of labor? Once a positive interest rate exists in an economy, that interest must be added to costs as a reflection of the distortions of utility. It is the *source* of that money cost that Böhm-Bawerk seeks to clarify in this argument. To use an analogy we might say that continuous inflation in an economy is caused wholly by a 'demand-pull' factor rather than a 'cost-push' force, but once the fact of inflation is established, those who deal with cost must take it into account in computing costs over time.

The Second Ground leads one to underestimate consistently the utility of future goods, and therefore the instruments of their production. The distinction between the statements that man prefers equal present pleasures to equal future pleasures, on the one hand, and that waiting for future pleasures is a sacrifice increasing the real costs of future goods, on the other, is important to Böhm-Bawerk's purposes. We have noted his overriding desire to ground capital and interest theory in utility, thereby uniting them with value theory, and to overthrow the cost-of-production theory then current. It was, therefore, crucial to eliminate the idea that abstinence was a real cost co-ordinate with that of the disutility of labor, and to substitute for it the concept of opportunity cost (itself, following Menger, a utility concept), with the time element entering solely as a distortion via the frailties of man of the true value of future goods. In rejecting such real cost theories Böhm-Bawerk sided once more with Karl Marx against the abstinence theorists.

The Third Ground: The Technological Superiority of Present Goods

Böhm-Bawerk, as we have seen, denied that capital had a productivity in the straightforward sense of the word. Further, it was his intent to demonstrate that the phenomenon of interest arises from a value premium that noneconomic forces place upon present goods. By virtue of this pair of negative and positive reasons, therefore, he was bound to interpret advantages yielded by capital over natural forces in terms of a technically generated value superiority of present goods over goods not yet in existence. This was his Third Ground for the existence of interest, and it opened the door for the productivity aspects of capital to enter his theory – surreptitiously to be sure.

Fundamental to this notion of a value superiority of present goods through their usage in production processes, but not identical with it as Böhm-Bawerk himself frequently represented it as being, is the concept of *roundabout processes* of production. A roundabout process is a method of employing inputs to obtain outputs that is time-consuming; that is, the method involves the building of those intermediate forms of products called capital goods between the initial investment of primary resources and the enjoyment of the final product. Such processes – as contrasted with direct methods of production in which primary factors are applied directly to the immediate derivation of consumable goods – are employed because their usage gives man a control over natural forces that enhances the physical productivity of the primary factors, albeit only after some lapse of time. Indeed, frequently this magnification of man's power over nature through the use of such methods is infinite, in the sense that their employment is a *sine qua non* of getting any product at all. Thus, while the use of fruit-picking machinery in the orchard enhances the powers of the invested primary factors over nature by yielding a larger return over what unaided factors would yield, direct methods are completely impotent in transporting people or goods through the air.

The enhanced physical productivity of roundabout processes Böhm-Bawerk declares to be an empirical fact, and he generalizes concerning it in a most important assumption. By increasing the degree of roundaboutness in the production of most goods, he says, it is possible to obtain some increase in quality or quantity if one increases this degree by the better means known. That is to say, not all longer processes are more productive; but it is always possible to find a better method that obtains a larger output if one may employ processes that interpose larger amounts of capital goods between the primary factors and the final goods. The size of such increments of product decreases as the processes become indefinitely longer, but they remain positive.

But how can this productivity aspect of time be introduced into value theory in such wise as to help explain how present goods acquire a greater *value* than equivalent amounts of future goods? That is, how can it be worked into the fabric of value over time? It is the failure of Jevons, Rae and others to integrate the elements of psychological time preference with the facts of productivity that Böhm-Bawerk criticizes. His general failure to grasp the vision of simultaneous determination of variation leads in this instance to an unwillingness to permit the psychological and the technological to enjoy separate but interdependent existences. Productivity implications of capital must be permitted to influence the interest rate only via their impact upon the demand for present goods over future goods. The admitted physical powers of roundaboutness, transmitted through capital, must directly enhance the attractiveness of present goods in terms of future.

Böhm-Bawerk attempted the accomplishment of this feat by subtly linking the advantages of roundaboutness to what he termed the 'technological superiority of present goods'. We may best illustrate his argument in the following way: assume at $t = 0$ that the economy receives a gift of additional present goods which permit it to invest in a roundabout process an increment of current primary factor service formerly used to produce present goods. To pose the problem in its simplest form, we suppose the new roundabout processes are 'point-input – point-output' in nature, so that after the primary resources are invested in $t = 0$, the capital so created embarks upon a self-maturing process in the nature of aging wine or growing timber, and no further additions of primary factor services are required.

Alternatively, suppose the economy were to receive exactly the same gift of current goods to be used for exactly the same purposes in $t = 1$. Böhm-Bawerk attempted to show that on bases completely independent of the First and Second Grounds the economy would prefer the goods receivable in $t = 0$ to the goods receivable in $t = 1$, and thus to demonstrate the technological superiority of the earlier goods. That is, if we assume the stationary state (absence of the First

Ground) to exist and the existence of zero time preference (absence of the Second Ground), the value of the goods in $t = 0$ is greater than the value of the same goods in $t = 1$.

In Böhm-Bawerk's discussion it is tacitly assumed that the goods obtainable from these processes are consumed immediately in the period during which they ripen. Suppose, now, that consumption goods available from the process expand in value (utility) without limit but by decreasing amounts as the primary resources remain invested for a longer and longer time span. Then, Böhm-Bawerk argued, the process beginning in $t = 0$ will always be preferable to that beginning in $t = 1$, because in any future period $t = f$ the process that began earlier will always be preferred by society to the one that began later because the former yields a larger value of goods were the society to harvest the final goods in $t = f$. Therefore, wholly on technological grounds, goods available earlier are superior to goods available later.

I shall criticize this argument in Section V, but my present desire is merely to prove that the argument concerning the technological superiority of present goods is not equivalent to the existence of the advantages of roundaboutness. To accept the latter as a valid proposition is not to accept the former. Consider the following counter-example: to produce a unit of wine requires three units of primary factor services and four periods of time, no more and no less. If the three units of factor services were given to the society in $t = 0$, through a gift of present goods that freed them for direct processes, the wine would accrue to the society in $t = 4$. On the other hand, if the resources were obtained in $t = 1$, the same quality and quantity of wine would be available in $t = 5$. Since we are abstracting from λ time preference and time preference proper, psychological factors do not dictate a preference for goods in $t = 4$ over those available in $t = 5$. So, with the First and Second Grounds neutralized, greater returns to roundabout processes do not translate into technological superiority of earlier goods.

The productivity of roundaboutness, therefore, not the technological superiority of present goods, is the Third Ground, and we have penetrated its thin disguise to recognize simply the marginal productivity of investment. Böhm-Bawerk's attempt to escape the concept has ended in failure. If the gift of present goods were made in either $t = 0$ or 1 or in both periods, a demand for them would exist by virtue of their ability to permit the society to benefit marginally and physically from the extension of the time structure of production.

Even in the absence of the First and Second Grounds, therefore, the interest rate *could* be positive if the marginal physical increment also were a positive value increment.

Indeed, after all the emphasis he placed upon the psychological bases for the existence of interest, Böhm-Bawerk pays great attention to this Third Ground as an overriding explanation of interest, and in the last chapters of *Positive Theory of Capital* comes very close to presenting an exclusively productivity theory of the determination of the rate of interest, despite his criticisms of the productivity theorists who preceded him. It is, he says, this force that acts directly upon the initiators of capital formation, and the capitalists are viewed at times as exerting positive pressures on the interest rate in an environment otherwise neutral in attitudes to present goods.

Beyond this, Böhm-Bawerk indicates that the Third Ground may be responsible for the existence of the First, so that in this sense also the two grounds may not be independent. Where no predominance of those who value present goods above future exists initially, the operation of the Third Ground can absorb present goods from the economy to the point where such an underprovision in present goods arises to require that a premium exist on them.

The Integration with Value and Distribution Theory

In Böhm-Bawerk's broadest vision, the demand for and supply of present goods are determined by the present supply of present goods versus their prospective provision in the future (First Ground); the psychological attitudes toward the present and future (Second Ground); and the existence of a stock of capital and a state of the arts that together determine the rate of transformation of present goods into future goods (Third Ground, somewhat altered). This supply and demand interact within primary factor service and loan markets to determine the premium on present goods which is the interest rate. The explanation of this determination and the imputation of the enhanced value produced through the instrumentality of roundabout processes must, therefore, be subsumed in the grander exposition of the operations of the market economy.

This attempt not only to build a satisfactory theory of capital and interest grounded in natural and psychological forces favoring present goods, but to integrate it with the whole of value and distribution theory must be emphasized: Böhm-Bawerk must be given credit for an analytical sally of the most impressive ambition. It was nothing short of an attempt to formulate a general equilibrium model inclusive of the economy's capital and interest variation.

He did not succeed in doing so, in great part because his crudity of technique led him to oversimplify and distort the value and distribution theory (absorbed largely from Menger) that was his point of departure. We may signal two shortcomings of peculiar importance in these respects and discuss each singly.

The Lack of Understanding of Interdependent Causation

The most punishing deficiency of his analytical vision, given the very nature of the task he set himself, was that Böhm-Bawerk never grasped the concept of a set of functional relationships that associate the dependent variables and independent variables of a problem in an interdependent way, setting up flows of causation among the dependent variables that are omnidirectional. Unlike Cournot, for example, he had no extensive preparation in mathematics that would have permitted him to understand the idea of simultaneity of variation; indeed, he believed models which featured it to be presenting circular arguments. Further, his deficiencies in technique misled him into believing that such simple expedients as introducing discontinuities into functions permitted the analyst to escape the circularity and identify a direction of causation in the system. In this way he tried to establish the thesis that marginal utility set product prices, and they set factor prices.

Böhm-Bawerk was determined to demonstrate that, given the amounts of the primary factor services in an economy, all value springs from the marginal utility of the goods produced, and that the law of costs is merely an indirect reflection of the law of marginal utility. From Menger he borrowed the idea of imputing marginal social utility to the services of factors of production. The value of a primary factor service was determined by the marginal utility of its least valuable use; the costs of production of every good into which these factor services entered reflected these marginal utilities by virtue of competition; and thus the statement that all goods exchange at their (in his construction, marginal and average) costs of production is merely another way of saying that all value in the economy springs from the marginal utility of consumer goods.

Suppose that a given resource used in the equilibrium production of two goods yields marginal utility products in the production of those goods of 10 and 8 respectively, unequal by virtue of the discontinuities of the marginal product curves. Suppose, also, that each of the goods absorbs one unit of the factor service per unit of product. Were the factors and the products perfectly divisible, the last units of the factor services employed in both processes would have yielded the same amount of utility; prices of the goods would have reflected the common marginal utility of the resource in every use; the price of the input service would be this imputed marginal utility product; and it would have been impossible to determine the flow of causation between product and factor service prices. Are product prices the projection of the marginal products of the factor services? Are resource prices the creation of marginal utility of product values? Or are both forces jointly the result of underlying factors and thus simultaneously determined?

The question can be answered definitively in the discontinuous case, Böhm-Bawerk believed, for here it is possible to determine the direction of causation. The 'true' marginal utility product of a unit of factor service in the example above is 8 units, the least valuable contribution to production. Therefore, even though the last unit of factor devoted to producing good 1 yields a product that gives 10 units of satisfaction, that good's effective marginal utility is only 8, and the good's price will reflect this opportunity cost. Therefore, Böhm-Bawerk argued, the marginal utility product in good 2 determines the price of good 1 via the pricing of the factor service. The duality of the two price systems is not sensed.

The introduction of time into this schema and with it the marginal utility of future goods – both the 'true' marginal utility and its Second Ground distortion – is a simple extension of this static formulation. The demand for primary factors to produce future goods is merely added to the demand for primary factors to produce present goods in the competition for the fixed amount of current primary factor services. The technology of roundaboutness and of direct processes determine the technical rates of transformation of current services into present and future consumer goods. Primary factor services will be distributed over all present and future good production possibilities by allocating each service unit to the highest remaining utility alternative, where the marginal utility of future goods is distorted negatively by the Second Ground. The utility of the least important use for each factor service determines the value of the service, and these are projected forward to determine all prices. Interest arises because a larger marginal physical product is required to get a unit of marginal utility in the future than in the present. When the future becomes the present, however, that marginal product will sell at a price that reflects the undistorted marginal utility, and the capitalist receives the difference.

In this manner Böhm-Bawerk relegated productivity to a subordinate position, whose hidden impact was transmitted via its influence on the equilibrium level of marginal utility. The 'technological superiority of present goods' was an attempt to subsume technology in the demand for goods by converting entrepreneurial demand for current factor services for future goods production (capital goods production) into a derived demand for present goods to support the owners of those factors. A symmetrical treatment holds for the demand for primary factor services to use in producing present goods by direct methods. Both sets of entrepreneurs buy the same factor services at the same rates in terms of present goods, but the user of direct processes employs them to the point where they produce a marginal utility product equal to their rate of pay, while the user of roundabout processes buys them until their marginal utility product in undervalued terms equals that cost.

All of value and distribution, therefore – for distribution hinges upon the pricing of the factor services and this depends upon the marginal utility of their products – both when abstracting from capital and interest as well as when including them, is grounded firmly in the subjective relations of the consumer to the goods he consumes.

Of course, modern theorists of the closed economic model object to this solution as a peculiarly distorted depiction of the functioning of the market mechanism. To link prices and marginal utilities so exclusively is one-sided, and ignores important technological relationships inherent in the pricing process. In $t = 0$, the state of technology and the quantities of factors (including, indeed, capital goods) are as much *data* or *independent variables* in the market economy as the psychological preferences of consumers among goods. The definition of marginal utility depends upon the scarcities of consumer goods, and these scarcities mirror the productive limitations set directly by technology and the society's legacy of primary factors. To cite preferences as the 'ultimate' determinant of value is to ignore co-ordinate data, much as Marx defined away the productivity of capital by merely viewing it as an enhancement of the 'ultimate' productivity of labor. Böhm-Bawerk attempted to overthrow a theory which accepted the existence of goods' utilities only to ignore their further implications – the cost-of-production theory of value codified by John Stuart Mill – by acknowledging the existence of technological and resource constraints only to ignore *their* interdependent implications.

Modern theory emphasizes the co-ordinate roles of consumer preferences, resource endowments and technology in the determination of prices. Indeed, Wicksell, who did more than any succeeding theorist to refine Böhm-Bawerk's ideas, overcorrected this emphasis on the psychological by flying to the other pole and stressing the determinative role of the technological. Finally, Irving Fisher in his later work achieved a simultaneity of determination of investment and the interest rate by the mutual interaction of consumer preferences, technology, and primary factor constraints of co-ordinate importance.

Rigid Factor Proportions

The tools that are necessary to integrate value and distribution theory into a whole were at hand in Böhm-Bawerk's day. On the one hand, one could assume that goods were produced in activities requiring rigid proportions among the factors. In a cookbook-like technology, if it takes one cup of flour and four eggs to make one cake, it is assumed that *X* cakes require *X* cups of flour and 4 *X* eggs, regardless of the value of *X* or the types and amounts of other goods produced in the economy. If one makes these assumptions concerning technology it is impossible to substitute one factor for another in any particular activity. The structure of factors is rigidly fixed in the activity.

In order to impute the value of the product to the various factors entering its production, therefore, and thereby distribute the value of product jointly produced by them among the various factors, the theorist must step outside the single production process. It is necessary to consider alternative activities available for producing the same good, if these exist; and/or activities for producing other goods. In brief, one must consider the productive activities for the economy as a whole. For if all factors are to be priced at their incremental contributions to value, in order to obtain substitutability among factors in such an economy, one must permit substitution to occur among activities. This allows the rigid little groups of factors to retain their exclusivity within the activity, yet by changing the activity levels of operation of these activities permits the opportunity cost of each factor service to be determined.

For example, the marginal value product of labor is the contribution to social product that would result were one more laborer to become available to the economy (say by immigration). That laborer enters an economy, however, in which all other primary resources are employed in the various activities of the technology, and in which those activities are being operated at determinate levels. How does the economy determine the marginal contribution of the new laborer when he can be used only with rigidly complementary factor services all currently employed? The mechanism consists of a wage rate decline, which makes labor-intensive activities more attractive, affecting product prices, and which results ultimately in a shifting of the composition of the activities of the economy until a new full employment equilibrium has been achieved and the new labor absorbed.

In this model, then, with the requirement that demand not exceed supply for each factor service and with *activities* substitutable in the above manner for the economy as a whole, it can be assumed that all factor services will be priced at amounts equal to their marginal incremental social value. Although there are a number of difficulties in this approach – for example, the output of the firm is indeterminate unless capacity restrictions are included, and some factors may be incapable of complete absorption and become free goods – modern economic theory has become increasingly enamored of approximating reality with it, and it is an acceptable approach to imputation. Indeed, it is the route taken by Walras when he accomplished the first satisfactory formal integration of microeconomic value and distribution theory, and it is the modern method of shadow pricing in the dual of programming. Incidently, thus dies the notion of causal flow from product to factor prices.

A second approach to this integration is via marginal productivity theory, by assuming that the factor services are finely divisible and substitutable for one another within a given process. Menger had already accepted the variability of factor proportions in production, the example of divisible and substitutable consumer goods were features of the new utility theory, and crude approaches to incremental reasoning concerning factor services had been current in economics at least since von Thünen and Ricardo. When this step is taken, of course, the integration occurs simply. The marginal value product functions of the factor services form the bases for demand functions for such services in each usage; these can be aggregated to obtain a total demand function for each factor; and the interaction of this demand with the fixed supply sets prices for the factor services which enter into the values of goods. Reverse flows of causation from product value to factor service prices register in changes in the marginal value product curves.

Except for his rejection of all notions of the productivity of capital, this approach seems well suited to Böhm-Bawerk's dynamic analysis, for it permits time to be an independent factor of production and primary factor services of different periods to have different marginal products because of their position in the time sequence of the production process. The direct marginal contribution of time as a factor of production results from its usage without complementary factor services; efficient production requires that the value of this marginal product of time be equated to the price of time – its interest cost. The indirect contribution of time to roundaboutness may be established by the same type of equality between the properly discounted marginal factor service costs and revenues. The purest case of time as an independent variable is the point-input–point-output process, exemplified by the well-worked cases of maturing wine and growing timber. The pure case of the second type of time process is the continuous-input–point-output example, in which time serves as a vehicle to carry a continuous sequence of factor services into the production process.

While theorists like Walras, Barone, J. B. Clark, Edgeworth and Marshall were developing these new approaches, or perhaps more accurately rediscovering them, Böhm-Bawerk's techniques were too crude to permit him to operate on the frontiers. As a consequence, he fell back upon the assumption of fixed proportions for his production theory, and because he never placed his imputation theory within the context of the whole economy, he did not truly integrate distribution and value theory, particularly in so far as capital and interest are concerned. Moreover, his desire to eliminate any independent productivity of capital biased him against an approach that gave a marginal product to capital goods, although he might have welcomed the concept of a marginal productivity of time and its equation to a marginal revenue of time. In fact, however, it was Wicksell who made this advance in Böhm-Bawerk's theory.

Böhm-Bawerk's method of dividing the value of product among the factors of production was, therefore, primitive, and, indeed, a regression from Menger's variable proportion treatment. Some of it is wrong: in the case he treats as frequent, for example, he concludes that if a group of factors with zero opportunity costs taken separately is used in fixed proportions to produce a good, the 'last' factor service hired gets the entire product. Of course, the concept of 'last factor hired' is not easily understood, but, even more fundamentally, under these conditions the whole group of factors is a unit, and imputation within it is impossible on grounds other than those of bargaining. Here is a measure of the trouble entailed by his failure to treat the factors against the whole economy as a backdrop in order to obtain the opportunity costs of the factor services.

But despite its primitiveness, Böhm-Bawerk's distribution theory did permit him to see one truth: the insufficiency of productivity considerations alone, without taking psychological attitudes into account, to explain a positive interest rate. Others had isolated both aspects of the problem; Böhm-Bawerk tried to disguise the productivity content; and Fisher was to excel Böhm-Bawerk in the neatness with which these phenomena were integrated. But the Austrian theorist did insist upon the proposition clearly and steadily, and he saw it as his own major contribution to economic theory. At the very least, for merely providing the starting point for Wicksell's and Fisher's more incisive analyses, his work is a major contribution.

IV THE PERIODS OF PRODUCTION, INVESTMENT AND ENJOYMENT

The Measurement of Capital

One of Böhm-Bawerk's more unfortunate analytical sallies was his attempt to quantify the degree of roundaboutness of production processes. The attraction of the goal is understandable: were it possible to find a time concept that was perfectly correlated with the 'amount' of some homogeneous essence called 'capital' embodied in the heterogeneous capital goods of a process, each production process could be ranked on a scale of roundaboutness in a manner independent of the prices ruling for primary factor services and of the interest rate. Such an index would not only establish the temporal substance of all capital and provide an invariant measure of the capitalistic depth of each process, but, properly weighted over several or all ongoing processes, it would permit the calculation of the capitalistic dimension of groups of activities up to and including the whole economy.

Let us first present the concepts involved for the simplest production process with which Böhm-Bawerk dealt – indeed, the one in which he interprets much of his system – and then move into the more complicated processes he discussed. For the single production process the simplest case was the point-input–point-output activity. All inputs except time and the free gifts of nature are invested during $t = 0$, and the immature product permitted to ripen until it is ready and is totally consumed at the start of $t = h$. Output is unambiguously related to input and time structure equally clear.

Böhm-Bawerk's nomenclature for the various periods he discusses was not conducive to clarity of analysis, and he confuses them himself at times. We shall, therefore, try to establish a clearer set of periods and use the terminology defined consistently. The *period of construction* is the time that elapses between the first devotion of primary inputs to the production of the good and the harvesting of the first finished goods in h; that is, it is the h-period time span between $t = 0$ and $t = h$. The *period of production*, on the other hand, is meant to measure the average time interval between the expenditure of primary factors and the start of product flow. Assume that we define a land-plus-labor unit in which these primary factor services are combined in fixed amounts, and assume that two such units are used in $t = 0$ to lay down grape juice. This grape juice is undisturbed until it is sold and consumed as wine at the start of $t = 4$. Then let us imagine the time periods to be equally spaced along a straight line, with their identifying number having an origin at $t = 0$.

The period of construction of the wine is 4 – all of periods 0, 1, 2 and 3 – in general a rather uninteresting period from the viewpoint of measuring roundaboutness. If we define the period of production as the length of time between $t = 0$ and the center of gravity of the primary factors invested over time, we may compute it by renumbering the periods with the fruition period $t = h$ as origin, multiplying each such number by the units of primary factor service added at that point, summing these products, dividing the sum by the total of primary inputs added over the time span, and, finally, multiplying the quotient by -1. Suppose F_t is the number of primary factor units entering the process at t, and let T be the length of the period of production. The weights applied to the factor doses are obtained by setting the origin at $t = h$ and using the new indices so obtained. In general, Böhm-Bawerk measures the degree of capitalization in a process as

$$T = \frac{-[(-h)F_0 + (-h+1)F_1 + \ldots + (-1)F_{h-1}]}{\sum_{t=0}^{h-1} F_t} \qquad (14.1)$$

$$T = \frac{-[(-3)2 + (-2)0 + (-1)0]}{2 + 0 + 0} = 3. \tag{14.2}$$

It is in the nature of the point-input–point-output case that the period of construction equals the period of production.

The *period of investment* is the period during which the typical primary service input unit remains in the production process. In the point-output case, where all product accrues at $t = h$, this period is identical with the period of production, since all inputs emerge as finished output at the same period. The period of investment may extend beyond the period of construction when point-output does not rule, so the final product accrues over a succession of periods h, $h + 1$, $h + 2$, etc. The period of production ends with the first unit of finished output obtained, but, of course, all inputs cannot be envisaged as emerging in h, and so the period of investment may differ from the period of construction and production.

In this distinction between the periods of production and investment we have encountered the concept of the *period of enjoyment*, or the period over which the flow of finished output occurs. A greater-than-unity value for this period was an afterthought for Böhm-Bawerk, to recognize the possibility that increased roundaboutness of processes might make products more durable as well as more plentiful. Further, it recognizes the possibility that the nature of a process may not be to present a lump of final goods in a single terminal period h.

Beyond a few brief references, this period of enjoyment never really entered into Böhm-Bawerk's analytical scheme of things. He declined to join it onto the period of production because the period of enjoyment, he said, is similar only superficially to the period of production. For example, while the period of production is an onerous period for the economy, the period of enjoyment is a pleasurable period. Also, he says, in the case of durable capital goods, the production of more durable capital goods can be handled by increasing the period of construction of the consumer goods involved. As a consequence of these arguments, and from the evidence of the cases he discusses in his work, it is safe to conclude that Böhm-Bawerk thinks consistently in terms of a point-output in $t = h$, and consumption of the product during that period. For example, in a famous example involving the investment of resources to pipe water into a house, Böhm-Bawerk's consideration ends when the period of construction ends: with the start of the flow of water.

But he does deal frequently with an extension of the case of point-input to the case of continuous-input, or, more exactly, to the case of multi-point inputs. Suppose, for example, that the period of construction is three periods long, with investments of 10 units of factors *each* period. Then, following 14.1 we should obtain a period of production as follows:

$$T = \frac{-[(-3)10 + (-2)10 + (-1)10]}{30} = 2. \tag{14.3}$$

In this example it is assumed implicitly that the consumption goods are ready at the start of $t = 3$, and that the investments of 10 input units are made at the beginnings of $t = 0$, 1 and 2. The point or points within the period at which the investments are made can make a difference in the measurement of the period of production. For example, let us choose a new analytical time period with half the length of the old, so that their new number within the time span is $2h = 6$, and let us now invest five units of resource at the start of each of the new periods. In terms of the old period we may interpret this as investing half the inputs at the start of the period and half in the middle. Then, the production period in new time units is

$$T = \frac{-[(-6)5 + (-5)5 + \ldots + (-1)5]}{30} \tag{14.4}$$

If we convert this to the unit length of the initial period defined for 14.3 the period of production is 1.75, which is less than 2, as calculated in 14.3 above.

If we now divide the length of the analytical period by 2 once more, and assume a distribution of investment at the start of the $4h = 12$ time periods of 2.5 input units, we should find that $T = 6.5$. In terms of the original time period this calculation yields a period of production of 1.625. We may generalize the results we are obtaining in these operations through the use of the following formula:

$$T = 0.5nh + 0.5, \tag{14.5}$$

where h is the terminal period of the original time span and n is a positive number whose reciprocal is the length of the new period as a fraction of the length of the old. Since we add the constant value 0.5

to $0.5nh$ as n grows larger, clearly that constant becomes relatively smaller and smaller. As we allow n to grow without limit, and therefore permit the size of the period to become smaller and smaller, we obtain in initial period units

$$\lim_{n \to \infty} \frac{0.5nh + 0.5}{n} = 0.5h + \frac{0.5}{n} = 0.5h. \tag{14.6}$$

Thus, where a constant rate of investment occurs over a period of construction and the period of enjoyment has a unit length, as the time periods t are permitted to shrink without limit, which is to say that the instantaneous rate of investment is equalized over the period of construction, the period of production approaches a length equal to one half the period of construction. It should be noted again that the period of production is the average period of time between the investment of the primary inputs and the receipt of the consumption goods, or the average time period spent by primary inputs between their investment and the sale of the product. Since all product emerges at the start of $t = h$ in the point-output case, the period of investment of the primary inputs equals the period of production.

Now, under the idealized conditions of Böhm-Bawerk's analysis – that is, most simply, for a point-input–point-output process or a continuous and uniformly distributed input–point-output process – T is an index of the amount of capital in the process when we view the economy as moving through time in a stationary manner. We may illustrate this in two ways, one of which is due to Böhm-Bawerk and the other to Boulding and Dorfman. In Böhm-Bawerk's example, an analogy is obtained by setting up a stationary flow with a point-input–point-output analogue model. A bank messenger is dispatched from a bank to the clearing house every hour; therefore, one primary input unit is 'invested' at $t = 0$ and emerges as 'output' at $t = h$ at the destination. At $t = 1$ another input is invested that emerges at $t = h + 1$, and again at $t = 2$ another input unit is invested to emerge at $t = n + 2$, and so forth. Thus, inputs are invested and outputs received in an unchanging flow. The period of construction (equals period of production equals period of investment) in this example is h, and the rate of flow of input (equals rate of flow of output) equals one per period. At any particular point of time, t, there are h messengers en route – one just beginning and $h - 1$ at the different intermediate stages of the process.

We may describe the relationship after the steady state has been established by the equation

$$T = K/X, \tag{14.7}$$

where T is the length of the period of production, K is the number of messengers (unfinished product) *en route*, and X is the rate of input into (equals rate of output from) the process. Of course, distance in the example is proportionate with time, and clearly for a given input rate of X the size of K – a stock of inputs suspended in limbo at any moment – is an index of the length of the process. In our specific example, $K = h$, and thus it varies exactly with the time to travel.

Let us appeal to the Boulding-Dorfman hydraulic analogy to convert the example to a continuous-input–point-output stationary state. Consider K to be a reservoir for the storage of water, and X to be the rate of inflow and outflow of water into and from the reservoir respectively. Then, when the system is in this state of flow balance, the volume of K reflects the cumulated input flows in limbo; indeed, T is the average period of time the typical molecule of water spends in the reservoir, so it is properly viewed as the period of investment of the average unit of water in the process. The economic analogue of this hydraulic example is complicated by the fact that primary factor inputs and product outputs do not have a common physical unit, but for illustrative purposes let us blink this problem.

Now, if we associate 'productivity' with the length of time the typical unit of X remains in suspension, then a rise in K (invested primary resources) increases the productivity of the fixed rate of input flow X, T rises along with the value of the fixed output rate and with the increase in the roundaboutness of the process. It was Böhm-Bawerk's assumption that if the entrepreneur desired and the primary resources were available he could always find a method that permitted an increase in K to be worthwhile in the sense of increasing the value of output by more than the value of input. If K is increased, however, after the adjustment to the new flow in the stationary state with X fixed, T must rise to reflect the change, as is clear from 14.7.

Therefore, Böhm-Bawerk argued (although he did not specify our stationary assumption), T rises and falls with the stock of capital K and is an exact measure of it; it is impossible to increase K without affecting the value of T. In this sense, then, the amount of capital reflects directly the depth of roundaboutness of a process.

Beyond this, Böhm-Bawerk indicates that by proper weighting it is possible to average over all processes to obtain the 'social period of production', which presumably measures the degree of roundaboutness and average period of investment of primary factors for society as a whole. However, he never indicated how this weighting would be determined, nor did he pursue very far the hope of constructing such a grossly summary measure.

Note that when the concept of point-output is dropped it becomes difficult to identify the time at which a specific input emerges from the reservoir as output. For example, when the period of enjoyment is not unit-valued, the period of investment of the primary resources is extended, but it may be impossible to obtain a period of investment in these circumstances for any specific input unit. However, it may still be possible in the stationary flow to obtain the period of investment for the average input.

A Rigorous Synthesis

In order to effect a synthesis of Böhm-Bawerk's ideas I shall present them within the context of his ambitious vision: a general equilibrium model. I shall simplify by dealing only with the stationary state, but the rigorous formulation will serve the purpose of presenting his ideas in an integrated and interrelated fashion.

I make the following assumptions.

Assumption 1 The final output of the economy is a single consumer good, and we shall use the good as the unit with which to measure the prices of all other goods and services. Thus, if the good is carrots, then all prices are quoted in some physical unit of carrots.

Assumption 2 The 'primary factor' is an amalgam of labor and land in a given proportion, and the number of available units of this composite input is fixed at N per period for the current and every future period.

Assumption 3 Production in the economy is a process in which doses of primary factor services are applied over a period of construction of length h at a constant rate of X per period. (Note that X is input per period over the period of construction.) At the end of the period a point output $\bar{X}$ of the consumer good is the culmination of the process. We may depict the technology formally as

$$\bar{X} = f(X, t), \tag{14.8}$$

and we assume further that $\partial f / \partial t > 0$, $\partial f / \partial X > 0$ everywhere; and that $\partial^2 f/\partial t^2 < 0$, $\partial^2 f/\partial X^2 < 0$ everywhere. That is, the function is continuous and at least twice differentiable, the marginal products of time and input streams are positive, but diminishing returns to both time and these primary input streams characterize the process.

Assumption 4 No capital good survives the period of construction of the consumer good.

Assumption 5 Let C be the rate of current consumption and F be the rate of future consumption *enjoyed in every period in the future*. We assume that there exists a preference function for the community, defined as

$$M = M(C, F), \tag{14.9}$$

which is an ordinal representation of the society's rankings of all baskets containing non-negative amounts of present consumption and future consumption flows in perpetuity. We assume that it is twice differentiable, does not reach a maximum for any basket of the two goods, and that it yields strictly convex indifference curves.

Assumption 6 The economy is in a stationary state. That is, in each period the economy repeats the previous period's operations exactly, with all primary inputs employed in exactly the same way to yield exactly the same results in terms of output and capital formation.

Input–Output Relations

In the point-output case, $\bar{X} = C$, which is to say that the output of each period exactly equals the consumption of that period. When the equilibrium period of construction of the stationary state has length θ, input into each phase of the production process is $1/\theta$ of N, the total input. That is, the stationary state requires that for a construction period of θ it is necessary that θ phases of the process be carried on simultaneously in every period t, each phase a different distance from completion, and each absorbing its proper proportion of the total input of the economy. The consumption of any particular period, therefore, is the result of N/θ of the factors being invested for θ successive periods, so that we may write from 14.8

$$\text{(I)} \qquad C = f(N/\theta, \theta). \tag{14.10}$$

Total Costs of Production

In each period of the stationary flow, N/θ units of primary factor service, receiving the equilibrium rate of return w in consumption goods and accumulating at the equilibrium interest rate i, are applied to each stage of the process. We may express the costs of output from one such process as

$$G = \frac{Nw}{\theta} \sum_{t=0}^{\theta} (1 + i)^t, \qquad (14.11)$$

but if we assume that the rate of inflow of factor service is instantaneous the cost can be approximated by

$$G = \frac{Nw}{\theta} \int_0^{\theta} e^{it}\, dt, \qquad (14.12)$$

or, upon integrating,

$$\text{(II)} \qquad G = \frac{Nw}{i\theta} (e^{i\theta} - 1). \qquad (14.13)$$

We may obtain an approximate value of G by expanding $e^{i\theta}$ by Taylor's theorem, and retaining only the first three terms. If we do so, we obtain

$$G = \frac{Nw}{i\theta} (1 + i\theta + 0.5i^2\theta^2 + \ldots - 1) \qquad (14.14)$$

$$\approx Nw(1 + 0.5i\theta) = Nw + 0.5\theta Nwi.$$

The first term in the last expression of 14.14 is the total primary factor service cost of the process, or N/θ units of the primary resource at the unit cost of w applied each of θ periods. The firm pays Nw/θ for primary factor services for θ periods, or Nw over the period of construction. Since these resources are applied continuously we may approximate the average amount per period on which interest must be paid as 0.5 Nw, so that average interest payments per period are approximately $0.5Nwi$ and total interest payments $0.5Nw\theta i$. If we define capital as the capitalized value of the interest payments, we obtain

$$K = 0.5Nw\theta i/i = 0.5Nw\theta. \tag{14.15}$$

Producers, therefore, may be seen from 14.14 to be paying primary factor and interest costs on the product.

The Capital Stock

In any period t of the stationary flow the whole set of processes will be operating side by side, each with different amounts of accumulated capital. The total cross-sectional capital at the end of the analytical period in the whole economy will be the value of capital stocks in every stage of maturation, or, at the end of period h

$$K = \int_0^h \frac{Nw}{i\theta}(e^{it} - 1)\, dt. \tag{14.16}$$

Integrating, we obtain for the equilibrium period of construction,

$$\text{(III)} \qquad K = \frac{Nw}{i^2\theta}(e^{i\theta} - 1 - i\theta), \tag{14.17}$$

which is the capitalized value of the interest income of the period.

Let us use a Taylor's series expansion again to derive an approximation, neglecting all terms but the first three:

$$K \approx \frac{Nw}{i^2\theta}(1 + i\theta + 0.5i^2\theta^2 - 1 - i\theta) = 0.5Nw\theta. \tag{14.18}$$

It can be seen that this yields the capital accumulated for the economy as a whole and duplicates 14.17. Note, however, that in this analogy we view the stationary state rate of outflow of the economy as Nw – the value of consumption goods going to primary factors only – as opposed to C, the period's total value of production of consumption goods which includes interest payments. Thus, the above formulation ignores consumption out of interest payments, and is only an approximation to the total rate of consumption goods outflow. Further, we should note that 14.18 can yield very poor approximations to 14.17, as indeed it does for the example we shall present below, whenever $i\theta$ is large, since in that event the remainder term in the series expansion converges slowly to zero.

Profits

In the stationary state, we require that profits π be maximized with respect to the period of construction, or that the value of period output less the value of period costs be a maximum. But because we are using the consumption good as a value unit, the value of each process output is $f(X, t)$, when costs are defined in 14.13. We may, therefore, define profit as

$$\text{(IV)} \qquad \pi = f(X, t) - \frac{Nw}{it}(e^{it} - 1). \qquad (14.19)$$

To maximize profits we differentiate with respect to time, remembering that X will be a function of t in the stationary state, and set the resulting expression equal to zero. Further, we recall that in the stationary flow we have θ as the equilibrium period of construction, and so we substitute this for t throughout:

$$\text{(V)} \qquad \frac{\partial f}{\partial X}\frac{dX}{dt} + \frac{\partial f}{\partial t} - \frac{1}{\theta}(Nwe^{i\theta} - G) = 0. \qquad (14.20)$$

These are the marginal conditions that require the additional revenue ascribable to time to equal the additional cost of time in the stationary equilibrium.

However, it is also a condition for a stationary equilibrium that these maximized profits be zero: that is, that the value of consumption goods produced in each period be equal to the primary factor plus interest costs of production. We therefore specify that

$$\text{(VI)} \qquad C - G = 0. \qquad (14.21)$$

Taken in conjunction with 14.20 this condition specifies that the marginal cost of time equal the marginal revenue of time where total cost equals total revenue.

Wealth

We define the price of a flow of one unit of consumption good per period in perpetuity as P_e, or the price of a unit of wealth. It is easy to demonstrate that

$$\text{(VII)} \qquad P_e = \frac{1}{i}, \tag{14.22}$$

or that the price of a unit of wealth equals the reciprocal of the interest rate. The value of a stream of receipts of one unit of consumption good per period for n periods is the sum of the discounted elements in the stream:

$$V = (1 + i)^{-1} + (1 + i)^{-2} + (1 + i)^{-3} + \ldots + (1 + i)^{-n} \tag{14.23}$$

Now, multiply 14.23 by $(1 + i)^{-1}$ and subtract this product from V. This yields

$$Vi = [1 - (1 + i)^{-n}], \tag{14.24}$$

and if we allow n to grow without limit it follows that

$$\lim_{n\to\infty} V = P_e = \frac{1}{i}, \tag{14.25}$$

which explains 14.22.

The total number of units of wealth, Q_e, yielding an unending stream of Q_e future units of the consumer good for every period of time, may be used as the measure of this income flow:

$$Q_e = Nw + Ki, \tag{14.26}$$

so that the wealth of the society, being the capitalized value of this future earning capacity, is

$$\text{(VIII)} \qquad Q_e P_e = Nw P_e + K. \tag{14.27}$$

It may be noted that the condition of a stationary state implies the constancy of consumption through time and its exhaustion of income: that is, that $Q_e = C$. From 14.17 and 14.27:

$$Q_e P_e = Nw P_e + \frac{Nw}{i^2\theta}(e^{i\theta} - 1 - i\theta), \tag{14.28}$$

from which it follows by steps

$$Q_e = Nw + \frac{Nw}{i\theta}(e^{i\theta} - 1 - i\theta)$$

$$= Nw + \frac{Nw}{i\theta}(e^{i\theta} - 1) - Nw$$

$$= G - C. \tag{14.29}$$

The Constrained Maximum Social Welfare

From 14.27 we derive the value of social wealth, and from 14.13, 14.17, and 14.21 we derive the definition of consumption as

$$C = Nw + iK. \tag{14.30}$$

Wealth plus income is, then,

$$C + Q_eP_e = Nw + iK + NwP_e + K, \tag{14.31}$$

which is the expression of the constraint of wherewithal on consumption in the present and future. We set up the Lagrangean form

$$L = M(C, Q_e) + \lambda(C + Q_eP_e - Nw - iK - NwP_e - K), \tag{14.32}$$

and, assuming that the maximum is an interior maximum, we set the partial derivatives equal to zero:

$$\begin{aligned} \partial L/\partial C &= \partial M/\partial C + \lambda = 0 \\ \partial L/\partial Q_e &= \partial M/\partial Q_e + \lambda P_e = 0 \\ \partial L/\partial \lambda &= C + Q_eP_e - Nw - iK - NwP_e - K = 0, \end{aligned} \tag{14.33}$$

where $d^2L < 0$ by Assumption 5.

If we eliminate the Lagrangean multiplier from the equations of 14.33 we obtain the familiar marginal rate of substitution conditions:

$$\frac{\partial M/\partial C}{\partial M/\partial Q_e} = i. \tag{14.34}$$

That is, the marginal rate of substitution of present consumption for perpetual flows of future consumption must equal the interest rate.

Equations (I) through (IX) form a system to determine C, G, K, π, θ, P_e, i, w, and Q_e, within a closed context – that is, in a general economic stationary equilibrium.

A Numerical Example

We suppose that we are given the following data, which include the preference function, the production function, and the endowment of primary resources available to the society:

$$M = (C + 100)^{0.1} (Q_e + 5)^{0.8} \tag{14.35.1}$$

$$\bar{X} = 40X^{0.2}t^{0.7} \tag{14.35.2}$$

$$N = 100. \tag{14.35.3}$$

Because we are specifying that the equilibrium we seek must be stationary, we know that a steady flow of consumption each period requires that θ stages of the process be operated simultaneously, each such stage employing the same production function and the same input arguments, but being at different phases of completion. That is, over a longitudinal slice of time each process begins in an initial period and terminates at the start of the period θ units from that initial period, passing through θ stages of production. From a cross-sectional view, however, in the stationary flow each of these θ stages will exist simultaneously, and an equal number of primary resource units will be added to each of the stages during the period.

Therefore, for the process whose function is depicted in

$$X = N/\theta, \tag{14.36}$$

and we may therefore write

$$\bar{X} = 40\theta^{0.5}N^{0.2} = 100\theta^{0.5}. \tag{14.37}$$

From (I) we have the requirement that consumption each period be the output for a construction period of θ:

(I) $C = 100\theta^{0.5}$.

From (II) we have defined costs as

$$\text{(II)} \qquad G = \frac{100w}{i\theta}(e^{i\theta} - 1)$$

$$\approx 100w + 50wi\theta,$$

and we shall use the approximation in the calculations below.

We define capital explicitly from (III), reducing it to terms involving C, w, and i by substitution from (II) and (VI):

$$\text{(III)} \qquad K = \frac{100w}{i^2\theta}(e^{i\theta} - 1 - i\theta)$$

$$= (C - 100w)/i.$$

We define profits

$$\text{(IV)} \qquad \pi = C - G$$

and specify their maximization

$$\text{(V)} \qquad 50\theta^{-0.5} = 50wi\theta$$

at zero:

$$\text{(VI)} \qquad C - G = 0.$$

We repeat the definition of P_e

$$\text{(VII)} \qquad P_e = 1/i,$$

and of wealth

$$\text{(VIII)} \qquad Q_eP_e = 100wP_e + K.$$

Finally, remembering that $C = Q_e$ by the demonstration of 14.29, the consumer equilibrium conditions for the society are

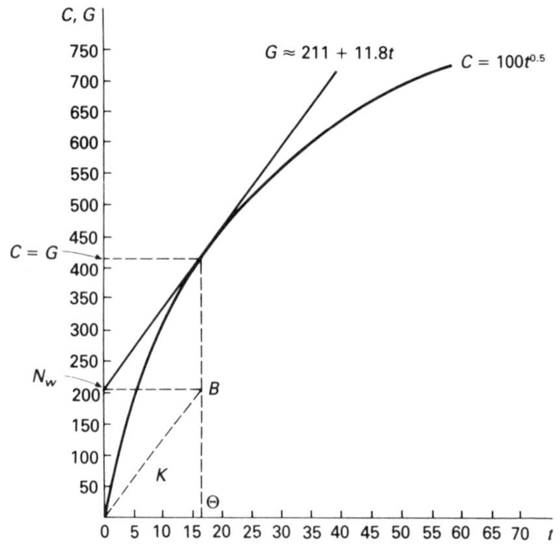

Figure 14.1 The approximate stationary solution for Θ, *C*, *K*, and *w*

$$\text{(IX)} \qquad \frac{0.125(C + 5)}{(C + 100)} = i.$$

When we solve this system by approximative methods, we obtain the following results: $\theta = 17.8$; $C = G = Q_e = 422$; $w = 2.11$; $K = 1{,}884$; $i = 0.112$; and $P_e = 8.93$.

On Figure 14.1 we have graphed the solution for C, G, θ, K, and w. The output is graphed as a function of t and, because all final output is consumption, that output is equated to consumption. The cost curve, G, is approximated by the Taylor series linearization of (II). The equality of the marginal revenue and cost of time imposed by (V) is represented by the equality of slopes of the functions at $\theta = 17.8$, and the no-profit constraint of (VI) is reflected in the tangency of the functions at this value of θ. The value of capital, K, is shown in 14.18 to be approximated by $0.5Nw\theta$, or half the area of the rectangle formed by the product of Nw and θ, graphed on Figure 14.1 as the triangle $OB\theta$. Finally, the solution Nw is the intercept of G with the y-axis.

On Figure 14.2 we have graphed the solution's values for C, Q_e,

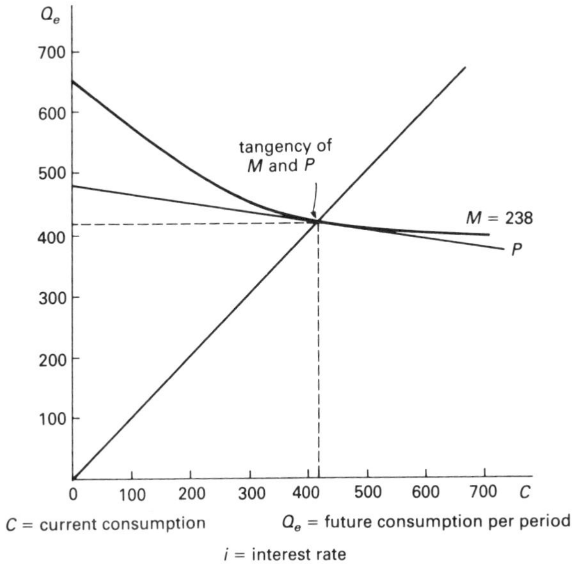

Figure 14.2 The approximate stationary solution for C, Q_e, and i

and i. Because the solution is stationary, the relevant indifference curve and price line must be tangent along the 45° ray from the origin at the point where $C = Q_e = 422$. The slope of the price line P is $-i$, and equals the marginal rate of substitution of present for future consumption, where the latter is perpetual flows of consumption.

It is difficult to analyze displacements of equilibrium in our model formally, but we may get some notions of the directions of movement of variables from Figures 14.1 and 14.2 by partial analysis. If N is held constant but w is increased with no change in the exponent of N in 14.37, the intercept of G with the y-axis in Figure 14.1 must rise and the slope of G must fall in order to re-establish the tangency with the production function. Because the slope of G is $0.5\theta Niw$, with N constant and w rising, it follows that θi must fall. But it is clear from the figure that θ rises, and so i must fall. Further, since $K = 0.5\theta Nw$, the rise in θ and w, with constant N, implies a rise in the capital stock K.

Figure 14.1 encompasses the productivity aspects of Böhm-Bawerk's theory, and it is clear that primary factors and time have become substitutable inputs. When w, the price of primary inputs, rises, a substitution toward the now relatively cheaper input, time, occurs, with a consequent reduction in its marginal value product and therefore *i*. The return to primary factors and the interest rate vary in inverse fashion with a constant amount of primary inputs.

On Figure 14.2, if we suppose as is customary that as the amount of current income and consumption grows, positive time preference becomes smaller, then, along the 45° line as one moves away from the origin, the slopes of the indifference curves become less and less steep. This implies that in the neighborhood of the line the loss of $1 in present consumption can be compensated by smaller and smaller amounts of future income flows. A fall in *i*, therefore, will yield a price line *P* which is less steep and that will find its stationary state tangency farther out along the 45° line.

V CRITICISM OF THE THEORIES

Böhm-Bawerk's theories of capital and interest have not sustained the onslaught of their numerous critics well in detail. We shall conclude our work by considering some of the major weaknesses of the theories that have not been noted already in previous sections.

The Vision of Production

The extreme simplicity of Böhm-Bawerk's vision of production results in his model treating all capital as arising in and perishing with the period of construction of consumer goods. All consumption goods resolve into the primary, nonaugmentable factors of production, and this resolution is exact only if no capital goods exist at the start of the construction period and none survives that period. We may soften this somewhat, however, when capital goods do exist at the start of the construction period if the primary inputs that enter their production can be identified as to time period and if their sum, period by period, approaches zero as time moves toward $-\infty$. This extension of the simple case does seriously compromise the usefulness of the concept of a period of production, however, nor can it be viewed as a likely realistic possibility.

Further, production processes do not involve the investment of primary resources at uniform rates over time, but rather a whole spectrum of possible rates and time-phasing of inputs exists. At the other end of the process, ordinarily there is no one terminal date at which all final product falls out as neatly as a candy bar from a coin machine; rather there is a set of output streams over time from which the firm may choose. Typically, therefore, for any given pattern of inputs over time there exists a spectrum of output patterns; and for any given pattern of output over time, there exists a choice of input patterns. Moreover, even after the process has yielded 'all' of its product in some sense, durable capital goods remain to perform their function for future processes.

Böhm-Bawerk's vision of the role of capital is really at bottom an extension of the wages fund or advances-to-labor theories of earlier economists. *All* capital goods extant at *t* (not merely the stocks of subsistence goods in being) are in effect nascent consumer goods that will provide imminently the support of primary factors. They form a fund of advances to support these factors during the intermediate phases associated with roundabout processes. The theory fails to integrate such stocks of indefinitely durable goods as paper money or even long-lived if perishable capital goods. Capital becomes in this way of thinking a quasi-homogeneous fund of consumer goods with a determinate size correlated with the period of investment.

But in continuous-input–continuous-output processes, with variable patterns of input and output over time, with capital goods surviving particular construction periods, with capital goods producing capital goods, the neatly dated and discrete activities of the theory become increasingly vague as to time lapse, date of initiation, and date of termination. In the stationary state it may be possible to find a pulse for the process or the economy as a whole, by virtue of the repetitive nature of activities; but the cyclical input–output pattern thus discerned is a last refuge from conceptual vagueness.

The principal opposing concept of capital and explanation of interest at the same level of ambition as Böhm-Bawerk's is associated with the *productivity* theories of interest, formulated by such thinkers as Walras, John Bates Clark and F. H. Knight. Their fundamental departure from Böhm-Bawerk's schema is the overthrow of the distinction between augmentable and nonaugmentable, or producible and nonproducible, inputs. They take the alternative view that all factors are ultimately produced by the investment of all resources – land, labor and capital goods – taken jointly and indecomposably, and that all such factors require maintenance and amortization payments between which it is impossible to distinguish.

The production process, therefore, in the view of the productivity theorists, is best studied from the social viewpoint as an application of the *services* of the whole complex of factors in existence at *t* to production, such production at once covering the maintenance-amortization expenditures of the society, the net investment of the economy, and the current requirement to throw off current goods for the period. The economy in any given period is a mass of assets which must be maintained, amortized, fed services for current outputs, and perhaps enlarged. It is misleading to look for some factors that are capital and some that are primary: all are produced, in the sense that even skilled labor must be produced through education involving the application of all factor services, land must be cleared or created and needs a flow of services to retain its productivity, and so forth.

Moreover, in this view of the economy as an organic mass of economically indistinguishable factors absorbing factor services and throwing off consumer and investment goods each period, it is impossible to specify a cycle of investment for inputs, either for the individual process or the whole set of processes. Sequences of inputs generate sequences of outputs, and it is a fruitless task to attempt a determination of the point at which specific inputs emerge as specific outputs or the average period of investment of input flows. The rate of interest is grounded in the technological capability of longer construction periods to generate greater amounts of output; but there is no 'waiting' or distortion of value through the time perspective, because after the construction period is over output begins and continues forever as long as maintenance-amortization applications of factor services are made to the capital. This theory does not eliminate time as a productive element, but it does permit it to be abstracted from as an independent factor by merely dating the inputs and outputs of a process.

Modern capital theory has moved a long way toward the productivity theory outlook by viewing *the* problem of capital as that of determining technologically efficient or Pareto-optimal (technologically efficient and meeting certain weak criteria for consumer welfare) paths of final goods production through time. The approach, found in the work of Malinvaud, for example, concentrates attention upon the time sequences of consumption; after determining the set of such sequences that are efficient or Pareto-optimal, the inputs and gross outputs over time necessary to produce the sequences are examined and capital formation is traced out passively as the baskets of non-consumed goods surviving each period. Interest is a reflection of the technological relations available and the society's preferences among goods at different points of time, interest rates being determined by the relations of prices over time and the choice of a good in which to measure prices.

This modern productivity theory of capital permits the theorist to escape the creation of a homogeneous fund of subsistence (*à la* Böhm-Bawerk) or of capital (*à la* J. B. Clark). Each capital good retains its identity and is priced over its time span of existence; theoretically speaking, capital goods themselves become as uninteresting in the time dimension as intermediate goods frequently are in the cross-section analysis. Rather, the focus of the modern approach is upon the stream of final outputs of the economy and the best methods of producing and distributing them, where these methods imply the use of different inputs at different time periods which in turn imply that certain incomplete forms of goods are created within certain periods and survive over others. The role of prices over time in bringing about efficient patterns of final outputs can then be studied, the intertemporal implications in terms of inputs and gross outputs can be obtained, and the implied price sequences can be compared to reveal interest rates.

The Period of Production

Böhm-Bawerk's definition of the period of production assumed two crucial features of production: only one primary factor service was involved in investments, and compound interest did not enter as an investment cost of the primary factor. If more than one primary factor exists, however, the physical unit of factor measurement must be abandoned, and resort taken to the *value* of factors invested period by period. Indeed, even in the case of the single factor, the nature of *capital* as a separable concept from a catalogue of physical inputs depends upon recognizing it as value concept. To use a value measure seems a plausible solution.

When we take this step, however, Böhm-Bawerk's period of production encounters a fundamental difficulty that renders it useless as an index of roundaboutness. That nettle is compound interest, for when the necessity to treat invested inputs in a value unit is recognized, the period of production becomes sensitive 'internally' to the rate of interest. Böhm-Bawerk desired to distill a physical, value-free characteristic of production processes, on the basis of which an economy could choose among methods of production in response to varying interest rates. That possibility disappears when different interest rates yield different lengths of the period of production for the same process.

On the other hand, if we elect to eliminate compound interest from the computation of the period of production, as Böhm-Bawerk did, so that T is a constant in the face of interest rate change, compound interest destroys its usefulness. It is surely a fundamental necessity of any index of roundaboutness that a more capitalistic process be more penalized by higher interest rates than a less capitalistic process. We shall demonstrate below that this need not hold for Böhm-Bawerk's measure of capitalization.

Let us illustrate both sets of difficulties with a simple example. Suppose there exist two processes that produce a unit of the same good in $h = 4$, but which require the following value amounts (before interest) of primary resources in the periods indicated:

Process / Period	0	1	2	3
1	0	6	2	0
2	4	0	0	4

Let F_t depict these values, and let us generalize formula 14.1 to the following:

$$T = \frac{-[(-h)F_0(1+i)^h + \ldots + (-1)F_{h-1}(1+i)]}{\sum_{t=0}^{h-1} F_t(1+i)^{h-t}} \tag{14.38}$$

Böhm-Bawerk's T is the special case where $i = 0$, and if we compute the periods of production for the processes in our example for this case we obtain $T_1 = 2.75$ and $T_2 = 2.50$. Note that in 14.38 the denominator is the cost of production of the product for the process, and if we symbolize these by C_1 and C_2 we obtain $C_1 = C_2 = 8.00$ in a world we still consider to have zero interest. To Böhm-Bawerk, Process 1 is more capitalistic than Process 2, but in a world of no interest the difference does not register in a cost disadvantage for Process 1. Suppose now that $i = 0.05$. Then $T_1 = 2.76$ and $T_2 = 2.61$ on the new basis of 14.38, so that Process 1 is still indexed as more capitalistic than Process 2. Now, however, $C_1 = 9.16$ and $C_2 = 9.08$, so that in a world of positive interest the more roundabout process yields higher costs of production, as Böhm-Bawerk would predict.

Thus far, then, the extension of Böhm-Bawerk's measure performs as we should desire it to do.

Now let $i = 0.20$. We then find that $T_1 = 2.78$ and $T_2 = 2.90$. A higher interest rate has now reversed the ranking of roundaboutness. What has happened, of course, is that the interest rate is now high enough to make the $t = 0$ inputs of Process 2 dominate the index; hence, the measure is sensitive to the rate of interest and no longer represents any purely technological feature of the economy. But our surprises are not exhausted, for we find that $C_1 = 13.26$ and $C_2 = 13.12$, so that the *more* capitalistic process (as computed by our generalized method) yields the *smaller* production costs at *higher* interest rates than formerly. The interest rate has become high enough to tilt the roundaboutness measure against Process 2 because of its very early inputs, but not so high as to override the interest costs on Process 1. Surely, then, a measure of capitalization which moves in a direction opposite from costs when the interest rate rises must be of limited value. Be it noted, however, that Böhm-Bawerk's interest-invariant T measures do not suffer this deficiency.

But now let us set $i = 0.30$, and $C_1 = 16.58$ and $C_2 = 16.64$. The more capitalistic process under our generalized calculations is now more expensive, as we should desire. It is now Böhm-Bawerk's measure which is in trouble, for it is his *more* capitalistic process that is *cheaper* at *very high* rates of interest. This is a most perverse result which the generalized measure has permitted us to escape, but not without creating its own ambiguities in the middle ranges. Neither the original nor the generalized measure of roundaboutness, therefore, can be recommended.

Notice the important result that Process 1 is favored at low rates of interest, Process 2 at high rates of interest, but Process 1 is again favored at very high rates of interest. A 'reswitching' of techniques occurs as the interest rate rises, which we have shown cannot be explained on the basis of Böhm-Bawerk's measure of capitalization or its generalization to accommodate value and compound interest phenomena.

The Technological Superiority of Present Goods

Böhm-Bawerk's Third Ground is subject to criticism because it is not truly independent of the First and Second Grounds, or the time preference aspects in his theory. If it be assumed that increasing the construction period in a process ultimately yields a zero marginal value product of time, and if the community has neutral λ time preference and time preference proper, were it to be offered a gift of resources in $t = 0$ or $t = 1$ it would be indifferent to their receipt in either period. With neutral time preference the community would invest the additional resources for the period of construction whose marginal value product is zero, so that the consumption goods would accrue h periods from the investment of factor services. Whether this period is h or $h + 1$ is of no consequence to the community for the gifts described above, and if one of the two were preferred, time preference must be operating contrary to assumption.

Thus, Böhm-Bawerk's attempt to eliminate the direct effect of capital goods' productivity from his analysis was a basic failure: not the technological superiority of present goods but the productivity of roundabout processes is the fundamental technological basis, and it can only contribute to the result that *value* can be created by longer methods of production, and therefore capital goods.

Positive Time Preference

Böhm-Bawerk's belief that human beings reveal a positive time preference has also been the subject of a great deal of debate. In its clearest statement, the question is whether, given a stationary flow in which the consumer looks forward to a steady stream of consumption over time, if he were to be offered an increase in consumption today or an equivalent increase tomorrow, he would prefer the former.

Joseph A. Schumpeter – an Austrian in other respects – declined to accept this proposition. In terms of Figure 14.2, he argued that the slopes of the indifference curves along the 45° line were zero, so that if the economy reached a stationary flow it would be with a null interest rate and with every investment opportunity that paid a positive return used up.

But at least conceptually this question is an empirical one, not a priori. Does the typical consumer reveal such a preference for present goods over future or not? In the real world where technological change and taste changes upset the status quo and prevent the economy from approaching the stationary state, the question cannot be answered by observation of the tendency of the interest rate.

A Final Word

A last resemblance to Karl Marx is that in every detail of his analytical mosaic Böhm-Bawerk evoked able criticism: in his value theory, his attempt to integrate capital and interest theory with it, his imputation theory, his concept of time preference, his treatment of the technological contribution of capital goods, and his correlation of time structure with the amount of capital, his work has invited assault.

Moreover, modern theory has turned away from the very problems he raised, asserting them to be extraneous to the central economic issues. But here, surely, the glib operationalism of present capital theory indulges an unsatisfying shallowness: the nature of capital and the explanation of the rate of return on it are neither false nor uninteresting questions, nor is the fact that Böhm-Bawerk and others failed to provide final answers logical evidence that in some sense the questions are badly put. Böhm-Bawerk's total structure is admittedly not a complete answer to these conundrums of the economic workshop, but neither are they false answers to false problems.

And after all the point-by-point criticism of Böhm-Bawerk's analytical structure, frequently the student finds that the kernel of the performance still makes a great deal of sense to him. Time *is* intimately involved in the concept of capital and in the receipt of interest; time *is* at once a productive element and a psychological deterrent; and the flow of consumption goods currently available *is* a limitation on the amount of saving and investment that can be undertaken in a healthy economy. Perhaps the attempt to measure periods of investment in simplistic ways is to oversimplify – but it is a striving to give definition to a phantom that will not down, a ghost that haunts the structure of economic theory. To treat capital goods wholly as factor services on a par with other resources is to neglect the fact that they are producible in a way that coal or petroleum or even agricultural land is not, and that a return to such goods cannot be viewed as merely a form of rent. The Böhm-Bawerkian vision remains in a meta-analytical sense a tempting one.

Perhaps we can do no better in closing on this note than to quote Schumpeter, who, as we have seen, had reservations about the theories. After an extensive presentation of his criticisms, Schumpeter wrote: 'But . . . if we recognize all this and if we introduce all these corrections, what is left of Böhm-Bawerk's capital theory and in particular of this period of production? Well, nothing is left of them except the essential idea. And this keeps on proving its vitality by every piece of criticism and every piece of constructive work it evokes.'

Select Bibliography

[1] Arvidsson, Guy, 'On the Reasons for a Rate of Interest', *International Economic Papers*, 6 (1956), pp. 23–33.
[2] Böhm-Bawerk, Eugen von, *Further Essays on Capital and Interest* (1909, 1912), Libertarian Press, South Holland, Ill. (1959).
[3] Böhm-Bawerk, Eugen von, *History and Critique of Interest Theories* (1884), Libertarian Press, South Holland, Ill. (1959).
[4] Böhm-Bawerk, Eugen von, *Karl Marx and the Close of His System* (1896), A. M. Kelley, New York (1949).
[5] Böhm-Bawerk, Eugen von, *Positive Theory of Capital* (1889), Libertarian Press, South Holland, Ill. (1959).
[6] Böhm-Bawerk, Eugen von, *Shorter Classics of Böhm-Bawerk*, Libertarian Press, South Holland, Ill. (1962).
[7] Böhm-Bawerk, Eugen von, 'The Positive Theory of Capital and Its Critics', *Quarterly Journal of Economics*, 9 (1894), pp. 113–31, 235–56; 10 (1895), pp. 121–55.
[8] Boulding, Kenneth, 'Professor Knight's Capital Theory: A Note in Reply', *Quarterly Journal of Economics*, 50 (1935), pp. 524–31.
[9] Boulding, Kenneth, 'Time and Investment', *Economica*, NS 3 (1936), pp. 196–220.
[10] Clark, John Bates, 'The Origin of Interest', *Quarterly Journal of Economics*, 9 (1895), pp. 257–78.
[11] Conard, Joseph W., *An Introduction to the Theory of Interest*, University of California Press, Berkeley (1963).
[12] Dorfman, Robert, 'A Graphical Exposition of Böhm-Bawerk's Interest Theory', *Review of Economic Studies*, 26 (1959), pp. 153–8.
[13] Dorfman, Robert, 'Waiting and the Period of Production', *Quarterly Journal of Economics*, 73 (1959), pp. 351–72.
[14] Fisher, Irving, *The Rate of Interest,* Macmillan, New York (1907), ch. 4.
[15] Fisher, Irving, *The Theory of Interest*, Macmillan, New York (1930).
[16] Gifford, C. H. P., 'The Concept of the Period of Production', *Economic Journal*, 43 (1933), pp. 611–18.
[17] Hayek, Friedrich A., *The Pure Theory of Capital*, University of Chicago Press, Chicago (1941).
[18] Hayek, Friedrich A., 'Mythology of Capital', pp. 355–83 in W. Fellner and B. F. Haley (eds), *Readings in the Theory of Income Distribution*, Blakiston, Philadelphia (1946).
[19] Hicks, J. R., *Value and Capital*, Oxford University Press, London, 2nd edn (1946), ch. 17.
[20] Hirshleifer, J., 'A Note on the Böhm-Bawerk-Wicksell Theory of

Interest', *Review of Economic Studies*, 81 (1967), pp. 191–9.
[21] Kaldor, Nicholas, 'Annual Survey of Economic Theory: The Recent Controversies on the Theory of Capital', *Econometrica*, 5 (1937), pp. 201–33.
[22] Kaldor, Nicholas, 'On the Theory of Capital: A Rejoinder to Professor Knight', *Econometrica*, 6 (1938), pp. 163–76.
[23] Knight, Frank H., 'Capitalistic Production, Time, and the Rate of Return', pp. 327–42, in *Economic Essays in Honor of Gustav Cassel*, Allen & Unwin, London (1933).
[24] Knight, Frank H., 'On the Theory of Capital: In Reply to Mr. Kaldor', *Econometrica*, 6 (1938), pp. 63–82.
[25] Knight, Frank H., 'The Quantity of Capital and the Rate of Interest', *Journal of Political Economy*, 44 (1936), pp. 433–63, 612–42.
[26] Knight, Frank H., 'The Theory of Investment Once More: Mr. Boulding and the Austrians', *Quarterly Journal of Economics*, 50 (1935), pp. 36–67.
[27] Kuenne, Robert E., *The Theory of General Economic Equilibrium*, Princeton University Press, Princeton (1963), ch. 4.
[28] Kuenne, Robert E., 'The Stationary State and the Technological Superiority of Present Goods', *Quarterly Journal of Economics*, 76 (1962), pp. 648–52. Reprinted as Chapter 15 in this volume.
[29] Kuenne, Robert E., 'The Technological Superiority of Present Goods', *Zeitschrift für Nationalökonomie*, 22 (1962), pp. 271–7. Reprinted as Chapter 16 in this volume.
[30] Malinvaud, Edmond, 'Capital Accumulation and Efficient Allocation of Resources', *Econometrica*, 21 (1953), pp 233–68.
[31] Neuberger, Egon, 'Waiting and the Period of Production: Comment', *Quarterly Journal of Economics*, 74 (1960), pp. 150–3.
[32] Samuelson, Paul A. 'A Summing Up', *Quarterly Journal of Economics*, 80 (1966), pp. 568–83.
[33] Schumpeter, Joseph A., *History of Economic Analysis*, Oxford University Press, New York (1954).
[34] Stigler, George, *Production and Distribution Theories*, Macmillan, New York (1941).
[35] Taussig, Frank, 'Capital, Interest, and Diminishing Returns', *Quarterly Journal of Economics*, 22 (1908), pp. 333–63.
[36] Wicksell, Knut, *Lectures on Political Economy*, Routledge, London (1934), I, pp. 144–95.
[37] Wicksell, Knut, *Value, Capital, and Rent*, Rinehart, New York (1954).

15 The Stationary State and the Technological Superiority of Present Goods*

In an illuminating article on Austrian capital theory published recently in this *Journal*[1] Professor Robert Dorfman repeats with seeming approval Böhm-Bawerk's 'Third Ground' for the existence of interest: an alleged technological superiority of present over future goods. He asserts, even more strongly than Böhm-Bawerk, the independence of this Third Ground from the prior two, which we shall categorize jointly as 'time preference'. Dorfman says, 'The validity of this dictum [the technological superiority of present goods] does not depend on time-preference or the perspective underestimation of future wants or any such subjective and perhaps irrational consideration.' He continues: 'It depends on the ability of present (consumer goods) to finance an accumulation of (capital goods) that will enhance the flow of future (consumer goods).'[2] It is not immediately apparent that future consumer goods do not possess this same capacity or that they possess it consistently to a lesser degree.

Dorfman has not in fact adopted the Böhm-Bawerkian outlook on the technological superiority of present goods. Indeed, implicitly he has redefined the concept to make it formally correct. This is a contribution to Austrian doctrine which should not go unmarked, and, because it is buried in the counterpoint of Dorfman's article, a short note is warranted to point out this achievement. Moreover, it points the way to a reconstruction of Böhm-Bawerk's own concept which will make it less vulnerable to criticism.

The interdependence of time preference and the productivity aspects of capital goods springs from two distinct sources. The first of

* Published originally in *Quarterly Journal of Economics*, 76(1962), pp. 648–52, and reproduced with permission.

these, pointed up clearly by Wicksell and discussed more recently by Knight, Metzler and Lerner,[3] is the essential fact that capital and interest are in their nature value concepts ultimately. In the simplest case, a net return to capital in a physical sense might well be nullified in a value sense given peculiarities of consumers' preferences. This first type of interdependence remains to haunt the capital theorist as ineradicable, but fortunately its thorniest implications may be abstracted from by use of the stationary state. It is this characteristic of capital which explains most of the fascination of the capital theorist with that construct. The determination of capital magnitudes and the theory of interest are, then, in a most fundamental sense, problems in general equilibrium theory, and any partial approach to them must lead to difficulties which can be dispelled only by placing the partial solution in a general equilibrium framework. We do not intend to deal with this first type of interdependence of consumer preferences and productivity considerations, but shall abstract from them.

The second source of the interdependence between these two elements of capital theory is a quite different one, and was injected by Böhm-Bawerk in his classic presentation of capital theory. It is much more subtle and rarely understood. Although it was the subject of a controversy between such capital theorists as Böhm-Bawerk, Irving Fisher, Bortkiewicz and Landry in the early part of this century, the issues were never clearly defined or resolved, and the dispute and its implications remain largely unknown.[4]

Briefly, the problem may be stated in the following form, which follows Böhm-Bawerk's original presentation. Assume as Böhm-Bawerk did that in some way it is possible to impute to the input of primary factor services of a given year the marginal value contribution of time acting upon the initial investment over a series of years. The classic case, of course, is the point-input–point-output example of the maturing of wine. If we assume that one unit of labor's worth of grape juice is laid down in Year 0 and allowed to age with no expenditure of other scarce resources upon it, and if we assume that the value of the wine in every year over a domain of years is given to us, we shall have the Böhm-Bawerkian example at its simplest.

In his numerical examples Böhm-Bawerk assumed that the function depicting the value of the wine over the span of years was strictly monotone-increasing: i.e. the marginal value product of time was always positive. This being true, at any Year n in the future, grape juice laid down in Year 0 was always more valuable than grape juice laid down in (say) Year 1. Since the curve never approached a maximum value, at least in the relevant time domain, technological superiority of goods in Year 0 over goods in Year 1 seemed evident (the very definition of the table providing the abstraction from the first type of interdependence discussed above).

Böhm-Bawerk's examples were challenged by Bortkiewicz, Landry and, most particularly, by Irving Fisher. They pointed out that if the curve of value in function of time approached a maximum, Böhm-Bawerk's technological superiority of present goods vanished. For the case in point, if wine became vinegar at the end of six years then at the end of Year 4 presumably grape juice will have attained its maximum value. The marginal value product of Year 5 would then be negative. This holding true, grape juice laid down in Year 0 and allowed to mature in Year 4 would have a value of (say) $2.50. But grape juice laid down in Year 1 and allowed to rest until Year 5 would also have a value of $2.50. Treating resources in Year 0 as present resources, and resources in Year 1 as future resources, we see that present resources would have no technological superiority over future resources since invested in processes of the maximum rational length they yield the same value-product. Faced with the choice between the consumption of $2.50 of wine in Year 4 or in Year 5, the decision-maker would prefer one to the other only on nontechnological grounds: i.e. his time preferences for consumption in different periods.

In response to Fisher's argument Böhm-Bawerk admitted that the curve of value in function of time, although it might not turn down, did in fact approach a finite limit.[5] However, he argued, the point in time at which this limit was approached was so distant normally as not to be relevant, since the scarcity of capital available to society would always assure that roundabout processes would never be taken that far.[6] Capital scarcity, therefore, gave to Böhm-Bawerk a positive marginal value product of time at the equilibrium period of investment. In the case of wine, the decision-maker would always stop the aging process well before the six-year limit, so that in fact that limit was not relevant.

This argument, however, leads merely to a revised statement of the original objection. Where such a limit exists and is approached we have seen that the decision-maker will favor a given amount of labor service in Year 0 over Year 1 only if he has a positive (marginal) time preference, assuming that the wine will be consumed in some time pattern rather than sold for future investment. 'Technological superiority' of present over future goods does not exist although the productivity of roundabout processes does. Böhm-Bawerk's interpretation preserves the 'technological superiority' of grape juice laid down in Year 0 over that in Year 1, since in Year n (always before the limit of the productivity curve is reached) wine laid down in the former year will always be more valuable than that laid down in Year 1. But the major reason why society does not go to the limit of such curves – capital scarcity – reflects in largest part consumers' time preferences. Thus, to an uncertain degree, Böhm-Bawerk's interpretation introduces the same type of interdependence of time preference and technological superiority objected to by his critics.

It follows from the arguments of both sides that technological superiority of present over future goods can be defined clearly only after every care has been taken to define and introduce into the decision-making 'neutral time preference'. It must be injected into the analysis in order that it can be excluded. We shall not take the space to do this, but suffice it to say that some plausible definitions of neutral time preference rob Böhm-Bawerk of his technological superiority of present goods even under his assumption.[7]

In Dorfman's example of a gift of net investible resources to an underdeveloped nation one straightforward interpretation of technological superiority may be objected to on the Fisher-Bortkiewicz-Landry grounds. If the net investment is used to raise income from a stationary level of Y to a stationary level of $Y + \Delta Y$ beginning in Year n, and if only the annual level of income is considered, the investible resources in Year 0 are no more desirable than those of Year 1, unless the community possesses positive time preference. The availability of the resources in Year 0 yields the higher income in Year n, while the availability of the resources in Year 1 yields the higher income in Year $n + 1$. But to argue in this way is merely to project the Böhm-Bawerk case of immediate terminal consumption of goods into a case where the goods are not totally consumed at the end of the initial period of investment. By the assumption of stationarity the stream of additional consumption repeats itself indefinitely into the future. Stationarity was not an assumption of the Böhm-Bawerk-Fisher utility analysis, of course, and terminal consumption of (for example) the wine was taken for granted.

Dorfman's definition of technological superiority implicitly assumes a perpetuity of ΔY, in which case in Year 0 a productivity curve which never approaches a limit in function of time is attained. In some Year $x \geqslant n + 1$ it will always prove more advantageous to the community to have begun the approach to a new stationary equilibrium earlier. In so interpreting the concept, Dorfman assures himself the analogue of Böhm-Bawerk's effectively unbounded curve of value in function of time, but needs only to assume nonsatiety of the consumer as a restriction on preferences.

It also points the way in which Böhm-Bawerk should have shaped his own reasoning to avoid the criticism leveled against it. Had he merely allowed the *possibility* of reinvestment of matured goods, fully or partially, rather than insisting upon their consumption, it may be shown for the more plausible definitions of neutral time preference that present goods will have a technological superiority in the sense that more goods will emerge in a cumulation of the stream of annual product over time from the former than from the latter.[8]

But there is a deeper truth that Dorfman as well as others who have followed Böhm-Bawerk in interpreting Austrian theory have sensed. The productivity of capital springs from the roundaboutness of indirect processes: this is the fundamental productivity element in the theories of interest and capital. The 'technological superiority of present goods over future' is an idiosyncratic construction of a theorist who was bound and determined to reject the concept of a simple productivity for unfinished consumer goods. To state the concept in such form as to make it formally correct is to reveal its triviality: if a series of annual payments is to end in Year x, it is preferable to begin their receipt in Year 0 rather than Year 1 only if Year x is indefinitely distant. However, under these conditions, the additional payment is negligible in relation to the stream of payments in perpetuity. The greater fruitfulness of roundabout production introduces into capital theory a simpler and primordial technological basis for that theory and for the theory of interest.

Notes

1. 'Waiting and the Period of Production', [3]. My appreciation is due Professor Haberler for extended criticism of earlier drafts of this paper and valuable suggestions to clarify points in the present one.
2. [3], 359. We have substituted the words in parentheses for the expressions used by Dorfman and having meaning only in the specific context of his presentation.
3. K. Wicksell [11], I, p. 180; F. Knight [6]; L. Metzler [10]; and A. Lerner [9].

4. L. von Bortkiewicz, [2]; I. Fisher, [4], pp. 53–74, and [5], pp. 473–85; and A. Landry, [8], pp. 197ff.
5. [1], pp. 215, n. 27; 170, and 176.
6. [1], p. 162.
7. This point is discussed at length in my article [7].
8. This point is discussed further in the reference in note 7.

References

[1] Böhm-Bawerk, Eugen von, *Capital and Interest*, III, Libertarian Press, South Holland, Ill. (1959).

[2] Bortkiewicz, L. von, 'Der Kardinalfehler der Böhm-Bawerkischen Zinstheorie', *Jahrbuch für Gesetzbegung Verwaltung und Volkswirtschaft, 1906*, Duncker & Humblatt, Leipzig (1906), pp. 61–90.

[3] Dorfman, R., 'Waiting and the Period of Production', *Quarterly Journal of Economics*, 73 (1959), pp. 351–72.

[4] Fisher, I., *The Rate of Interest*, Macmillan, New York (1907).

[5] Fisher, I., *The Theory of Interest*, Macmillan, New York (1930).

[6] Knight, F. H. 'The Quantity of Capital and the Rate of Interest, I,' *Journal of Political Economy*, 44 (1936), pp. 433–63.

[7] Kuenne, R., 'The Technological Superiority of Present Goods', *Zeitschrift für Nationalökonomie*, 22 (1962), pp. 271–7. Reprinted as Chapter 16 of this volume.

[8] Landry, A., *L'Interêt du Capital*, V. Giard & E. Brière, Paris (1904).

[9] Lerner, A., 'On the Marginal Product of Capital and the Marginal Efficiency of Investment', *Journal of Political Economy*, 61 (1953), pp. 1–14.

[10] Metzler, L., 'The Rate of Interest and the Marginal Product of Capital', *Journal of Political Economy*, 58 (1950), pp. 289–306.

[11] Wicksell, K., *Lectures on Political Economy*, Macmillan, New York (1934).

16 The Technological Superiority of Present Goods*

I THE PROBLEM

It is a notable feature of the interest theories of Böhm-Bawerk's successors that they did not adopt an integral tenet of his beliefs: the technical superiority of present goods. Workers such as Wicksell, Schumpeter and Hayek chose instead to ground the technological basis for the existence of interest in the greater 'productivity' of roundabout processes or in innovation. There exists a temptation, however, to believe either (*1*) that the greater productivity of roundabout processes is identical with the proposition that present goods are superior technically to future goods, or (*2*) that while these two are not identical propositions the technical superiority of present goods may be shown to follow in simple fashion from the roundaboutness postulate conjoined with certain assumptions which are independent of time preference.

The first of these temptations claimed Böhm-Bawerk himself as a victim at times. For example, in answering Fisher he wrote:

> When a timber dealer simultaneously clears a 100-year-old and an 80-year-old forest, he will value the older forest higher simply because it actually yields more timber. And when we drink 'old' wine and young wine, we value the older wine higher simply because it is superior in quality. Is this or is this not a real technical superiority?[1]

In this passage Böhm-Bawerk identifies falsely (*1*) the preferability of the results available today of longer processes to those of shorter

* Published originally in *Zeitschrift für Nationalökonomie*, 22 (1962), pp. 271–7, and reproduced with permission of Springer-Verlag, Vienna.

processes available today, and (*2*) preferences between the results of an equally long process available in Year n or in Year $n + 1$. Also, in his earlier works, there is some evidence that Fisher felt that since he had revealed the falsity of the technological superiority dictum as presented by Böhm-Bawerk, he had destroyed the entire productivity base of Böhm-Bawerk's theory.

The independence of the doctrine from time preference assumptions has been explicitly challenged by Bortkiewicz, Fisher, and Landry,[2] and Böhm-Bawerk's response to the first two authors' criticisms forms the famous Excursus 12 in his *Further Essays on Capital and Interest*. In reading this famous debate one has the feeling that the issues were left hanging in midair, with Böhm-Bawerk yielding the major ground but unconvinced that in so doing he had surrendered the field. This paper seeks, therefore, a definitive solution to the problem, or at the least to secure many of the loose ends left dangling fifty years ago.

In Section II we shall seek to define exactly the problem in hand and to give rigorous expression to the concept of 'neutral' time preference. In Section III we shall attempt the statement of a solution to the conditions under which present goods may be said to have a technical superiority for a point–input, point–output process. A brief summary and conclusion is provided in Section IV.

II DEFINITIONS

We shall say that present goods are technologically superior to future goods if, after abstracting from time preference in manners to be discussed shortly, a nation decides in Year *0* that it prefers a gift in Year *0* to an equivalent gift in Year *1*.[3]

In order to neutralize the impact of time preference on the discussion of the technological aspects of capital and investment, it is necessary to make explicit alternative definitions of neutral time preference. It would seem that three of these might qualify as implying a type of neutrality:

Definition 1 The time of consumption is completely ignored in the utility function, as is the possibility of satiation, more goods at any time always being preferred to fewer goods at any time. This might be the outlook adopted by an underdeveloped nation's policy-makers were they to put the maximum social emphasis upon growth in the uses of an *increment* of investible resources with which we shall deal below. One difficulty with this definition is that its pattern of investment may be approximated in given situations by patterns of *negative* time preference with the discount factors for future goods falling continuously through time. Thus, what appears to be neutral time preference has a close kinship with negative time preference.

Definition 2 Let us assume that it is possible to approximate any pattern of social time preference we might wish to deal with by the discount factor $d_t = (1 + \theta)^t$, where t is the time period and θ the psychological rate of time discount. Then, if the preference function for the alternative receipts of consumer goods at the different time periods, $t = 1, \ldots s$, which could arise from the increment of investment, is defined as $U = U(X_1, \ldots X_s)$, where $U(X_i) = U(X_j)$ when $X_i = X_j$, for all i, j, we may consider the society to maximize its welfare by choosing a set of annual consumptions X_t. We may regard time preference as neutral if $\theta = 0$ for $0 \leqq t \leqq n$, where n may approach infinity. That is, if the society attempts to maximize its satisfaction from the investment gift over a possibly infinite time period, we shall consider time preference to be neutral in the second sense. This definition of neutrality seems to be 'purest' in that its affinities with negative or positive time preference are the least of the three alternatives. This is not to say that it is the most 'useful', 'realistic', or 'natural'.

Definition 3 Under the same conditions as those of Definition 2 we shall merely recognize the possibility that the society may choose to maximize satisfactions over a finite time period, so that $\theta = 0$ for $0 \leqq t \leqq m$, where m is finite, and θ approaches infinity for $t > m$. In this definition, of course, we may view the finite cutoff point as being dictated by a break in the regime of neutral time preference for the institution of a regime of absolute positive time preference. In this sense it resembles Definition 1 in its non-purist nature.

III BÖHM-BAWERK'S ANALYSIS

Suppose an underdeveloped nation were offered either (*1*) a unit of wheat at the beginning of Year *0* or (*2*) a unit of wheat at the start of Year *1*. On the basis of technological superiority alone, will the nation prefer (*1*) to (*2*), (*2*) to (*1*), or be indifferent between them? If it is assumed that the unit of wheat will purchase a unit of labor which, working with free land will, in one year's time, produce a given quantity of capital goods, we have an elementary capital problem. To be concrete, let us assume further that these capital goods, after maintenance and replacement expenditures, if any, can produce an annual net output of (say) 0.1 unit of wheat indefinitely.

Suppose the nation had a unit of labor in Year *0* and faced the alternatives of using it to produce wheat for consumption in Year *0*, or Year *1*, or Year *2*, and so forth. If the country elected to use the labor in processes whose investment period was greater than one year, it would have to add labor in future years to that invested in Year *0*, but we shall imagine that it is always possible to impute to the Year *0* labor its share of the ultimate product.

In Böhm-Bawerk's presentation[4] the amount of product imputed to a unit of Year *0*'s labor invested for increasing lengths of time rose without limit within the time span illustrated in his tables.[5] This being the case, at whatever year one wishes to halt for consideration (say Year *12*), the product imputed to a unit of labor of Year *0* is always larger than that imputed to a unit of labor of Year *1* for a period of investment ending in Year *12*. Consequently, he argues, present goods are technologically superior to future goods.

A fundamental assumption of Böhm-Bawerk's analysis was that of 'capital scarcity'. By virtue of it he obtained an assurance that the marginal product of time at the point chosen for the length of investment period would be positive. Böhm-Bawerk argued that the case of technological superiority of present goods held only for those realistic cases where the community's stock of present goods prevented it from approaching the limit of the curve.[6]

A difficulty with this assumption is that it may be viewed as introducing, to a debatable degree, a non-neutral time preference, since one of the major reasons why sufficient present goods do not exist to finance such investment periods is found in the existence of positive time preference, as Böhm-Bawerk himself agrees. Further, while such an assumption may be acceptable when the totality of investment is under consideration, it hardly seems appropriate to the present example where only a small portion of the total labor supply is in question.

We shall therefore generalize the Böhm-Bawerk case in the following way. Suppose the underdeveloped nation were told by a wealthy nation of the latter's willingness to underwrite the use of a unit of labor in a process of any length the former nation might wish to undertake. That is, if it chose to invest a unit of Year *0* labor for five years, the underdeveloped nation would receive all the wheat necessary to finance the total labor necessary to sustain such a period of investment. However, at the end of the period, the lending nation would reclaim all of the product except the amount imputed to one unit of labor invested for the whole of the period. We should therefore be dealing with a 'quasi-point input–point output' case, and Böhm-Bawerk's tables could be considered to be the net returns to the underdeveloped nation from the investment of one unit of labor for variable lengths of time.

It is assumed that this offer is made in Year *0* and that the underdeveloped nation is forced to make an irrevocable decision in that year concerning the amount of the gift it will consume and the amount it will invest, as well as the length of the period of investment. Further, we shall consider two relevant circumstances: (*1*) whether the returns to a unit of labor through time rise without limit or approach a limit, and (*2*) whether the underdeveloped nation is constrained to consume the wheat at the end of the investment period or may commit itself, in Year *0*, to reinvest it in a new process at the end of the initial and subsequent investment periods.

For each of the four cases obtained by pairing these two sets of circumstances, and for each of the definitions of neutral time preference, would the recipient nation prefer the initial delivery of wheat to begin in Year *0* rather than in Year *1*, or would it be indifferent between these alternatives? In the first case we shall speak of present (Year *0*) goods being technologically superior to future (Year *1*) goods; in the second case no such technological superiority exists. Let us consider each case in turn.

Case 1 Returns to Time Rise Without Limit, Enforced Terminal Consumption

Definition 1 Goods in Year *0* would be technologically superior to goods in Year *1*, since all of the gift would be invested in an indefinitely long time period. The amount of the goods available from an investment in the earlier year would always be larger and therefore preferable to the amounts available from investment begun in the later year.

Definition 2 The investment period would be terminated in Year p when the returns to the units of labor invested would satiate the community and the marginal satisfaction derived from a lengthening of the period to Year $p + 1$ would be zero. In this case the marginal productivity of time would still be positive. Further, no technological superiority of present goods would exist since the same amount of wheat would be obtained from the process, whether begun in Year *0* or Year *1*, the only difference being the receipt of the product in Year p or Year $p + 1$ – of no importance to the community.

Definition 3 In this case the investment period will terminate in Year m if started in Year *0*, the fruition year being dictated by the 'cutoff' point in the community's pattern of discount rates. Were the process to be started in Year *1* the cutoff point would be Year $m + 1$. But, in Year *0*, when the offer is made, the community prefers the first option, since goods receivable in Year $m + 1$ possess negligible utility for the society. Therefore, the logic of our framework leads us to declare present goods 'technologically superior' to future goods, but no doubt will leave many disturbed about the true neutrality of time preference in obtaining this result.

Case 2 Returns to Time Rise to an Upper Limit, Enforced Terminal Consumption

Definition 1 For an implicit assumption of this definition of time preference neutrality, Bortkiewicz and Fisher demonstrated that present goods would not be superior to future goods, since the investment period would terminate at Year n when the marginal product of time was zero. The community would be indifferent to the receipt of the same quantity of goods in Year n or Year $n + 1$. Böhm-Bawerk admits his critics are correct but denies the relevance of the case by asserting that Definition 3 is alone valid and, moreover, that under this definition the cutoff point will always occur before Year n, the year of zero marginal product.[7]

Definition 2 The terminal year will occur at the earlier of two possibilities: Year n or the year at which satiation occurs. In either case present goods will not possess a technological superiority.

Definition 3 In this case the terminal Year m will be greater than, equal to, or less than $n + 1$, the number of years' investment before the zero marginal product point is reached. If $n + 1 < m$, then goods receivable in Years *0* through $m - n + 1$ are perfect substitutes, but goods received thereafter are inferior 'technologically'. If $n + 1 = m$ or if $n + 1 > m$, goods in Year *0* will be technologically superior to those available in Year *1*, in the first case for the same reason as in Case 1, Definition 3, and in the second because fewer goods will be available in the cutoff year.

Böhm-Bawerk's examples belong to Case 2 and assume time preference to be neutral in the sense of Definition 3 when $n > m$. Consider the following statement:

> Therefore, it certainly was neither 'arbitrary' nor 'unsuitable for the problem', but rather required by its very nature, that I mentioned the case in question in my suppositions and that in the unavoidable 'interruption' of my series I assumed the border of the higher productivity zone to lie beyond and not before my point of interruption. For in reality the period of time for whose wants we plan to provide through the use of our means of production and which can exert an influence on our valuations of producers goods never and nowhere extends beyond the zone of higher productivity of longer roundabout methods. We will observe that in economic calculations we never make investments for the sake of wants that lie beyond this zone.[8]

Case 3 Returns to Time Rise Without Limit, Reinvestment Possible

In those cases where the nation may commit itself to reinvest the net returns at the end of investment periods certain modifications in these conclusions occur.

Definition 1 No change from Case 1.

Definition 2 A succession of investment periods will be selected such that in Year t, for t sufficiently large, a satiation output is produced with sufficient surplus to produce, in a one-year investment period, a satiation output in Year $t + 1$ and a similar surplus. Since goods in Year *0* will allow this state to be reached a year earlier than goods in Year *1*, and therefore will produce a greater flow of goods over any time period selected, they will be superior technologically to goods available in Year *1*.

Definition 3 A pattern of investment and consumption will be selected to maximize satisfaction over the period from Year *0* to Year m. Abstracting from satiation, we conclude that goods in Year *0* will be preferred to goods in Year *1* by virtue of the community's discounting in Year *0* of goods receivable in Year $m + 1$.

Case 4 Returns to Time Rise to an Upper Limit, Reinvestment Possible

Definition 1 If returns to successive investment were constant, the society would select the optimal investment period, reinvest the gross proceeds upon its completion in an investment period of similar length, and enjoy an indefinite rise in product. However, the ultimate limitations on non-producible factors will lead ultimately to a point where reinvestment in a period of any length will yield no enhancement of product. This is the point of maximum product and it will be produced and maintained indefinitely. Since a beginning in Year *0* would allow its attainment one year earlier than one in Year *1*, goods in the earlier year will be superior technologically.

Definition 2 No change from Case 3.

Definition 3 No change from Case 3.

IV SUMMARY

The existence or nonexistence of a 'technological superiority of present goods' under the three definitions of neutral time preference suggested is more complicated than Böhm-Bawerk, Bortkiewicz and Fisher imply by their analyses.

All of this principally remained with the cases of enforced terminal consumption, all three ultimately agreeing upon the reality of the bounded-returns-to-time condition. Thus, Case 2 was relevant to their analyses, and disagreement centered upon the acceptance of Definition 1 by Bortkiewicz and Fisher and of Definition 3 with $n + 1 > m$ by Böhm-Bawerk. Both of these definitions of neutral time preference are tinged with non-neutral impurities, but perhaps these are strongest in the latter case, implying as it does a cutoff point in the planning period. Be it noted that if the 'purest' definition of time preference neutrality – Definition 2 – be adopted for Case 2, Fisher and Bortkiewicz would still be correct in maintaining the non-superiority of present goods.

On the other hand, if Böhm-Bawerk had chosen to extend the analysis to the cases of reinvestment, his position would be correct for all definitions of neutral time preference whether or not time were limitlessly productive.[9]

Notes

1. E. von Böhm-Bawerk [1], p. 178.
2. L. von Bortkiewicz [3]; I. Fisher [5], pp. 53–74, and [6], pp. 473–85; and A. Landry [8], pp. 197 ff.
3. This case is suggested by the recent article of R. Dorfman [4].
4. E. von Böhm-Bawerk [2], pp. 260–8.
5. While he denies the practical existence of a point on this curve where it turns down (the marginal product of time becomes zero and then negative), Böhm-Bawerk does admit that the curve approaches a finite limit. For example, in somewhat ambiguous language, he says: 'Fisher remarks correctly that "the limited earth can never produce unlimited products", and that, therefore, these do not grow to an infinite maximum by an infinite process of increasing productivity, but can only infinitely approach an ultimate maximum. The latter situation would be fully compatible with my opinion that higher productivity shows a successive tendency to lessen. However, I have neither an inclination nor a theoretical interest in asserting an "infinite" continuation of my law in this physical and mathematical sense' (p. 215, [1], note 27. See also pp. 170, 176).
6. [1], p. 162. Hence the lack of 'theoretical interest' in whether or not the period-of-investment curve approaches a limit recorded in the quotation of note 5.
7. [1], pp. 155–7.
8. [1], p. 156. See also G. Haberler [7].
9. Dorfman [4] has taken this step implicitly in his analysis of technological superiority.

References

[1] Böhm-Bawerk, Eugen von, *Capital and Interest*, III, Libertarian Press, South Holland, Ill. (1959).

[2] Böhm-Bawerk, Eugen von, *The Positive Theory of Capital*, Stechert, New York (1923).

[3] Bortkiewicz, L. von, 'Der Kardinalfehler der Böhm-Bawerkischen Zinstheorie', *Jahrbuch für Gesetzbegung Verwaltung und Volkswirtschaft, 1906*, Duncker & Humblatt, Leipzig (1906), pp. 61–90.

[4] Dorfman, R., 'Waiting and the Period of Production', *Quarterly Journal of Economics*, 73 (1959), pp. 351–72.

[5] Fisher, I., *The Rate of Interest*, Macmillan, New York (1907).

[6] Fisher, I., *The Theory of Interest*, Macmillan, New York (1930).

[7] Haberler, G., 'Irving Fisher's *Theory of Interest*', *Quarterly Journal of Economics*, 45 (1931), pp. 510–11.

[8] Landry, A., *L'Intérêt du Capital*, V. Giard & E. Brière, Paris (1904).

Part IV

Money in a General Equilibrium Framework

Introduction to Part IV

Two discussions have occurred over the last forty years concerning the presence of money in static general equilibrium systems that incorporate insufficient uncertainty to give rise to precautionary or speculative motives for holding it. The first arises over preconditions for the existence of money in such frameworks, and the second questions the manner of its incorporation and the consequences for the determination of money prices. For the theorist, both issues are important in that they challenged classical interpretations of the functions of money and the manners in which demands for it were determined; they inspired in the field of monetary theory new investigations of its foundations, very much as Austrian theory did for capital and interest theory.

Part IV deals at length with both issues, with an attempt to clarify the first and a denial of the existence of the alleged problem of the second in Walras's system, but finds an analogous problem in Keynes's model. In the course of the discussion of both topics I have explicated my interpretation of Walras's extremely complicated method of integrating money into his theory and of unifying money with capital and interest, modifying it in directions that I feel improve it. It is my contention that by distinguishing rigorously between money as a stock of assets and money services as a flow from that stock, Walras made a substantial contribution to both the foundations of monetary theory and to the means of integrating it with capital and interest theory.

Chapter 17 deals primarily with the first of these issues. Marget and Hicks, among others, argued that if a static, essentially riskless economy were functioning, money could not exist in the absence of some friction which led to transaction costs for investing over short periods. In this environment transactors could perceive with certainty the times at which they would require cash balances to finance the purchases for which cash receipts up to that time would not suffice. They would, therefore, seek to lend cash balances between those dates and, were such loan markets frictionless and costless in terms of entry and exit, all would seek to dispose of money balances and no one would desire to hold them more than instantaneously. Money, therefore, could not exist.

The notion that money would be needed at any point constitutes a retreat from the assumption of zero uncertainty, for it indicates the risk that individuals will default on the obligation to take from the social stockpile only the value of their contributions, past and present. This paper features the costs of clearing accounts among agents which would arise with this minimal amount of uncertainty, even in the stationary state, and that money may have a cost advantage as an impersonal clearing instrument that obviates record keeping. That is, I assume that there is a need to 'identify' transactions to ensure entitlement, and that this is a precondition for money's existence to prevent *potential* default. A second precondition is that the indicated cost advantages over keeping accounts, or using personal credit instruments ('bills'), or barter, make it the preferred instrument.

To these conditions must indeed be added the inability to lend or undesirability of borrowing money balances between transactions when such transactions occur continuously or between discrete points throughout a period. Marget and Hicks, among others, found these in prohibitive transactions costs, and, at one point, Patinkin introduces the rather strained assumption that times of payment and receipt are randomized. He also follows Schumpeter in accepting the possibility that stocks of cash must be held over periods without investment possibilities because of institutional arrangements as in Walras's 'fulfillment week' economy. The latter seem a more useful and realistic assumption to me, though perhaps these can be construed as a form of transaction friction. The time necessary to seek out loans, or the periods during which such markets are open, may effectively force the agent to hold stocks of money idle.

Therefore, with very little uncertainty introduced into the stationary state, concerned with the potential dishonesty of individuals, and with realistic market or institutional frictions, it is not difficult to establish the existence of money in that environment. The price of its service and hence the money interest rate will be positive. Because real assets are an alternative to holding stocks of money as assets yielding saleable services, the rate of return on real assets cannot fall below that on money. It is difficult to accept Schumpeter's contention, therefore, that the interest rate would disappear in the stationary state.

Chapter 18 deals with two criticisms of neo-classical general equilibrium theory in terms of its success or failure in integrating the monetary sector with real sectors. The first was Oskar Lange's contention that it embraced Say's Law, which asserted that demand for and supply of money were identically equal: economic agents never attempted to change their demand for money by substituting against goods and services. Excess demand for money, therefore, was identically zero, so that a change in the price of money had no impact upon its own excess demand and hence upon any real good excess demand. Any price level could rule so that absolute prices were indeterminate. The monetary and real sectors are rigidly separated and money, depending upon a price for its existence, cannot perform its exchange function.

I am sympathetic to Lange's criticism with respect to Walras's models, but it must be stated in somewhat more sophisticated manner in those systems. Walras divided his money markets into one for the exchange of cash balances and one for the exchange of money as an asset. I have attempted to reconstruct Walras's model to eliminate this fault, and that effort will be discussed in later chapters, especially Chapter 20.

Patinkin's criticism explicitly excludes Say's Law as the problem and, in the final version of his criticism, focuses upon the *tâtonnement* process of reattaining general equilibrium when the system is disequilibrated. His contention is that Walras and other neo-classicists argued as if equilibrium was first attained in the real goods markets and then in the money market, so that no feedback from one such sector to the other was recognized. Consequently, the real and monetary sectors were dichotomized, not formally or structurally as in the case of Say's Law but effectively in the re-equilibration process.

The chapter contains my defense of Walras, but indicates that because Patinkin's criticism is a more sophisticated one than Lange's, being based upon *procedural* rather than *structural* considerations, it is more difficult to establish the case. I feel that one must concede, however, that the *structure* of Walras's system does not contain the Invalid Dichotomy, so that, despite Walras's discussion of the system out of equilibrium, a valid depiction of the re-equilibration cannot display Patinkin's Invalid Dichotomy. The internal logic cannot permit it.

Chapter 19 constitutes an attempt to place Keynes's *General Theory* in the stream of neo-classical general equilibrium theory and points to a 'partial dichotomy' of his monetary and real sectors which is vital to the theorems he deduces from his theory. Keynes's distinctive contribution is the construction of a general system in which the flexibility of the money wage rate cannot eliminate an excess supply of labor, and the major questions concerning the system concern the manner in which it departs from 'classical' models to obtain this result and the credibility of that departure.

His work contains an irony, in that he 'abolished' Say's Law which, as seen in Chapter 18, dichotomizes a system containing it into monetary and real sectors, only to introduce a 'partial dichotomy' which has similar effects. That is, Keynes must block off real-balance effects from an increasing real money stock as the wage level falls from exerting impacts upon real goods sectors, and must eliminate substitution and income effects from the rising prices of securities on real goods sectors. This he does by assuming that the real-balance effect is exhausted in the securities markets and that the substitution and income effects that spring from the rise in security prices are prevented by the liquidity trap. The real and monetary sectors are dichotomized, and another assumption concerning the fixity of money wages prevents prices from continuing to fall towards zero.

I have always felt that the analysis of this paper presents the distinctive features of the Keynesian system within the mainstream of general equilibrium systems, and gets to the core of the substantive changes Keynes made in the postulates of those systems. That he substituted one dichotomy for the more stringent one of Say's Law, and that rigid money wages anchored the price level, not the unemployment level, are rather clearly brought out by the comparisons of the systems. I have felt some disappointment that the paper did not have more impact upon Keynesian analysis than it has had.

Chapter 20 presents a reconstruction of Walras's theory of money, designed to eliminate his implicit adoption of Say's Law and to integrate money clearly and unambiguously with his savings and investment sector. A major contribution in this area was his clear separation of money as an asset and money as a flow of cash-balance services. No theorist made the distinction as clearly as Walras, and, in my opinion, it is a major step in understanding the relations between money-as-flow and money-as-stock – a problem of some interest in post-Keynesian economics.

A difficulty is Walras's ambiguous and confusing treatment of money capital and its relation to real capital goods, and the relations of both of these to securities markets. Moreover, Walras's adoption of Say's Law refers to money as a stock: the agent never attempts to change holdings of money *assets* over a period, but is free to vary amounts of cash *balances* used. Nowhere does Walras discuss the possibility of agents purchasing more money assets over the period, and that urge for a 'neutral' money led him to sacrifice one of the potential advances his model permitted. The whole logic of his system, with its insistence on identifying markets in which each type of good was traded, argues for a 'market' in which the price of money as an asset – the 'price level' – is determined. That market is the obverse of every other market and money is purchased or sold in it by excess supplies and demands for other goods and services.

I have, by adding four new variables and equations in the model, allowed agents to seek to change the holdings of money stocks, thereby determining a 'price' of money and integrating money with real investment. These changes are wholly within the logic of Walras's economics, and indeed complete it. The reconstruction makes his monetary sector determinate by elimination of Say's Law, provides a distinctly different presentation of monetary theory, and escapes the Invalid Dichotomy.

Chapter 21 is a short comment on a paper by Professor E. M. Mishan concerning Say's Law and Walras's Law which, I think, makes several valid points about the usage of identities in theory and their relation to equilibrium conditions. The first concerns Mishan's implication that only equilibrium conditions in a model could qualify as behaviour assumptions, and that identities cannot reflect choice behaviour. My point is that the role of identities takes its meaning from the model in which they are employed, and that they can very well reflect agent choice. In Lange's treatment of Say's Law, in which the relation is treated as both an identity and a mode of behavior, the economic interpretation is that decision-makers act in such manners as never to seek to increase their money holdings. That certainly is a legitimate usage of a behavioral assumption, whatever one may think of its realism. A more familiar usage might be the quantity theory of money: if economic agents act behaviourally to hold a proportion k of their transactions value in money balances, then the identity MV=PT holds in the aggregate. Does the latter have no behavioral content?

The treatment of identities proposed by Mishan reveals its difficulties when he treats Walras's Law as a nonidentity, but a refutable equilibrium condition. But that condition – that in an economy with n commodities and with $n-1$ markets in equilibrium the nth must be also – is an identity and cannot be refuted. If the nth good is money, then the value of aggregate demand of the $n-1$ other goods is identically the aggregate supply of money. The nth 'market' is simply the sum of the obverses of all other markets, and if, in those other markets excess demands are zero, excess supply of money must be zero. But the identity holds true only if every amount demanded is an amount of money supplied, which implies that each consumer conforms to a budget constraint. If every consumer conforms behaviourally to a budget constraint, then this value of aggregate demand and supply holds identically, in or out of equilibrium. Walras did assume behaviourally that each consumer did so conform, and hence his 'Law' was an identity. Is it not also behavioural, resting so fundamentally upon individual choice to act nonfraudulently?

Mishan also attacks Dorfman, Samuelson and Solow as having imposed Say's Law upon their production model when they did not include money in their system. In addition, he asserts that Walras, whom he interprets as permitting agents to change their holdings of money stocks, introduces an equilibrium condition for zero excess demand for such stocks, making Say's Law an equilibrium condition eliminating the need for identities. But Walras's Law would continue to hold in the aggregate, now containing money as another good, assuming consumers adhere to budget constraints.

Lastly, as indicated in Chapter 20, I have rejected the notion that Walras did include the potential for changing money stocks in his budget constraint, and hence included Say's Law as an identity. But I have every sympathy with any economist who attempts to plumb Walras's extremely unclear monetary theory, as I have made clear in other selections above, and Mishan is not the first theorist to have concluded the opposite is true.

Chapter 22 contains a summary of recent work and my own continuing interest in the integration of money into general equilibrium theory and obtaining a clearer understanding of its relation to capital goods and interest. My feeling about the usefulness of general equilibrium theory is spelled out at the beginning: it is a metaphorical depiction of complex interdependence whose involuted patterns we will never disentangle fully, but which gives a vision of economic reality, or a guiding paradigm. It is in this spirit that I construct a series of simple dynamic capital models with the intent of casting light upon the nature of interest and the intertemporal, as opposed to the intratemporal, functions of money in such systems.

That role is primarily to provide a single rate of interest for an economy through usage of its 'own-rate' in a current *numéraire* economy. A different insight into the concept of interest is obtained if we use a constant external *numéraire* economy. Then it is clearly seen that interest and capital are value concepts, own rates are different for each good and become the same only in artificial constructs like the stationary state or the von Neumann balanced growth economy, and that *real* interest is zero. The role of interest in such an economy is to keep the intertemporal values of a given stock of capital constant over time when the value of real goods is being affected by productivity and time preference. Hence, interest is not truly a net return to capital ownership.

The paper concludes with my views, developed rather fully in the preceding chapter of Part IV, about the problems of integrating money and real sectors, but with, perhaps, some new insights. The partial frameworks of Marshall inhibited this integration in two ways. First, as Patinkin's work highlighted, it failed to deal fully with the substitutability and complementarity relations the money sector had with the real. This affected the Keynesian attempt at a general system as well. And, second, it failed to distinguish clearly between money services and money stocks.

Clower and Leijonhufvud have reinterpreted the Keynesian contribution as that of an informal disequilibrium model in whose development Keynes presented the static model as a preliminary. This admits, at least implicitly, that the latter has the fundamental faults summarized in Chapter 19. The Clower-Leijonhufvud version of Keynes, as I understand it, is given in brief summary, and some of its policy implications indicated.

In revisiting the selections of Part IV I find very little that I would retract or do differently today, had I the chance to rewrite them. One characteristic of them I think will be clear to the reader: my interest in general equilibrium theory in general and in monetary theory in particular is dominated by a concern with the primitives, or fundamentals underpinning the structures. It is the hope to gain insights into the intratemporal roles of money, the preconditions that must exist for its emergence and continued existence, its manner of integration with capital markets and capital goods, its intertemporal functions, and its relation to interest, that sustains my interest in general equilibrium theory despite its formal analytical poverty. The future of general economic equilibrium theory, I am convinced, lies in the willingness of economists to employ it in more informal ways to win perspectives that can be formally studied more fruitfully in less ambitious models. To continue to build increasingly arcane formality into it which contributes no useful insights into its foundations or points to no new paths of research into real economic systems is to assure its future irrelevance.

17 On the Existence and Role of Money in a Stationary System*

I INTRODUCTION[1]

This brief paper seeks to examine the basis upon which the existence and use of money rests in a stationary framework. Its major innovation is to uncover a neglected cost in the circulation of goods and services – the costs of clearing accounts between economic units – and, ultimately, to ground the use of money under stationary conditions in an advantage money possesses inherently on this count over personal credit.

Our 'vision' of economic processes will be Walrasian. It is assumed that there exists a large community of economic units which holds, at the start of a market-day, given stocks of factors whose services can be appropriated by firms. The latter enter markets to buy these services for one period at the equilibrium prices which emerge at the end of the market-day. The consuming units, which own the services, arrive at understandings with firms for the future delivery, at equilibrium prices, of goods produced by the firms. All markets are purely competitive and the pairings of buying and selling units are purely random, market-day after market-day. An external condition imposed arbitrarily upon the economy is the constancy of the capital stock from one period to another, but markets exist in which all types of assets are traded, either as newly produced commodities to replace depreciated goods, or as the means *individuals* have of consuming more than their current incomes or acquiring larger stocks of earning assets.

An important feature of the economy is the lapse of time between the absorption of inputs by the productive units and the emergence of

* Published originally in the *Southern Economic Journal*, 25 (1958), pp. 1–10, and reproduced with permission.

outputs. The delivery of productive services, their transformation into outputs, and the delivery of these goods to consuming units, occur during a 'fulfillment-week', in which the understandings arrived at in the final equilibrium of the market-day are carried out. Finally, the assets from which the hired services were obtained, as well as all new assets, are delivered to their owners at the end of the fulfillment-week.

Under these conditions, the economy faces a problem of synchronizing inputs and outputs in the only way it can: by the accumulation of inventory capital. That is, inventories of finished goods and raw materials (we shall abstract from semifinished stocks), extant at the start of the fulfillment-week, can eliminate the need to wait until the flow of goods begins from current production. Therefore, under any direct or indirect type of exchange, the services of inventories will be saleable – let us say to firms initially – and the prices of these services will enter into the prices of final products. These real costs are inescapable if a society wishes to acquire the means with which to overcome the inconveniences of time. In a Walrasian system, for example, they are pooled by society, in that every unit of product produced during the week bears its proportionate share of these synchronization costs, regardless of the time of its delivery during the week. Thus, the consumer pays no more for a good delivered Tuesday than for one delivered Friday. We shall, however, assume that other types of 'costs', such as inconvenience of storage, lead economic units to distribute purchases over the length of the week.

We shall define money as a stock of good with no inherent utility, acceptable in the discharge of obligations, *whose use does not involve the recording and clearing of debit and credit accounts for economic units*. This last characteristic of a true money has been little stressed in the literature, but it forms a major distinction of an impersonal means of transacting from any type of personal instrument for so doing. Throughout the paper we shall accept the hypothesis familiar to us all: barter, in the sense of an exchange of spot goods for spot goods, is so inefficient a method of transacting that its costs effectively rule it out of the realm of relevance.

Demand deposits may be seen to fall somewhere between money and personal credit instruments, but since they require clearing between units we shall rule them to be credit instruments. Although we shall treat them as a sort of riskless bill purchased by units at par, and although we shall assume that they circulate at no discount, we shall consider the demand depositor as receiving earnings in the form of clearing services to a full or partial extent of their costs.

Given these conditions, then, we ask: under what circumstances will there arise a stock of money, a market for money services, and a market for the asset, money, in this circular flow? What conditions must be fulfilled to assure the continuance of this stock and these markets?

II THE CASE OF PERFECT KNOWLEDGE

We shall suppose that every productive and consuming unit in the economy will engage in 'understandings' only to the extent that its stock of wealth plus the willing loans of others allow, and that it will honor all 'undertakings' as to quantity and time of delivery. This is true, we shall assume, because the state of knowledge in all markets at all times is such that each unit foresees that any such inconsistency in its overall weekly plan would be apparent immediately to every other unit in the economy, it would be deprived of access to markets for the week, and severe penalties of other types would follow. The risk, therefore, that any individual will default in any respect is nil.[2]

Under such conditions, it may be argued that the market-day becomes superfluous, since, if foresight is sufficiently acute to discern the preceding, it should be true that equilibrium prices and quantities are immediately apparent to all in the economy.[3] Nevertheless, 'understandings' between individual units will be required to assure the co-ordination of deliveries of inputs and outputs, if one refuses, as we shall, to admit the possibility of foreseeing the purely stochastic variation in the economy. In the absence of these undertakings between units, all economic units will tend to move together immediately to correct imbalances, since the cognizance of such defects would be instant and action immediate. We should be in the situation, to use Schumpeter's analogy, of inhabiting a frictionless world in which *all* heat moves immediately to correct the slightest departure from an even distribution, and thereby renders the world uninhabitable at the first lapse from the ideal state. Of course, we could assume a pluperfect foresight in which all units foresee that all other units will move to create further disequilibrium, and, therefore, no unit will move. However, there is no end to this ponderous gavotte, and we shall not begin it.[4]

But despite the need for undertakings so that co-ordination of units can occur, they are not necessary to insure against the default of these units. We may look upon this economy as a series of open stockpiles upon which units place their supplies at agreed-upon intervals and from which they take a socially condoned value of desired services and goods. No need for any medium of exchange will arise: there is no use for money or open-book credit, checks, bills of exchange, receipts, etc.[5]

A medium of exchange (of whatever type) arises only when there exists a need to 'identify' transactions – to keep track in some way, or to make a note of, economic units' deposits upon and withdrawals from society's stores. Under the state of perfection of knowledge *here* assumed, no 'personal' or 'impersonal' type of instrument (distinguished by the existence or nonexistence of an economic unit's pledge) will arise for 'identifying' transactions.

III THE CASE OF IMPERFECT KNOWLEDGE

Let us now move to an economy in which potential default by economic units is not ruled out by perfect foresight. Suppose it to be possible in the conditions of 'open stockpiles' for units to withdraw goods without ever replacing them. Under these circumstances, explicit notice must be taken of transactions: withdrawal must be accompanied by a token which has the function either of (*1*) explicitly identifying the person who obtained the goods, or (*2*) noting the nonnecessity of identifying the unit by assuring the deliverer of the goods that the receiving unit has a valid claim to them.[6]

In this case, however, it is assumed that personal credit instruments arise to deny the possibility of default, but that *once the identification of units has been made, default once more becomes nil.* In such circumstances, the credit instruments would be needed in order to deter default, but no clearing between units at the end of the period would be necessary. If it be assumed that the costs of such credit instruments – let us say simple receipts signed by agents of the economic units receiving goods – are negligible, once more we find that no additional burden to the real costs of synchronization is borne by society. These credit instruments would be kept by the forces of competition at face value, and any attempt to exchange via a stock of impersonal instruments could not succeed if a *sine qua non* of its continued existence was a positive price for the services of these instruments.

This type of credit instrument has an important kinship with money as defined in this paper. Although it is literally a personal credit instrument, the fact that it is not cleared makes it impersonal in a *de facto* sense. It is one of the important features of money, as will be apparent later, that it serves as a token of transactions that requires no clearing between individuals. It would not be farfetched, therefore, if we stepped outside the rigorous definition of money and looked upon this type of economy as one in which the price of the services of 'money', and therefore the price of 'money' itself, is zero.

IV THE CASE OF IMPERFECT KNOWLEDGE AND CLEARING

But the mere existence of uncleared personal credit instruments could not for long guarantee the nonexistence of default under conditions of imperfect foresight. Sooner or later, individual units would take advantage of this lack of clearing to draw more from the stockpile than they deserved, and a need for clearing these instruments would arise.[7] We may suppose that all transactions are conducted by the signing of receipts, and assume further that a prior cost-minimization decision by society has led to a central clearing agency where individual units' accounts are cleared at the end of each fulfillment-week. Assume, finally, to keep the new institution Walrasian, that it is forced to price at average cost for its clearing function in an as yet unspecified manner. At last, then, we have arrived at conditions in a circular flow in which the services of a 'transactions identification' instrument become a real charge against the economy – a real cost necessitated by the need to obviate the possibility of default by clearing credit instruments.

As soon as the need for clearing comes into being, and its cost becomes greater than zero, there is need to consider the possibility of substitutes as methods of 'identifying' transactions. Specifically, an *impersonal* credit instrument whose mere possession is sufficient evidence of the validity of a claim upon society's goods can (*1*) eliminate the cost of clearing while it (*2*) identifies transactions. If we assume that if provides these services costlessly, then holders of a given stock of it may sell its services period after period at prices that just entice the marginal economic unit to transact with it rather than use the credit mechanism.[8] The upper limit to the price of the service rendered by money must be the cost of transacting a unit of *numé-*

raire's business with the cheapest personal credit instrument requiring clearing.

It must not be forgotten that there was *no* default on the credit instrument before we introduced money. *Actual* default is zero and is recognized to be such by every unit in the economy as long as regular clearing is carried on. Thus, in our economy, the costs of actual default do not enter into the price of credit instruments, although, of course, the basic reason for their existence (as well as that of money) and for their being cleared is to prevent *potential* default. The distinction is vital in understanding the reason for the existence of money, for the individual, in deciding to transact by purchasing the services of money in the market or availing himself of the use of his personal credit pledge, is absolutely certain that payment in both will be accepted, although a discount will rule on the latter.

Although it is an awkward approach that leads its user into difficulties, let us follow Hicks in assuming the existence of money before we establish that it will arise in a personal credit-instrument economy. Assume that money is borrowed by the delivery of personal credit instruments (bills) which are riskless in the credit sense and on which the interest rate cannot change since they are redeemable at the end of the fulfillment-week. Then,

> If bills stand at a discount, and consequently earn interest, is there anything to stop any individual from investing all his surplus funds in bills, and holding them during the week in that form? If there is nothing to stop him, then money has no superiority over bills, and therefore cannot stand at a premium to bills. The rate of interest must be nil.[9]

But, he continues,

> . . . If people receive payment for the things they sell in the form of money, to convert this money into bills requires a separate transaction, and the trouble of making that transaction may offset the gain in interest. It is only if this obstacle were removed, if safe bills could be acquired without any trouble at all, that people would become willing to convert all their money into bills, so long as any interest whatsoever was offered. Under the conditions of our model, it must be the trouble of making transactions which explains the short-rate of interest.[10]

But while this explanation can be used to show why money can continue to exist, once born, it cannot answer the question of why it came into existence.[11] This can be seen more clearly if, unlike Hicks, we start with an economy where only a credit instrument of a personal type is being used and ask if money will come into being. The cost of converting money into bills cannot now be used to explain the interest rate, for the 'bills' are being used as circulating media already without conversion to money. Hicks comes to this point and says:

> If this were to happen generally, there would be no cost of investment, and, therefore, so it would appear, no reason for bills to fall to a discount . . . If default risk is so generally ruled out, that all traders reckon and are known to reckon, a particular bill as perfectly safe, then there is no reason why that bill should stand at a discount, for the obstacle of cost of investment can be circumvented . . .[12]

Thus, the existence of money, and with it the interest rate, cannot be explained by the costs of converting personal credit instruments to impersonal ones. By the logic of Hicks's own arguments, in the absence of default risk, money cannot exist in these circumstances. But he never comes to grips with the question of why credit does not replace money. He admits the possibility, but shrugs if off by refusing to accept the feasibility of its being generally acceptable:

> But this general acceptability is something different from the mere absence of default risk, which we assumed previously. A class of bills may be regarded as perfectly safe by those who actually take them up, and yet these persons may be different from those to whom the borrower has to make these payments. These latter would not accept his bills, so he has to pay *cash*; the former are perfectly willing to lend, but require interest to compensate for their cost of investment.[13]

Thus, when the question is finally posed by Hicks's analysis: when, under conditions of a regime of personal credit instruments which are riskless methods of identifying transactions, will money come into existence? Hicks backs away from answering it. As we shall see in the next section his investment impediments are a separate condition for the existence of money, but he has omitted a prior one.

That condition we have found in the need to clear personal credit instruments which circulate to prevent potential default (when actual default is zero). But, of course, our singling out of this expense because of our belief that it would be major should not blind us to the disadvantages inherent in such an instrument as money: the need to transport it physically, risk of loss or theft, and so forth. Only if the costs of using an impersonal instrument which obviates the need for clearing are cheaper than those of using a personal instrument can money come into existence: but this cost advantage may be complete, in which case money would completely replace credit instruments, or it may be only partial, in which case credit instruments may be used along with money.

But what of the twilight case: bank deposits? Why should economic units which can transact with money or pay the costs of clearing their receipts purchase *stocks* of personal-impersonal credit instruments which require clearing? If the cost advantages of a non-clearing instrument were absolute, they could not exist in these circumstances. Similarly, if the cost advantages of a personal instrument requiring clearing were absolute, no such *stocks* of instruments need arise, for surely the need to clear current transactions and account for stocks as well cannot be any cheaper than merely clearing transactions. They *can come* into existence, therefore, only when neither of our pure methods has an absolute advantage, so that economic units may desire to use clearing or non-clearing at will. But will they? It would seem to the writer that our assumptions allow no middle ground: bank deposits, which possess the disadvantages of both media, would always be at a disadvantage to one or the other, whatever the nature of the transaction. Bank deposits cannot exist.[14]

Next, let us assume that the existence of a credit instrument plus clearing, and the provision of societal penalties, do not eliminate the occurence of default entirely. If the clearing agency does not assume the responsibility and charges the loss to the seller of the goods, then the *numéraire* value of the credit instrument will be reduced further by the percentage charge that will just cover the individuals' defaults. If the clearing house sets up an insurance fund, the risks must enter into the costs of clearing in order to keep profits zero. Therefore, the existence of *actual* default among the units of an economy does not introduce novel elements into our analysis, for the roots of the latter continue to lie in the costs of successful elimination of *potential* default of much greater magnitude: the former merely adds an additional burden.

V THE CASE OF IMPERFECT LENDING FACILITIES

Given these two preconditions for the holding of money – the need to identify transactions and the existence of cost advantages in its use for this purpose – two relevant considerations arise:

1. Economic units will seek to hold it, and to purchase it for holding, between the end of the fulfillment-week and the beginning of the following market-day, as they would seek to purchase and hold any other asset that yields as saleable service;
2. Economic units will seek to economize their use of its services.

From the first of these springs the assurance that individuals will not, even in a stationary state in which the good money has no direct utility for economic units, seek to reduce their stocks of it to zero at the end of the market week. This is true for money as for any other stock of asset: inventories of consumption goods, for example. If no reason exists for the unit to liquidate its holdings of other stocks, no reason exists for it to seek to reduce its stock of money to zero.

The second consideration leads to an objection many economists have raised. Since money is a good whose service is required only at a set of discrete points for each economic unit in the fulfillment week, individuals will seek to lend it to other units between payment dates, or even to make contracts to rent durable consumption goods with cash balances between such payment dates. Imagine, first, that the distribution of the *numéraire* value of transactions to be made throughout the fulfillment-week is a uniform one at discrete points, so that, for example, 1/10 of the total value of transactions occurs at each of 10 time points in the week. Between these points, demand for the services of money will be zero, so that units will seek to lend or rent goods until the next payment date. Thus, its velocity tends toward infinity, and the price level rises with it until it becomes useless as a medium of exchange.

If payment dates may occur continuously through the week, the same result will occur. Some method of preventing the use of cash balances between payment dates must be devised as a third condition for the holding of money balances, and, therefore, the continued use of money as a medium of exchange.

Two solutions have been advanced. First, the Hicks-Marget investment impediments, which assume that the costs of making extremely short-term investments outweigh their returns, can be looked upon as forcing the economic units to hold cash balances instead of securities between payment dates. Baumol has worked out a model based upon the impediments of brokerage charges in the broadest sense of the term, and Marschak has generalized the concept to one of perfect markets for goods and discusses such frictions for a pure exchange economy.[15]

A second method is that developed by Patinkin by which the payment dates for economic units are fixed in time, but the amounts of goods received and delivered, thereby leading to payments and receipts of money by economic units at any given payment time, are randomized. The hope is that such a scheme will force economic units to hold cash if the drawings for receipts and payments are made just before the payment time, insofar as economic units will be unable to foresee accurately their net requirements of cash until it is needed. But, the mere fact that contracts of all types are made only on the market day does not force units to hold cash balances. Such randomization will lead units to form some expectation of a needed cash balance, and then seek to make contracts to lend it between payment dates, obtaining control of it just before *every* payment date. Unless impediments exist to prevent this, randomization has had no effect upon the desire of units to lend cash between payment times. Seemingly, the times at which payments in excess of receipts may occur would have to be randomized.[16]

Some have objected that under the circumstances, where money is held by default, there cannot truly be a demand for cash balances. For example, Schumpeter wrote:

> If people get their incomes' each Saturday and spend them on consumers goods each succeeding Monday – transactions between firms being excluded – then the money will lie about in the vaults of firms from Monday to Saturday, not because there is any demand for cash holdings, but because the institutional arrangements wills it.[17]

Taking a somewhat similiar tack, Patinkin argues that in the Walrasian system the individual holds money, not because he maximizes his utility in so doing, but because the nature of the system requires him to.[18] The present writer has expressed himself as believing that the services of money have a utility to economic units in a stationary state that can be handled best outside of a marginalistic framework.[19] The performance of the necessary identification of transactions most cheaply gives money a utility, and the mere fact that it is not in continuous use – like the tractor on the farm – does nothing to lessen this.

VI CONCLUSION

Given (*1*) the need to identify individual units' withdrawals and deposits of goods and services, (*2*) the cheaper cost of doing so with a non-clearing type of instrument than with one that must be cleared, with or without actual default occurring, and (*3*) some inhibition or inducement to hold it between payment dates, money will exist in a stationary economy.

Its services will sell for a price which can be no higher than the cost of the cheapest alternative method requiring clearing. Therefore, it becomes an asset in which units can invest to obtain a yield in futurity, and a market for it as an asset will exist in which it will be priced.

From the price of its services and its price as an asset can be reckoned the 'rate of interest' which is, as for any other asset, the yield of a good, durable and limited in quantity, which provides a saleable service. In this case it is the ability to perform most cheaply the necessary task of identifying transactions. Since the price of its service will always be above zero in the conditions envisioned, the interest rate will be positive and less than infinite as long as future income has a utility. Since an equilibrium condition in asset markets is the equality of all net rates of return, the rate of return on real assets can never drop below that on money. Therefore, as long as money exists in a stationary flow, the interest rate and return on goods can never, *à la* Schumpeter, decline to zero.

Lastly, we have not argued the probability of the preconditions for the existence of money, but merely the necessity of their existence in order for money to come into being and remain extant.

Notes

1. I am grateful to my colleague, W. J. Baumol, for reading the manuscript and suggesting several improvements. Errors, however, are jealously guarded as the author's property.
2. See S. P. Chambers [3], p. 39, and J. C. Gilbert [4], pp. 150–1. This

seems to be the degree of perfection of knowledge implied by Professor F. H. Knight when he discusses the nonexistence of money in a circular flow. See [8], p. xxii. This would tend to rule out the difficulty he noted in the original work, however: ' . . . Attention may be directed to another tendency fatal to free competition under theoretical conditions. This is the matter of the inflation of credit. With all forms of friction eliminated there would seem to be hardly a limit to the substitution of credit for any sort of commodity as a medium of exchange and a stable value-standard would apparently be impossible to establish' (1921 edition, pp. 193–4n).

3. See O. Morgenstern [14]. This paper considers the game-theoretic aspects of the assumption of perfect foresight and demonstrates the impossibility of its existence.
4. The step is a familiar one, made famous by Sherlock Holmes and Professor Moriarty, except that in the present case the final bow by the former to the latter is not a logical consequence of the choreography. *Ibid.*
5. The literature reveals that others have discussed the nonexistence of money in a 'perfect foresight' context which was not 'perfect' enough to rule out other media of exchange. Marget, for example, argues that no cash balances would exist, but that units would have to convert all goods into some acceptable commodity before obtaining the goods they desire. He suggested that units might go to a 'public treasury' to obtain a 'certificate' to spend immediately, so that 'money' would not be absent. Quite apart from the fact that this is not 'money' in our sense (since it is not a stock of good), no such need would arise under the conditions we hypothesize. See. A. Marget [11], p. 159.

   Similarly, Rosenstein-Rodan puts much stress on the absence of money rather than any transactions medium, making his 'certain foresight' ambiguous: ' . . . it is inconsistent to assume at the same time a state of general certain foresight and the existence of money: they are mutually incompatible. Money (as cash balance) exists only and insofar as general foresight is not certain, it is a function of the individual's feeling of uncertainty, a means of melting it: a good satisfying the want for certainty. Since certain foresight is assumed in static equilibrium, money and static equilibrium are incompatible.' See [18], pp. 271–2.
6. Pigou asserts that in one-good production economy no need for money arises, since all factors can be paid in the good. With imperfect foresight assumed, this solutions neglects completely the need to 'identify' deliverers of services and receivers of the good when such deliveries are not perfectly synchronized. See A. C. Pigou [17], p. 70. Pigou seems to combine money and credit instruments in his considerations, and therefore abstracts from the problems that concern us. See, for example, his conclusions about money on pp. 74–5, which are phrased in terms of barter or no barter, not money vs. credit instruments.
7. This need may not be continuous. If penalties for default are made severe enough and the probability of being caught high enough, randomized clearing-or-no-clearing decisions at the end of each period may be sufficient, but costs may not be reduced greatly under those of continuous clearing. We shall assume that clearing is necessary period by period in this paper.

8. Hicks has indicated that no monopoly bank could have a free hand in the pricing of its services, since substitutes as media of exchange would arise to limit its pricing power. See 'Gleichgewicht und Konjunktur', [5], p. 448n.
9. [7], p. 164.
10. [7], pp. 164–5.
11. See F. Modigliani [13], pp. 82–3, for the first discussion of this point. But he substitutes for Hicks's reasoning an unsatisfying theory of interest: '. . . for certain categories of people (entrepreneurs as well as spendthrifts) it is worthwhile to pay a premium to obtain spot cash against a promise to pay cash in the future' (p. 83). Here, too, the reason why credit instruments cannot be used and money dispensed with is begged. Later, in his attempt to show that the disappearance of money in a stationary state, and the substitution of existing securities for money, would not lead to a zero interest rate, Modigliani limits his analysis to bonds and stocks. But if the 'securities' used were the 'receipts' we have been dealing with (because, for example, securities of the bond-type do not exist beyond the fulfillment week in a Walrasian system), no need to use bonds and stocks would arise, and the interest rate would decline to zero. Modigliani comes to the conslusion that this is true in the case of *indefinite* credit expansion, but does not connect it with the stationary state.
12. [7], p. 165.
13. [7], p. 166.
14. Therefore, the defense that the following writers were thinking of money in terms of demand deposits rather than paper cannot excuse their neglect of the costs of clearing personal credit instruments.

    Patinkin has failed to provide an explanation of why credit cannot drive money from his own model in [16]; G. Becker and W. Baumol, in [2], p. 367, make the point that if no investment frictions exist money and securities will circulate at par, neglecting the costs of clearing, and Baumol repeats the assertion in [1], p. 550; Rosenstein-Rodan stresses repeatedly that it is the 'store of value' function of money – the 'want for certainty' – that explains the existence of money [18]; and J. C. Gilbert in [4], pp. 146–54, distinguishes the 'inconveniences of time' from uncertainty as a cause of the existence of money, but does not attempt to prove that money furnishes a cheaper method of coping with these inconveniences than credit. Similarly, J. Marschak, in [12] rules out barter as a means of transactions, but does not discuss credit substitutes, although his provision that only money has a perfect market in the sense of zero transactions costs contains our concept of clearing costs implicitly. See pp. 87–100.
15. J. R. Hicks, [6], pp. 6–7, and [7], p. 165; A. Marget, [11], pp. 159–61; W. J. Baumol, [1]; J. Marschak, [12], pp. 87–100. See also D. Patinkin, [16], p. 136 and [15], pp. 62–95; G. Becker and W. J. Baumol, [2], p. 367; Rosenstein-Rodan [18], pp. 269–72; F. Modgliani [13], p. 51.
16. [16], pp. 86–95. I may be doing Patinkin an injustice but I do not find evidence that the *payment times* themselves are randomized.
17. [19], p. 547.

18. [16], p. 393.
19. See pp. 332–3, and [10], pp. 631–6.

References

[1] Baumol, W., 'The Transactions Demand for Cash: An Inventory Theoretic Approach', *Quarterly Journal of Economics*, 66 (1952), pp. 545–56.

[2] Becker, G. S. and W. J. Baumol, 'The Classical Monetary Theory: The Outcome of the Discussion', *Economica* (NS), 19 (1952), pp. 355–76.

[3] Chambers, S. P., 'Fluctuations in Capital and the Demand for Money', *Review of Economic Studies*, 2 (1934–5), pp. 38–50.

[4] Gilbert, J. C., 'The Demand for Money: the Development of an Economic Concept', *Journal of Political Economy*, 61 (1953), pp. 144–59.

[5] Hicks, J., 'Gleichgewicht und Konjunktur', *Zeitschrift für Nationalökonomie*, 4 (1933), pp. 441–55.

[6] Hicks, J. 'A Suggestion for Simplifying the Theory of Money', *Economica* (NS), 2 (1935), pp. 1–19.

[7] Hicks, J., *Value and Capital*, Oxford University Press, London, 2nd edn (1946).

[8] Knight, F. H., *Risk, Uncertainty and Profit*, London School Reprint, London (1933).

[9] Kuenne, R., 'Walras, Leontief, and the Interdependence of Economic Activities', *Quarterly Journal of Economics*, 68 (1954), pp. 232–54. Reprinted as Chapter 1 in this volume.

[10] Kuenne, R., 'Walras, Leontief, and the Interdependence of Economic Activities: Comment', *Quarterly Journal of Economics*, 69 (1955), pp. 631–6.

[11] Marget, A., 'The Monetary Aspects of the Walrasian Theory', *Journal of Political Economy*, 43 (1935), pp. 145–86.

[12] Marschak, J., 'The Rationale for the Demand for Money and "Money Illusion",' *Metroeconomica*, 2 (1950), pp. 71–100.

[13] Modigliani, F., 'Liquidity Preference and the Theory of Interest and Money', *Econometrica*, 12 (1944), pp. 45–88.

[14] Morgenstern, O., 'Perfect Foresight and Economic Equilibrium', *Zeitschrift für Nationalökonomie*, 6 (1935), pp. 337–57.

[15] Patinkin, D., 'Relative Prices, Say's Law, and the Demand for Money', *Econometrica*, 16 (1948), pp. 135–54.

[16] Patinkin, D., *Money, Interest and Prices*, Row Peterson, Evanston, lll., 1st edn (1956).

[17] Pigou, A. C., *The Economics of Stationary States*, Macmillan London (1935).

[18] Rosenstein-Rodan, P. H., 'The Coordination of the General Theories of Money and Price', *Economica* (NS), 3 (1936), pp. 257–80.

[19] Schumpeter, J. A., *Business Cycles*, II, McGraw-Hill, New York (1939).

18 Patinkin on Neo-Classical Monetary Theory: A Critique in Walrasian Specifics*

I INTRODUCTION[1]

The recent publication of Professor Don Patinkin's *Money, Interest, and Prices* may be regarded as the appearance of a summary statement of the thought of its author as it has evolved over the last decade. Together with the allied criticism of neo-classical monetary theory by Lange,[2] Patinkin's challenge to the determinateness of that body of analysis and to its successful integration with the 'real goods' equations of neo-classical systems constitutes the most formidable attack ever made against it. As such it deserves to be examined in detail.

The bibliography of this controversy is already an extensive one.[3] Further contributions to it, however, will be forthcoming because of the marked evolution of views apparent between the earlier articles and the book. Patinkin's present analysis makes the categories of 'right' and 'wrong', provable 'existence' and provable 'non-existence' of the Invalid Dichotomy, in which terms much of the previous discussion took place, largely outmoded. Concomitantly, future appraisal and discussion must turn much more than past around the detailed analyses of neo-classical systems in the book.[4]

This paper, therefore, has much narrower aims than earlier critiques of Patinkin's analysis. In Section II it seeks to relate the well-known Lange criticism of neo-classical theory to the Patinkin analysis as the writer understands the latter. Section III contains a generalization of the Lange-Patinkin analysis from the specific case of

* Published originally in the *Southern Economic Journal*, 26 (1959), pp. 119–124, and represented with permission.

money to that of any good in a general equilibrium system. Section IV uses the frameworks of Sections II and III to examine in detail Patinkin's analysis of neo-classical theory as it relates to Walras's general system. And Section V contains a brief summary and conclusion.

Much of the following will be mildly critical of Patinkin's analysis, but the writer does not wish to slight the solid contribution to the effective integration of monetary variables in a static general system made by Patinkin in the major themes of his work. Our concern will be with certain elements in the counterpoint, much of which is wholly acceptable.

II THE INVALID DICHOTOMY

The most convenient starting point for an understanding of Patinkin's current position is Lange's analysis of neo-classical systems.[5] In this analysis (as in Patinkin's) neo-classical economies are interpreted as those in which consumers and entrepreneurs adjust their desired holdings of cash balances by changing their excess demands for goods (if, indeed, the system allows them to). Should one of these decision-making units desire to increase its balances, it does so by decreasing its demand for and/or increasing its supply of real goods. In the model, goods and cash balances are (or should be) substitutes – a point which becomes important in Section IV, since the Walrasian model does not conform to it.

Lange shows that the existence of Say's Law – or, more accurately, Say's Identity in the Becker-Baumol terminology[6] – is an effective bar to the existence of money as a medium of exchange. Such an economy is provably equivalent to a barter economy, and if an equation specifying that demand must equal supply in the money sector is added, such an equation will contradict the identity of demand for and supply of money which is implicit in Say's Identity.

The validity of Lange's criticism of neo-classical theorists hinges upon the acceptance by the theorist of Say's Identity. But this is not true of Patinkin's, or, rather, the latter goes beyond Lange in his criticism of the neo-classicists, including the latter's area of criticism as a sub-case of his own. Patinkin explicitly sides 'with those who deny that [Say's Identity] is a basic component of the classical and neo-classical position.'[7] What, then, is the 'Invalid Dichotomy' present in neo-classicism if it is not the rigorous segregation of money and real sectors (to the extinction of the first) implicit in Say's Identity?

To Patinkin, the problem is that these theorists implicitly accepted the outlook that the homogeneity postulate – demand and supply equations for real goods are functions only of relative prices – was necessary to a rational economy and did not consistently relate it to their attitudes (explicit or implicit) toward Say's Identity. Therefore, they were led to assert that relative prices were determined in real goods markets and, independently, the price level was fixed in the 'money market', thereby dichotomizing their models, without explicitly recognizing the interdepence necessary to avoid monetary indeterminateness. Or, less strongly, if they did not assert this dichotomy, they did not deny its possibility:

> This is the crucial point. The dynamic groping of the absolute price level toward its equilibrium value will – through the real-balance effect – react back on the commodity markets and hence on relative prices. And it is precisely the constant failure to find this point explicitly recognized – and, indeed, the constant sensation of being just on the verge of having it explicitly contradicted – that is the basis of our original contention that the roots of the invalid dichotomy are to be found in the neoclassical analyses of Walras, Fisher, Pigou, and Cassel.[8]

Patinkin's focus, therefore, is less upon the logical conditions of some rigorous model – as Lange's is, since his criticism is directed at the existence or nonexistence of Say's Identity – and more upon the implications of these authors' discussions of how equilibrium is attained in real and money markets. Walras becomes an ideal case in point as a major target for Patinkin: Lange's analysis would lead us to ask the question, 'Does the rigorous Walrasian model contain Say's Identity or not?' It might permit a second question: 'Apart from Walras's specific assumptions, does the internal logic of Walras's system lead to a contradiction of Say's Identity, or does it not?' Although Lange did not ask these questions of Walras's analysis, they spring logically from his paper.

Patinkin's criticisms lead to some such question as the following: 'Whatever the conclusion about Say's Identity as posed in the two questions above, when Walras steps outside the equilibrium conditions of his model to discuss the *tâtonnement* process by which equilibrium is reached, does he tend to dichotomize such market processes into real and monetary?'

III A GENERALIZATION OF THE INVALID DICHOTOMY

We shall seek to answer all three of these questions for the specific case of Walras. Before we do, however, it will be valuable to generalize the Lange-Patinkin line of criticism in order to gain greater insight into the nature of a general equilibrium system.

Assume that we are describing the general equilibrium of a system with $n - 1$ real goods and money, in which all prices are measured in units of some *numéraire* outside the system. We shall begin then with n excess demand equations which are equated to zero (one of which is not independent), where excess demand is obtained by subtracting the supply function from the demand function, good by good. We shall say no more about the characteristics of the system, but we will assume that a price vector (P^0) is found whose elements fit the model.

Next, let us relax the conditions of the model sufficiently to permit us to lift the variable P_i from the system and, similarly to a government policymaker or administrator, arbitrarily increase its value to $P_i^0 + dP_i$. *If we temporarily abandon all of our preconceived notions of how well-behaved, rational, logically constructed economies should behave*, three cases will hold an interest for us:

Case 1 No excess demand in the system departs from zero, and therefore no other price in the system is affected. We shall term this the case of complete non-integration, for when a price change has no impact upon a system, in the sense of failing to impose the need for a new *tâtonnement* process, it is not truly a part of that system.

Case 2 The excess demand for the good whose price changes is affected, and departs from zero, but in such a way as to have no effect upon any other excess demand or price in the system. (Once more the reader is enjoined to suppress his Euclidean economic notions). We shall term this the case of partial integration.

Case 3 At least one other excess demand besides that for the good whose price is disturbed departs from zero, in which case full integration of the price, P_i, is achieved.

Assume that we now assert the general impossibility of Case 2 in a rational economic model: that is, that in a system containing all possible alternatives for expenditure, any one excess demand can change without affecting at least one other. If, in spite of this general restriction, we assert that there is one good whose price does have this property, we are involved in a contradiction. Either the structure of the system does not in fact deny the possibility of Case 2 and we have a partially integrated variable, or, contrary to appearances, Case 2 must not exist and Case 1 or Case 3 must be operative. If we have structured the system specifically to prevent Case 3 – taking care that the price of the good under discussion does not affect any other excess demand – then we must in fact have Case 1. That is, in spite of the appearance of affecting only its own excess demand, the logic of the system must dictate that it is completely non-integrated.

What is it in the nature of the relationships among most goods which makes of their price changes Case 3 examples? In a simple exchange economy, the *substitution* and *income effects* are the paths of interdependence that give the system coherence. Since our concern is not with the stability of the system or with gauging the strength or net direction of changes in variables, given the *existence* of a non-zero level for either or both of these effects, even should they cancel out at the end of the *tâtonnement* phase, our condition for full integration will have been met.

The essence of Say's Identity is that the inability of changes in the price of money to lead to substitution or income effects implies that the demand for and supply of it are independent of its price and identically equal.

Patinkin's Invalid Dichotomy asserts that the neo-classicals argued as if, when the price of money departed slightly from the equilibrium value, with no compensating changes in its quantity, the *excess* demand for *real* balances remained unchanged while other excess demands were undisturbed. Therefore, they implicitly treated money as a Case 1 good in real terms. As far as nominal balances are concerned, Patinkin argues, a change in the price of money was interpreted as affecting only the excess demand for nominal cash balances. But, since interdependence of the system denied this possibility, this implies that the price of money does not affect the excess demand for cash balances, so that any price level could rule. Again, money is shown to be a completely nonintegrated commodity.

Lastly, Patinkin's stress on the necessity for an effect which he rechristens 'the real-balance effect' is the demand that the income-substitution effect for money be built into the system so that Case 3 integration can occur.

IV PATINKIN ON WALRAS

As a preliminary to the discussion of Patinkin's criticism of Walras, the following sketch of the latter's definitive monetary theory is presented.[9] Two specific functions of money are presented in Walras as (1) the cash balance function, allowing payments to be made as delivery contracts are fulfilled even if receipts fall short of claims, and (2) the asset function, allowing the owner of money a return for lending it to others to serve in the role of function (1). The price per unit of money *service* – that is, of "leasing" from the owners of money stocks one unit of money for use during one period of contract-fulfillment with return of the unit to its owner at the end of that period – is p_{u}' (using some good other than money as *numéraire*), and it is determined in a promissory note market where owners of money stocks meet demanders of cash balances as lessors confront lessees. In the budget equation of the lessor, on the income side, along with his sales of labor, land, and capital goods services, is the term denoting his earnings from his lease of money assets, $o_u p_{u'}$.[10] One important implication of this is that in Walras money must be introduced simultaneously with borrowing and lending. No Walrasian money economy exists without 'bonds' (in the present case, promissory notes).

The price of money as an asset – i.e. as a yielder of services selling for $p_{u'}$ per unit per period – is p_u (again, measured in terms of some other good). These units of money, purchased as any other asset for delivery before the *next* market-day for sale of services commencing the *following* period, like any other asset, must in equilibrium sell at a price equal to the price of the service capitalized at a common rate of return ($p_{u'}/i$). The price, p_u, in Walras, however, is identically equal to this ratio.

The key to Walras's monetary theory is this duality: two functions – transactions and asset; two prices – $p_{u'}$ and p_u; two markets.[11]

Given this brief summary, let us then pose the three questions suggested in Section II. First, does Say's Identity exist in the formal Walrasian model? The answer is unmistakably 'yes', although the construction of the model requires a different interpretation of the phrase. The existence of a promissory note market in the model means that economic units can adjust their holdings of cash balances to any desired level, but Say's Identity prohibited them from changing *their holdings of money as an asset at the end of the period*.

But if there is no market for cash as an asset it can have no price. And if p_u is indeterminate cash balances cannot yield any useful service, since, unlike the services of physical assets which can be priced even if the assets themselves cannot, the utility of money services depends upon money having a price. Therefore, Walras's obtaining of p_u by capitalizing $p_{u'}$ at the rate of return on physical assets is gimmickry, not economics. Money as asset is a Case 1 good in Walras. Thus, the indeterminacy is not strictly a Langean indeterminacy, but it exists.[12]

Second, does the internal logic of the Walrasian system admit of Say's Identity (in the above sense)? The author has argued elsewhere at length that the whole burden of logic in Walras requires that consumers be allowed individually to desire to change their holdings of money assets at the end of the period. He has introduced this possibility into the model by reconstructing it. With this reconstruction, money exists at a determinate price, its services are priced, and an essentially new framework is available to analyze monetary problems.[13]

But, as made clear in Section II, these two questions are used only as approaches to the third, or that which Patinkin's new analysis poses: if we have the reconstructed, formal Walrasian model referred to immediately above, so that we may abstract from Say's Identity in the manner of Patinkin, does the Walrasian discussion of the manner in which equilibrium is established in the real goods and money markets imply his acceptance of the Invalid Dichotomy?

Patinkin asserts its presence is 'highly probable'[14] or, at least, if Walras did not advocate it he certainly did not oppose it.[15] We may deal quite speedily with the formal framework of Walras, for Patinkin now accepts the integration of money as a Case 3 good in the model. Suppose $p_{u'}$ and p_u rise proportionately, so that the rate of return on money, as well as all other prices, are unchanged. Then, the aggregate budget equation term, $o_u p_{u'}$, will change, and via this change in income at least one other excess demand will have to change.[16] On the other hand, if only p_u were to change, direct substitution for or against goods and securities (from the capital goods market) would occur in addition to the change in the budget equation term. Ergo, in the Walrasian system, reconstructed to eliminate Say's Identity, money is a Case 3 good.

Thus, in spite of Patinkin's assertion that the 'real-balance effect' must enter at least one equation in order to yield a determinate price level, the Walrasian system obtains this result without linking the absorption of goods to the value of a stock of money wealth. Rather, the real value of an *income flow* – the earnings on money lent as working capital – provides the non-homogeneity required. Therefore, Patinkin's criticism of the absence of the real-balance effect is not relevant.[17]

But the main burden of Patinkin's present criticism has moved to Walras's discussion of the *tâtonnement* process. The primary basis for this criticism is the following passage from Walras which we must quote at some length:

> On referring back to the various terms that enter into the composition of (the value of inventories demanded in money form), we perceive that they are not absolutely independent of $p_{u'}$, since $p_{u'}$ figures in the term $o_u p_{u'}$ of the (budget constraint), which together with the equations of maximum satisfaction, enables us to deduce the quantities (of inventories desired in money form) for any one party to the exchange, and, consequently, the aggregate quantities (of such services) for all parties together. We must admit, however, that the dependence of these items on $p_{u'}$ is very indirect and very weak. That being the case, the equation of monetary circulation, when money is not a commodity, comes very close, in reality, to falling outside the system of equations of economic equilibrium.
>
> If we first suppose economic equilibrium to be established, then the equation of monetary circulation would be solved almost without any *tâtonnements*, simply by raising or lowering $p_{u'}$ according (as demand for money differed from supply), at a price $p'_{u'}$ which had been cried at random. If however, this increase or decrease in $p_{u'}$ were to change (the value of inventory services desired in money form) ever so slightly, it would only be necessary to continue the general process of adjustment by *tâtonnements* in order to be sure of reaching equilibrium. That is what actually takes place in the money market.[18]

After quoting this passage at somewhat greater length, Patinkin comments:

> This is the sum total of Walras' discussion of the *tâtonnement* by which the level of $p_{u'}$ and hence p_u is determined. It shows that Walras is quite willing to assume that, with $p_{u'}$ arbitrarily fixed at the level $p'_{u'}$, a *tâtonnement* in the commodity markets first achieves equilibrium there; that, once this is accomplished, a *tâtonnement* in the money market then determines the equilibrium

level of $p_{u'}$; *and that this latter tâtonnement can be carried out without reacting back and disturbing the equilibrium initially achieved in the commodity markets*. It fails to indicate any realization on the part of Walras that such a dichotomization of the pricing process contradicts the very conditions necessary for the existence of a monetary economy.[19]

The present author asserts merely that other evidence exists which tends to soften the impact of Patinkin's charges. First, the last sentence in his comment flies in the face of the statement by Walras above that the lack of strength of the dependence comes close to excluding the monetary equation from the system. Second, the longer passage quoted by Patinkin contains a statement that only if the demand for and supply of money balances were brought into equality could the demand for and supply of the *numéraire* good be equated (given equilibrium in all other markets). The required linkage to the real goods sector is here explicitly recognized.

Third, Patinkin's charge that Walras did not bother to trace out the reactions back upon the real sector's prices of the *tâtonnements* in $p_{u'}$ is true, but he does recognize that the 'feedback' case is what actually occurs in the market. Moreover, by this time, he had expressly dealt with the feedback in the multiple exchange economy, the production economy, and the capital goods economy, so that he might reasonably have felt it was unnecessary to repeat it. For example, on page 329 he refers to the price of inventory services as being determined by one equation (excess demand for them equal to zero) – in the midst of his treatment of money – yet it would be difficult to infer that he consistently envisions them as Case 2 goods.

Lastly, Patinkin himself quotes Walras's tracing of the impact of an increase in commodity money upon prices, in which the latter shows (imperfectly, as Patinkin points out) how reactions throughout the price system result in a final neutral effect upon the interest rate.[20]

These arguments seem rather stiff evidence against the dominant presence of the Invalid Dichotomy in Walras. At the least they serve to temper the indictment of Patinkin.

V CONCLUSION

The present writer concludes that Walrasian monetary theory – and, to the extent he typifies it, neo-classical theory as well – is more subject to the Lange criticisms than to the broader challenge of Patinkin, although, as indicated above, even the former must be modified for Walras. At the same time, however, Patinkin's present attack is a more sophisticated one than Lange's, and therefore less susceptible to 'proof' or 'disproof' in some final sense.

Notes

1. I am indebted to colleagues W. J. Baumol, S. T. Beza, E. O. Edwards and R. E. Quandt for reading an earlier version of this paper and giving valuable guidance.
2. O. Lange [6].
3. The reader is referred to Patinkin [7], pp. 477–8 and 488.
4. See, in particular, [7], Supplementary Notes B–H, and L, as well as chs. I–VIII.
5. See [7].
6. G. Becker and W. Baumol [1].
7. [7], pp. 119.
8. [7], p. 112. The preceding paragraph draws mainly upon the analysis on pp. 107–19.
9. Cf. L. Walras [8], Lessons 29, 30, and 31. The present writer's interpretation of this area of Walrasian theory is contained in [5].
10. Patinkin interprets this term as consumers' excess supply of cash used to purchase real capital goods times the rate of interest on these goods. This is a provably incorrect interpretation: new capital goods in the Walrasian model do not begin to earn a return until the period *following* construction and purchase. The term that does represent the purchase of new fixed and circulating capital goods, $d_e p_e$, does not introduce any interest payments into the budget equation. The term $o_u p_{u'}$ must be seen as the amount earned in current income by purchasing promissory notes as a means of lending entrepreneurs (primarily) their working balances of money. See Patinkin [7], pp. 399–400.
11. This interpretation is presented in the writer's [2]–[5].
12. These propositions are discussed at great length in [5]. Space limitations prohibit their inclusion here.
13. See [5].
14. Patinkin [7], p. 108.
15. See [7], pp. 455–6.
16. For Patinkin's recognition of this, see [7], pp. 403–4. See also the writer's [2], pp. 335–6. The element of nonhomogeneity upon which the price level is hung is this term in the budget equation. Patinkin states that the writer failed to see this as the real issue, but the reference cited will indicate that this was exactly his point.
17. Patinkin [7], p. 411.
18. L. Walras [8], pp. 326–7, with minor changes.

19. Patinkin [7], pp. 402–3.
20. See [7], p. 406, and Walras [8], p. 406.

References

[1] Becker, G. S. and W. J. Baumol, 'The Classical Monetary Theory: The Outcome of the Discussion', *Economica* (NS), 19 (1952), pp. 355–76.

[2] Kuenne, R., 'Walras, Leontief and the Interdependence of Economic Activities', *Quarterly Journal of Economics*, 68 (1954), pp. 232–54. Reprinted as Chapter 1 of this volume.

[3] Kuenne, R., 'The Architectonics of Leon Walras', *Kyklos*, 10 (1956), pp. 241–9. Reprinted as Chapter 2 of this volume.

[4] Kuenne, R., 'On the Existence and Role of Money in a Stationary State', *Southern Economic Journal*, 25 (1958), pp. 1–10. Reprinted as Chapter 17 of this volume.

[5] Kuenne, R., 'The Walrasian Theory of Money: An Interpretation and a Reconstruction', *Metroeconomica*, 13 (1961), pp. 95–105. Reprinted as Chapter 20 of this volume.

[6] Lange, O., 'Say's Law: A Restatement and Criticism', *Studies in Mathematical Economics and Econometrics*, University of Chicago Press, Chicago (1942), pp. 49–60.

[7] Patinkin, Don, *Money, Interest and Prices*, Row Peterson, Evanston, III., 1st edn (1956).

[8] Walras, L., *Elements of Pure Economics*, Allen & Unwin, London (1954).

19 Keynes's Identity, Ricardian Virtue and the Partial Dichotomy*

> Ricardo offers us the supreme intellectual achievement, unattainable by weaker spirits, of adopting a hypothetical world remote from experience as though it were the world of experience and then living in it consistently. With most of his successors common sense cannot help breaking in – with injury to their logical consistency.
>
> (J. M. Keynes, *The General Theory*, p. 192.)

What was the contribution of Lord Keynes to the development of general equilibrium theory? It is becoming possible now, with the passage of a quarter-century, to view Keynes's *General Theory* as an (important) episode in the continuous development of general neo-classical systems. From one vantage point, it is paradoxical that his work should have been accepted as the Grand Departure from its neo-classical progenitors – even though Keynes presented it in that light – since his interpretations of these systems are almost always kept visible for purposes of comparison. On the other hand, Keynes's impact on economic policy, on the methodology and analytics of quantitative economics, and on the working economist's 'vision' of reality, may be seen even today to be of that order of power rightfully termed 'revolutionary'.

From the restricted viewpoint of the pure theory of general economic systems, however, Keynes's fundamental challenge to neo-classical theory is quite clear: it lies in his construction of a model from which flexibility of the money wage rate cannot eliminate an excess supply of labour. In the sense of pure general equilibrium theory this is the core of the Keynesian performance, and his altera-

* Published originally in the *Canadian Journal of Economics and Political Science*, 27 (1961), pp. 454–60 and reproduced with permission.

tions to general neo-classicism to achieve to achieve such a theory of employment are the most interesting aspects of 'the Keynesian Revolution'.[1]

In the main current of doctrinal development, the overthrow by Keynes of Say's Law (in the Identity sense of Becker and Baumol[2]), if this in fact he did, was a necessary but pedestrian clearing of ground for the major work.[3] We say this because, even if it were shown that Say's Identity was a customary assumption of the neo-classical models of Keynes's day, it is now clear, by virtue of the work of Lange and Patinkin, that it is not a necessary projection of the logic of these systems.[4] Not only does Say's Identity effectively remove money as a good from models which retain it as a postulate, but it violates the logical symmetry which a proper neo-classical model employs to integrate money with the body of real goods equations.[5] This part of the 'Revolution' may now be seen, twenty-five years later, to be only deceptively disturbing to the postulates of neo-classicism, if indeed it was necessary at all.[6]

The essence of the Keynesian system is of quite different stuff. It consists in altering the neo-classical postulates to derive qualitatively different theorems about the employment of labour – a fundamentally new set of theorems derived from postulates not contained in the neo-classical model and violating its vision of economic reality.

It is the irony of the Keynesian Revolution that, while Keynes is viewed by his adherents as having laid the basis for his innovations by destroying Say's Law, it is necessary in rigorously depicting his own model to introduce into it an identity as intuitively objectionable as the one he rejects! Sketched out in the Marshallian vagueness of the *General Theory* is this rigidity which we have called 'Keynes's Identity', and it serves to 'semi-dichotomize' his model into real and non-real sectors in the same manner as the much maligned Say's Identity.

The present paper approaches Keynes's Identity by first discussing a neo-classical model and its dichotomization by Say's Identity in Section I. It then proceeds to an economy where the assumption of Say's Identity is dropped, but where the monetary effect of price changes is identically zero, in Section II. Against this it then derives and discusses, in Section III, Keynes's Identity and the partial dichotomization of the economy. A brief summary and conclusion are contained in Section IV.

A fundamental difficulty charged with great potential for misunderstanding must be faced at the outset: we must formalize 'the

Keynesian model'. The General Theory is many theories, the analysis bears the sophistication of genius, the qualifications genuinely meant to confront a recalcitrant reality rather than carping critics.[7] Any attempt to formalize such analysis is destined to be unfair to its subtlety and to be to some extent a distortion.[8] None the less, in gauging the theoretical contributions of Keynes, sooner or later one must seek the fundamental essence of his system. Not to do so is to deny him a major strength: the creation of a model co-ordinate with, but alternative to, the neo-classical general equilibrium model.

I A NEO-CLASSICAL MODEL AND SAY'S IDENTITY

We assume an economy in which there are two consumer goods, Y_1 and Y_2; two factor services, z_1 and z_2, the latter being labour services; one capital good, c_1, producing a unit of z_1 per period eternally; one type of security, e, whose unit is the riskless promise of the perpetual receipt of a net return of one unit of money per period; and money, U. Markets are purely competitive, expectations of future prices are unique values and of unitary elasticity in every case, and consumers and firms are viewed as adjusting their holdings of cash balances by either reducing their demand for the goods they purchase or by increasing their supplies of the goods they sell or both; that is, by reducing their excess demands for goods other than money.

Denoting market excess-demand functions by E, we may state the conditions for equilibrium in this economy as follows:

$$E_{y1}(P_{y1}, P_{y2}, P_{z1}, P_{z2}, P_{c1}, P_e) = E_{y1}(P) = 0 \tag{19.1}$$

$$E_{y2}(P) = 0 \tag{19.2}$$

$$E_{z1}(P) = 0 \tag{19.3}$$

$$E_{z2}(P) = 0 \tag{19.4}$$

$$E_{c1}(P) = 0 \tag{19.5}$$

$$E_e(P) = 0 \tag{19.6}$$

For the sake of completeness, we define i_k, the rate of return on capital, as:

$$i_k \equiv P_{z1}/P_{c1} \tag{19.7}$$

The rate of return on securities, i, is defined as:

$$i \equiv 1/P_e \tag{19.8}$$

For an equilibrium in the investment sector, it is required that

$$i = i_k \tag{19.9}$$

That is, the net rate of return on all assets must be equal.

$$-E_u - E_eP_e \equiv E_{y1}P_{y1} + E_{y2}P_{y1} + E_{z1}P_{z1} + E_{c1}P_{c1} + E_{z2}P_{z2} \tag{19.10}$$

is a statement of Walras's Law.[9]

For convenience we shall term the goods whose equations are found in systems 19.1 to 19.6 'included goods', and the equations 'included equations', while good U (money) will be denoted the 'excluded good'. The latter is merely the good whose value of excess demand is implied by Walras's Law from the value of excess demands for all other goods. It may be selected quite arbitrarily and need not even be the *numéraire* good. We have chosen money as the excluded good, however, since it is truly the obverse of other markets, reflecting their states of excess demand directly in a manner not true of other goods, given the peculiar role of money in the system.

For the moment the following interpretation is given the securities and capital good markets. Securities are supplied by middlemen who obtain money by their sale in order to purchase capital goods, and redeem their securities in the capital good at the end of the period. Consequently, at the beginning of the period, all wealth is held either as physical wealth (stocks of c_1) or in paper money which is the debt of an external agency (the government). No stocks of securities exist, and, further, the value of excess demand for securities is not identical to the value of excess supply for the capital good.

From a position of general equilibrium we suppose the economy to be jarred by an increase in the excess demand for money to a non-zero value, and the consequent dislocation of all other markets. Prices in all included markets but that for z_2 (labour) are adjusted to re-equilibrate those markets, leaving the excess demand for money (the excluded good) identically equal to the value of excess supply of labour.

If the model is to return to full equilibrium it will be necessary for price flexibility to exist in the labour market, for P_{z2} to fall, and, having fallen, to induce a rise in E_{z2} (excess demand for labour). We may depict the changes in the equilibrium values of all prices but P_{z2}, and the change in E_{z2} after all markets but the markets for z_2 and U (money) have been returned to equilibrium after a fall in the price of labour (P_{z2}), by the system below:

$$\begin{bmatrix} E^{y1}_{y1} & E^{y2}_{y1} & E^{z1}_{y1} & 0 & E^{c}_{y1} & E^{e}_{y1} \\ E^{y1}_{y2} & E^{y2}_{y2} & E^{z1}_{y2} & 0 & E^{c}_{y2} & E^{e}_{y2} \\ E^{y1}_{z1} & E^{y2}_{z1} & E^{z1}_{z1} & 0 & E^{c}_{z1} & E^{e}_{z1} \\ E^{y1}_{z2} & E^{y2}_{z2} & E^{z1}_{z2} & -1 & E^{c}_{z2} & E^{e}_{z2} \\ E^{y1}_{c} & E^{y2}_{c} & E^{z1}_{c} & 0 & E^{c}_{c} & E^{e}_{c} \\ E^{y1}_{e} & E^{y2}_{e} & E^{z1}_{e} & 0 & E^{c}_{e} & E^{e}_{e} \end{bmatrix} \begin{bmatrix} dP_{y1}/dP_{z2} \\ dP_{y2}/dP_{z2} \\ dP_{z1}/dP_{z2} \\ dE_{z2}/dP_{z2} \\ dP_{c}/dP_{z2} \\ dP_{e}/dP_{z2} \end{bmatrix} = \begin{bmatrix} -E^{z2}_{y1} \\ -E^{z2}_{y2} \\ -E^{z2}_{z1} \\ -E^{z2}_{z2} \\ -E^{z2}_{e} \\ -E^{z2}_{e} \end{bmatrix} \tag{19.11}$$

or

$$[E_{z2z2}][dV] = [E^{z2}_{i}] \tag{19.12}$$

where $E^{j}_{i} = \partial E_i/\partial P_j$, the subscript has been dropped from c_1 as unnecessary, and where, if we denote by $[E]$ the matrix obtained by substituting the column vector $[E^{z2}_{i}]$ for the fourth column of $[E_{z2z2}]$, the latter may be seen to be the matrix of the principal minor of $|E|$ formed by striking out the fourth row and fourth column of $[E]$.

From 19.10 we obtain

$$-\ dE_u/dP_{z2} \equiv E_{z2} + P_{z2}dE_{z2}/dP_{z2} \tag{19.13}$$

and, since we began from a position where 9.10 was reduced to

$$-\ E_u \equiv E_{z2}P_{z2}, \tag{19.14}$$

we may substitute for E_{z2} in 9.13 to get

$$P_{z2}dE_{z2}/dP_{z2} \equiv E_u/P_{z2}\,[1 - dE_u/E_u \cdot P_{z2}/dP_{z2}]. \tag{19.15}$$

And we may solve 9.11 by Cramer's method for dE_{z2}/dP_{z2}:

$$dE_{z2}/dP_{z2} = |E|/|E_{z2z2} \tag{19.16}$$

Now, it was assumed in neo-classical analysis: that P_{z2} was flexible; and that $dE_{z2}/dP_{z2} < 0$. In the Hicksian 'stability' analysis of neo-classical systems, for example, it is required only that the economy be imperfectly stable to ensure that the ratio of determinants in 19.16 will be negative, although, of course, perfect stability will guarantee it, *a fortiori*.[10] If all individuals are maximizing satisfactions and all firms are maximizing profits, and aggregate income effects may be neglected, the necessary and sufficient conditions for perfect stability will hold.[11] Thus, whatever one's feelings about the shortcomings of Hicksian stability analysis, if one is willing to make rather unheroic assumptions one can obtain a negative dE_{z2}/dP_{z2} as a theorem derivable from the postulate of maximizing behaviour in the neo-classical model, or, failing that degree of boldness, merely assume it as Keynes does in his presentation of rival models. From 9.16 and one of these two bases,

$$dE_{z2}/dP_{z2} = |E|/|E_{z2z2}| < 0 \tag{19.17}$$

It is obvious from 9.15 that a condition for effecting any shift downward in E_{z2} is that E_u be flexible, that is, that it be possible to change the excess demand for money. But Say's Identity states that $E_u \equiv 0$. Therefore, were Say's Identity postulated, dE_{z2}/dP_{z2} would not be defined for our initial position since such a position could not exist. If we assert that all but one included market is in equilibrium, and the excluded good is money, we must be incorrect under a Say's Identity regime; the inability of the market for money to be in disequilibrium implies that the remaining included market must be in equilibrium.

Or suppose Y_1 is chosen as excluded good and *numéraire*, and money is converted into an included good with price P_u. Then, when an excess supply of labour arises, it can have no effect upon E_u under the regime of Say's Identity, and neither can the price changes evoked throughout the system to re-equilibrate the relevant markets. Also, if one remaining condition is met – that changes in P_u have no effect upon the excess demands for any other good in the system – money is a completely non-integrated good.[12] That is, to generalize, if one of the included equations is an identity, it cannot be used to define an equilibrium condition. Were that good's price to have an impact upon other included goods' prices, we should have one more price to determine than equations to do the determining. If it has no such impact, we may remove it bodily and retain a determinate system of prices[13] in terms of the *numéraire*, Y_1. The economy is so effectively dichotomized into real and money sectors that money disappears and barter conditions prevail.

II THE ZERO MONETARY EFFECT AND LANGE'S RELATION

Let us now relax the assumption that $E_u \equiv 0$, so that the public may seek to adjust its nominal balances to fit its purposes with the money supply fixed. Then 19.15 begins to have some interest for us, since the excess demand for money may now depart from zero. It states that if P_{z2} (the price of labour) falls k per cent and all other included markets are returned to equilibrium, the excess supply of labour will fall if, and only if, the relative change in the excess demand for nominal cash balances is (*1*) negative, and (*2*) greater than k in absolute value.[14] Only in this case will the real value of excess demand for money decrease with a rise in the 'price of money' (the reciprocal of the price level), which means that the real purchasing power spent on included goods must rise. This influx of new purchasing power spent on such goods cushions the induced fall of prices of goods other than labour, preventing at least one of them from falling the full k per cent. The result must be an increased excess demand for labour by virtue of the induced substitution of labour for other factors, or the expansion of outputs by virtue of marginal costs falling more than the prices of all outputs, or a combination of both effects.

We may trace through the effects of a k per cent fall in P_{z2} by assuming that all other prices but P_e fall k per cent instantaneously, P_e (the price of securities) remaining constant. Then i_k (the rate of return on capital goods) remains the same, and the demand for and supply of securities will rise to keep real investment unchanged. The new set of prices will be an equilibrium set for all goods but z_2 and U, unless excess demand for the latter falls by more than k per cent. The case where the relative fall in excess demand for cash balances equals k per cent is the one that Lange has called the 'zero monetary effect.'[15]

In this case we may measure the decline in the 'price of goods' unambiguously as k per cent, so that we need not measure the real value of the decline in cash balances in wage units. Suppose, then, we had an economy which was characterized at all points by a zero monetary effect; that is, one where

$$dE_u/E_u \equiv dP_{z2}/P_{z2} \tag{19.18}$$

We shall term such an unrealistic economy a 'quasi-Say's Identity' economy. Such a system would be one in which wage flexibility would fail to eliminate unemployment, and prices would tumble toward zero as abortive efforts to reduce the real wage were made by labourers. If, instead, we began from a position of full employment equilibrium, and by relaxing our equilibrium conditions, effected arbitrarily a k per cent reduction in all prices but P_e, real-balance effects would be zero. We should have the special case of the zero monetary effect where $E_u \equiv 0$, and we should have a full-blown Say's Identity economy. Just as in the latter, a quasi-Say's Identity economy would be a dichotomized economy, if all prices were flexible, with an indeterminate price level.

But such a model would be an alteration of the neo-classical models in two important respects. First, the *excess* demand for money balances in neo-classical systems is not homogeneous of degree one in all prices but P_e, even if *demand* for such balances were assumed to be. The fixity of the money supply assured that a k per cent decline in all prices but P_e and (we assume for the moment) a k per cent decline in the demand for money balances would lead to a greater-than-k percentage decline in excess demand for money.[16] Under these conditions, then, $dE_{z2}/dP_{z2} < 0$.

The neo-classical models implied, if they did not state explicitly, a second quality of the excess demand for nominal cash balances. The quantity theory of money assumed that a k per cent increase in the supply of money, *ceteris paribus*, with a fixed output and velocity of money, would lead to a k per cent increase in all prices but P_e. The mechanism which projected the economy to the new price level was the operation of the real-balance effect of an increase in the money supply. For the individual, demands for and supplies of the included goods were functions of real balances. Were this 'real-balance effect' not to operate, so that the public swallowed increases in the money supply indefinitely, we should have a Say's Identity or quasi-Say's Identity economy, depending upon whether we began with an equilibrium or disequilibrium in the money market, and the economy would be effectively dichotomized.[17]

But a drop in all prices save P_e by k per cent, the supply of money being constant, is equivalent to a rise in the money supply by $100(k/1 - k)$ per cent. This enhancement of real purchasing power inherent in the public's stock of money wealth should lead to shifts in the demand for goods in the included sector. It may increase the demand for consumption goods, in which case the 'Pigou effect' operates to prevent all prices from falling k per cent in equilibrium, thereby decreasing the excess supply of labour in the manner stated above.[18] Should it increase the demand for securities we have the 'Keynes effect'.[19] Or, lastly, should it decrease the supply of labour (increase the demand for leisure) by virtue of the larger real balances held by labourers, we should have the 'Tobin effect'.[20] Any or all of these effects will lead to a fall in the excess supply of labour, or, identically, a decline in the nominal excess demand for money by more than k per cent. Continued declines in P_{z2} under such conditions will lead the economy back to full equilibrium.

Therefore, with the real supply of money increasing and the real demand for it decreasing as its price rises, the neo-classical system should yield the desired result. But, if we wish to make the assumption that the demand for nominal cash balances falls exactly $k(1 - S_u/D_u)$ per cent (where S_u is the supply of cash and D_u the demand for cash) with a k per cent decline in all prices but P_e, or the assumption that the monetary authorities reduce the money supply by k per cent while the demand for cash balances is homogeneous of degree one in all prices but P_e, we may have a quasi-Say's Identity economy, and thereby achieve Keynes's goal. Needless to say, it is difficult to find any virtue in such a *deus ex machina*.

III KEYNES'S IDENTITY

Of course, Keynes did not simply employ the above model with a zero monetary effect to obtain his results. He did state the problem as we have done:

> For, whilst no one would wish to deny the proposition that a reduction in money-wages *accompanied by the same aggregate effective demand as before* will be associated with an increase in employment, the precise question at issue is whether the reduction in money-wages will or will not be accompanied by the same aggregate effective demand as before measured in money, or, at

> any rate, by an aggregate effective demand which is not reduced in full proportion to the reduction in money-wages . . .[21]

And he was quite dogmatic in answering it for one set of postulated circumstances – his model in which changes in expectations and other shift-inducing factors were not allowed to occur, so that his functions did not shift in response to such factors:

> Does a reduction in money wages have a direct tendency, *ceteris paribus*, to increase employment, *ceteris paribus* being taken to mean that the propensity to consume, the schedule of the marginal efficiency of capital, and the rate of interest are the same as before for the community as a whole? . . . we have already answered in the negative . . .[22]

The one hedge on direct effects was that the interest rate be held constant, an assumption we shall discuss below.

In his discussion of a second model, where he allows shifts in the schedules to occur, he concludes that the impact upon the propensity to consume will probably be downward because of the redistribution effects of wage changes, and that shifts in the liquidity preference and marginal efficiency schedules are likely to be unfavourable to a reduction in the excess supply of labour. These conclusions are much less dogmatic than those of the 'first' model above.[23]

We shall return to this first model where distributional and expectational shifts are absent. How did Keynes obtain these results? One of the difficulties encountered in answering the question is that neither Keynes nor his critics possess the Ricardian virtue so admired by Keynes in the quotation at the beginning of the paper; common sense obstructs logical consistency at crucial points. Keynes's hedge concerning the constant rate of interest in his conclusion above is one such example. From the side of his critics the debate whether rigid money wages are or are not required in the Keynesian model to obtain an underemployment equilibrium is another example. Those who would argue that they are do so in this wise: if wages were not rigid downward, the decline in prices would lower the interest rate, whose impact upon investment would decrease unemployment, or, failing that, would bring about real-balance effects which would ultimately restore full employment. Therefore, the Keynesian model must rely upon rigid wages for its underemployment.[24]

Let us follow the Ricardian high road, arguing (*1*) that Keynes did construct a system which yielded an excess supply of labour with no tendency towards self-correction, (*2*) that it was not merely the existing neo-classical sub-case of rigid money wages, and (*3*) that it was not a naïve quasi-Say's Identity economy. We shall seek the conditions necessary to obtain the desired results within the Keynesian framework, taking care to guard against the intrusion of 'common sense' modifications of the conclusions following from logical consistency.

We alter our assumption about expectations in only one respect: by assuming that there is given for expectations of any future P_e (the price of securities) a schedule relating it to current P_e, and that this schedule is stable and invariant throughout our analysis. We also make one alteration in our premises concerning securities, assuming now that a stock of them exists at the beginning of the period, all of which are held by and are debts of private individuals. This assumption assures us that, if individuals and firms are similar in their reactions to gains and losses in real wealth, no 'real-indebtedness effect' will occur with price declines, since the gain in real wealth by the creditor will be exactly matched by the rise in the burdens of debtors.[25] The only other non-money stock of existing wealth at the start of the period is holdings of c_1, the non-depreciating capital good, whose value changes proportionately with P_1.

Now, let us start with a full general equilibrium and assume that the public suddenly desires increased real balances, obtaining them by reducing excess demands for consumption and investment goods. Let all markets but those for labour and money be re-equilibrated, so that 19.14 holds at the initial disequilibrium position in this economy as in our neo-classical economy.

As before we assume that P_{z2} (the price of labour) is bid down k per cent, and that all prices but P_e fall an equivalent percentage. This will not disturb the equilibrium of the markets for Y_1, Y_2, z_1 and c_1, nor will it change the excess supply of z_2, if no real purchasing power enters this real sector from the paper (money and securities) sector.

If we denote by dE_u^* the differential of cash balances before the securities market is returned to equilibrium, then, when only the markets for U, e, and z_2 are out of equilibrium, it follows from 19.10 that

$$- dE_u^*/dP_{z2} - P_e dE_e/dP_{z2} \equiv E_{z2} + P_{z2} dE_{z2}/dP_{z2} \qquad (19.19)$$

and, substituting for E_{z2} from 19.14

$$E_u/P_{z2}[1 - dE_u^*/E_u \cdot P_{z2}/dP_{z2} - P_e dE_e/E_u \cdot P_{z2}/dP_{z2}]$$
$$\equiv P_{z2} dE_{z2}/dP_{z2} \qquad (19.20)$$

Since we shall rule out negative real-balance effects, in order for $dE_{z2}/dP_{z2} = 0$, it is required that

$$dE_u^*/E_u + P_e dE_e/E_u = dP_{z2}/P_{z2}. \qquad (19.21)$$

That is, the money value of the change in excess demand for paper assets as a percentage of the excess demand for money formerly existing in the initial state of disequilibrium must be equal to the percentage price change. But since we assume constant stocks of money and may assume negligible changes in the stock of securities by virtue of current investment demand, changes in the *excess* demand for money and for securities are equal to changes in the *demand* for these paper assets. Therefore, we may state condition 19.21 as follows: the value of the change in demand for paper assets must equal $-k$ times the excess demand for money in the initial disequilibrium position.

If this condition holds, the direct impact of the wage decline on the real sector, the rate of interest held constant (which is equivalent to P_e being held constant by 19.8 above), will be nonexistent, in line with Keynes's conclusion. But, of course, the real value of money holdings will have increased by virtue of the price decline. Since P_e is 'the price which equilibrates the desire to hold wealth in the form of cash with the available quantity of cash',[26] or, in Metzler's apt phrase, P_e is determined 'largely by the decisions of asset-holders concerning the proportions in which they wish to hold money and securities',[27] the ratio of real money wealth to real security wealth will have risen, and, under Keynes's assumption that the desire to hold money rises as P_e rises, the latter must be allowed to restore equilibrium in the paper asset sector. Holding tenaciously to our goal of obtaining a model in which such adjustments do not affect the excess supply of labour, we must obtain this re-equilibration of the securities market without disturbing E_{z2}. Let

$$dE_u/E_u = k' \qquad (19.22)$$

where dE_u is the final change in the excess demand for nominal cash balances after all markets but that for labour have been re-equilibrated, and E_u is the excess demand for cash in the initial disequilibrium position. Then Keynes's Identity states that

$$dE_u^* - dE_u \equiv Q_e dP_e, \; k' \equiv \frac{dE_u^* - Q_e dP_e}{E_u} \tag{19.23}$$

where Q_e is the (nearly) fixed stock of securities. That is, Keynes's Identity states: *given a state of excess supply in the labour market and a fall in the money wage rate, this excess supply will never be affected if the monetary effect is confined to the securities market or is absent.*[28]

If the demand for money is less-than-infinitely elastic with respect to P_e in this 'Keynesian' model, P_e must rise and all potential effects of this rise (equals a fall in the rate of interest) on the real goods sector must be effectively nullified to preserve Keynes's Identity. The demand for consumption and investment goods must be completely interest-inelastic, or else 19.9 would no longer hold and net investment would occur to contradict Keynes's Identity. Otherwise we should be back to the neo-classical world where falling interest rates were changing investment (and savings?) to reduce the excess supply of labour.

This neo-classical model seems to be the economy Keynes envisages: Keynes's Identity holds, but substitution and income effects springing from the reduction in the interest rate act to reduce the excess supply of labour. In his constant-interest-rate hedge, above, and in the following quotation, it appears clear that we have not yet reached a distinctively Keynesian economy:

> in the extreme case where money wages are assumed to fall without limit in face of involuntary unemployment through a futile competition for employment between the unemployed workers, there will, it is true, be only two possible long-period positions – full employment and the level of employment corresponding to the rate of interest at which liquidity preference becomes absolute (in the event of this being less than full employment) . . .[29]

We may now relax our Ricardian rigour to assume that E_{z2} rises as the economy functions in a crippled neo-classical manner until the liquidity trap is reached. Keynes's Identity appears to be a peculiar encumbrance, keeping direct real-balance effects out of real goods markets, but income and substitution effects are operating in the familiar way to reduce unemployment.

But when we have reached the point where the demand for cash balances becomes infinitely elastic, say at P'_e, before E_{z2} shrinks to zero, we have reached the distinctively Keynesian model.[30] Further declines in the interest rate are arrested and shifts in the savings, investment and supply of labour schedules by virtue of the real-balance effect are prevented by Keynes's Identity. The latter now becomes Lange's zero monetary effect as an identity relationship, and the economy a quasi-Say's Identity economy. Substitution and income effects ceased with cessation of rises in P_e at the ceiling P'_e, and now the real-balance effect, the only remaining effect which can reduce unemployment, disappears.

Common sense will out! Haberler simply refuses to accept this Keynesian equilibrium as a resting place. If P_{z2} continues to fall, even in the liquidity trap the schedules must shift by virtue of the real-balance effect, he argues, and 'the truth is that what would happen in this case cannot be told within the Keynesian framework.'[31] A continuous rise in real wealth being wholly expended on money balances cannot continue, common sense argues. But it must if the contribution of Keynes to general equilibrium economics is to exist at all, and we shall retain the Ricardian virtue.

Going down the Keynesian road to its logical end, we are troubled by the appearance of Lange's zero monetary effect in a world without income or substitution effect possibilities to anchor us. P_e is consistent with any level of money prices which does not lower it from the liquidity trap value P'_e. Therefore, the price level is indeterminate, and, if P_{z2} remains flexible, it will fall towards zero and take all prices but P_e with it. We may determine a price level, of course, by arbitrarily setting a price floor on some one good, such as the money wage rate, as Keynes does. Thus, rigid money wages are necessary to determine a price level, not to obtain a less-than-full employment equilibrium in the Keynesian model.

But note what has been done to the neo-classical general equilibrium model in constructing this Keynesian model. Initially, money and real goods are substitutes in the sense that the public will seek to increase its excess demand for money by reducing excess demand for real goods, or, presumably, *vice versa*. But once this initial substitution has taken place, the economy is suddenly dichotomized to a partial extent.[32] Even while P_e rises it is the income and substitution effects which are holding paper and real sectors together; no real-balance effect is operating directly to hold them together. When this tenuous reed breaks, the dichotomy is complete; the public swallows increased real balances limitlessly.

This is what we mean by 'partial dichotomization'. The real value of excess demand in the real goods sector remains constant as that of the paper sector swells without limit. The price level is indeterminate and P'_e (the ceiling price of securities) ceases to exercise any effect upon the excess demands for real goods. The initial ability of the public to increase excess demand for cash at the expense of excess supplies of real goods indicates that the dichotomy is not complete. Why it is that, once crossed, the bridge ceases to exist is not clear, even given the expectations of future movements in P_e. If the public always sought to change its real balances by adjusting only its holdings of securities, we should have a complete dichotomy between paper and real sectors in this sense, even though income and substitution effects would hold both together. But, then, how Pyrrhic would be the Keynesian 'victory' over Say's Identity!

IV SUMMARY AND CONCLUSION

In an attempt to isolate the meaning of Keynesian economics for the body of general equilibrium theory we have sought to find in it a rigorous model which achieves the results its author sought. Certainly we have been rigorous and simple, perhaps overly so for a fair presentation of the kernel of theoretical Keynesianism stripped of its *obiter dicta*. Once more we plead the inevitability of the vice given the choice of the task, and leave the reader to decide how great the distortion is.

We have attempted to show the differences of Keynes's model from the neo-classical counterpart by bringing to the fore the Keynesian Identity – a statement of the exclusive nature of the real-balance effect permissible in the system. More importantly, we have pointed to the implications of this Identity, strictly adhered to in full Ricardian virtue, for the dichotomization of the economy into real and paper-asset sectors.

What remains of the Keynesian performance in this theoretical area if one's intuition leads one to view its theorems sceptically because of the presence of the Keynes Identity postulate? The writer believes that Patinkin's interpretation will grow to be more generally accepted in the future.[33] That is, involuntary unemployment is most readily – perhaps only – explainable in terms of the dynamics of adjustment of the market economy. To the extent that Keynes focused attention upon the slowness, or weakness, of these adjustments, his analysis contributes fruitful insights. But in the field of statics, where the existence of counteracting forces, however weak, is sufficient to yield full employment, his performance was essentially a failure.

Notes

1. In this sense, chapter 19 is the most important chapter of *The General Theory of Employment, Interest, and Money* for it deals with the effects upon the Keynesian model of changes in the wage rate most explicitly (but unsatisfactorily).
2. G. Becker and W. Baumol, [2], 356–7.
3. Cf. G. Haberler [6], pp. 173–5. For a recent contrary view, see E. O. Edwards [4], p. 420. Professor Edwards asserts that Keynes's major contribution was to emphasize the aggregate demand curve, although, in retrospect, he neglected factors which would eliminate excess demand for money at full employment. But to say this is to imply that a distinctively Keynesian static theory does not exist, and that Keynesian unemployment was due to the slowness of adjustments in a neo-classical world. In short, something more than the possibility of an excess demand for money is needed: namely, a set of postulates providing for its perpetuation.
4. O. Lange [15], pp. 49–68, and [16], and D. Patinkin, [21], pp. 119–21. See especially p. 119, where Patinkin explicitly sides with 'those who deny that [Say's Identity] is a basic component of the classical and neoclassical position'.
5. R. Kuenne [13], pp. 119–24.
6. To choose another illustration of our point: even were Patinkin's hypothesis proved that Keynes's speculative money holdings were not a function of the price level, not only is such a postulate unnecessary for his system, but it violates the logic of the system. Its removal cannot but improve the model and may constitute an important preliminary, if indeed the postulate is present at all. But such removal cannot be a major disturbance to the Keynesian model. Cf. Patinkin [21], pp. 173–4. (However, the equations on p. 173 have been reversed in their numbering through typographical error.)
7. Cf. R. Harrod [8], p. 69: 'No one had a greater sense than he of the complexity of economic adjustments and of the numerous reservations that had to be made to a generalization. In his case these reservations always sprang from a strong sense of their importance and not in the least, as in some writers, merely to safeguard himself from criticism – for on the latter point he was notoriously and signally indifferent.'

8. Cf. A. Smithies [25] p. 563.
9. Lange [15], p. 50.
10. J. R. Hicks [10] pp. 315–19, 325. The literary statement, pp. 62–77, contains an ambiguity which can be misleading. See R. Kuenne [14].
11. Except for those cases where the conditions for Hicksian perfect stability are necessary and sufficient for true dynamic stability – when only two goods exist in the economy, when the matrix $[E]$ is negative definite, when all goods are gross substitutes – we cannot derive the condition that principal minors of $|E|$ must be opposite in sign to the latter. Only the sign of $|E|$ is derivable from the Routhian conditions, which are necessary and sufficient for the real parts of the latent roots of the characteristic equation of a linear dynamic system to be negative.
12. Kuenne, [13], pp. 120–1.
13. Cf. Lange [15], and F. Modigliani [19], pp. 214–16.
14. This result was derived explicitly by Lange [16], pp. 20, but it was stated clearly by Keynes [12], pp. 259–60. Lange develops his analysis for an economy in which all real excess demands are homogeneous of degree zero in all prices but P_e (the price of securities), the supply of money, and the supply of securities, which is the homogeneity postulate implicit in neo-classical systems not characterized by Say's Identity.
15. See [16] pp. 5–20. The terminology in this area has become a bit overextended, so that a glossary may prove helpful. Let us postulate a k per cent fall in all prices but P_e (i.e., all prices but the rate of interest, i). Then:

    (*1*) Lange, beginning from a position of disequilibrium in two markets, defines a 'monetary effect of the price change' in terms of changes in the *excess* demands of the two markets out of equilibrium. Since the excess demand for money reflects the second disequilibrated market perfectly, he needs only to concentrate upon changes in the money market. This latter market must therefore register any changes in the *amounts demanded* in the equilibrium position of goods other than money. If dE_u/E_u is equal to $-k$, none of the amounts demanded of the goods whose markets are once more in equilibrium can have changed, and the monetary effect is defined to be zero. If dE_u/E_u is negative and greater than k in absolute value, some of the amounts demanded of these goods must have increased, and if dE_u/E_u is negative and less than k in absolute value, some of these amounts demanded must have decreased. Lange speaks of a 'positive monetary effect' in the first case and a 'negative monetary effect' in the second.

    (2) Patinkin, beginning and ending with *all* markets in equilibrium, defines a 'real-balance effect of the price change' in terms of changes in the *amounts demanded* of goods and of money balances. This definition is for a subcase of Lange's: where the excess demands in the money market and the market for the second good are zero.
16. [16], p. 14. With a fixed supply of money, S_u, the demand for money, D_u must expand by only $k\,(1 - S_u/D_u)$ if the excess demand for money is to fall by k per cent.
17. This work is, of course, that of Patinkin. See [21], pp. 39–45.
18. This effect was discussed by A. C. Pigou in [22], pp. 343–51; [23], pp.

241–51; and in ch. IX, pp. 123–34, of [24]. Pigou's analysis originally was stated in terms of the long run and for the effect of a drop in prices on the supply of money wealth. Kalecki pointed out that the effect could only operate on that portion of the money supply which was not offset by private indebtedness to the banking system. See [11], pp. 131–2. Pigou changed his concept in his later work to one of the effect of a fall in prices on physical as well as money wealth, but the concept remained a long-run concept.

This effect was stressed by Haberler in [5], pp. 242, 389, 403, 491–503, in the context of the short run. It was independently discovered by T. de Scitovzsky in [3], pp. 69–88, but he felt it to be of such long-run quality that it held true only of generations, not of individuals (pp. 72–3). His wealth concept is one of money balances in the special conditions of his non-capitalistic model.

The effect of wealth enhancements upon consumption has received extensive consideration since these early formations. Patinkin has refined Pigou's 'wealth' concept to one of net indebtedness of the government. See [20] and J. Tobin [26], p. 112. The effect of an increase in physical wealth upon consumption was noted briefly by Keynes. Cf. G. Ackley [1], pp. 154–61, and Patinkin [21], p. 464.

19. This will be discussed extensively below. The term is Haberler's used in [7] and is to be distinguished from Ackley's use of the same term for the long-run effect described in n. 18.
20. Tobin, [26], pp. 109–23. Should this effect operate, as Tobin points out, the supply of labour becomes a function of the money wage, lending credence to the familiar Keynesian assumption.
21. [12], pp. 259–60.
22. [12], p. 260.
23. [12], pp. 261–71.
24. Cf. Modigliani [19], p. 211; Pigou [23], pp. 241–51; Patinkin [20], pp. 278–9; and Haberler [6], p. 167.
25. This result springs from Kalecki's exception to Pigou's early analysis and is explicitly applied to securities by Patinkin in [21], p. 205.
26. Keynes [12], p. 167.
27. L. Metzler, [18], pp. 97–8.
28. Patinkin seems to have been the first to state this explicitly in [21], pp. 14, 163, and 463–71. Cf. also Metzler [18], p. 95.
29. [12], p. 191.
30. Or at least what Metzler has called the Keynesian system at its simplest [18], pp. 95–6. Cf. also Patinkin [20], p. 257.
31. Haberler [6], p. 169. See also [7].
32. Patinkin hints at this when, after discussing Keynes's limitation of the real balance effect to securities, he says: 'Looking back on the nature of these errors, we cannot but be struck by the irony that they should have emanated from the man who did most to demonstrate the fundamental inseparability of the real and monetary sectors of the economy . . .' See [20], p. 270.
33. [21], p. 237.

References

[1] Ackley, G., 'The Wealth-Saving Relationship', *Journal of Political Economy*, 59 (1951), pp. 154–61.
[2] Becker, G. S. and W. J. Baumol, 'The Classical Monetary Theory: The Outcome of the Discussion', *Economica* (NS), 19 (1952), pp. 355–76.
[3] de Scitovzsky, T., 'Capital Accumulation, Employment, and Price Rigidity', *Review of Economic Studies*, 8 (1940–41), pp. 69–88.
[4] Edwards, E. O., 'Classical and Keynesian Employment Theories: A Reconciliation', *Quarterly Journal of Economics*, 73 (1959), pp. 407–28.
[5] Haberler, G., *Prosperity and Depression*, League of Nations, Geneva, 3rd edn (1941).
[6] Haberler, G., 'The General Theory', in [8], pp. 173–5.
[7] Haberler, G., 'The Pigou Effect Once More', *Journal of Political Economy*, 60 (1952), pp. 240–6.
[8] Harris, S. (ed.), *The New Economics*, Dennis Dobson, London (1947).
[9] Harrod, R., 'Keynes, the Economist', in [8].
[10] Hicks, J., *Value and Capital*, Oxford University Press, London, 2nd edn (1946).
[11] Kalecki, M., 'Professor Pigou on the "Classical Stationary State" – A Comment', *Economic Journal*, 54 (1944), pp. 131–2.
[12] Keynes, J. M., *The General Theory of Employment, Interest and Money*, Harcourt, Brace, New York (1936).
[13] Kuenne, R., 'Patinkin on Neo-Classical Monetary Theory: A Critique in Walrasian Specifics', *Southern Economic Journal*, 26 (1959), pp. 309–15. Reprinted as Chapter 18 of this volume.
[14] Kuenne, R., 'On Hicks's Concept of Perfect Stability in Multiple Exchange', *Quarterly Journal of Economics*, 73 (1959), pp. 309–15. Reprinted as Chapter 3 of this volume.
[15] Lange, O., 'Say's Law: A Restatement and Criticism', in O. Lange (ed.), *Studies in Mathematical Economics and Econometrics*, University of Chicago Press, Chicago, 1942, pp. 49–60.
[16] Lange, O., *Price Flexibility and Employment*, Principia Press, Bloomington, Ind. (1944).
[17] Lutz, F. and L. Mints (eds), *Readings in Monetary Theory*, Blakiston, Philadelphia (1951).
[18] Metzler, L., 'Wealth, Saving and the Rate of Interest', *Journal of Political Economy*, 59 (1951), pp. 93–116.
[19] Modigiliani, F., 'Liquidity Preference and the Theory of Interest and Money', *Econometrica*, 12 (1944), pp. 45–88.
[20] Patinkin, D., 'Price Flexibility and Full Employment', reprinted in [17], pp. 252–83.
[21] Patinkin, Don, *Money, Interest and Prices*, Row Peterson, Evanston, Ill., 1st edn (1956).
[22] Pigou, A. C., 'The Classical Stationary State', *Economic Journal*, 53 (1943), pp. 343–51.
[23] Pigou, A. C., 'Economic Progress in a Stable Environment', reprinted in [17], pp. 241–51.

[24] Pigou, A. C., *Employment and Equilibrium*, Macmillan, London, 2nd edn (1949).
[25] Smithies, A., 'Effective Demand and Employment', reprinted in [8].
[26] Tobin, J., 'Asset Holdings and Spending Decisions', *Papers and Proceedings, American Economic Association*, 43 (1952), pp. 109–23.

20 The Walrasian Theory of Money: An Interpretation and a Reconstruction*

Economic theory offers few more interesting challenges than the monetary analysis of Léon Walras.[1] Its importance as a precursor of the better-known 'Cambridge-k' approach, its evolution in the writing of its author, and its importance as the first rigorous attempt to treat money in a fashion analagous to 'real' goods, have been treated elsewhere.[2] Less has been done in the study of its true relationships to the savings-investment equations of the model – relationships which differ significantly from their more famous Wicksellian counterparts – and this study will attempt to fill this gap to some extent. As presented by Walras, the analysis is highly condensed, overly mathematized, and often obscure.

A second aim of this paper is to demonstrate that within the meaning of the whole system as Walras himself conceived it there exists a fundamental contradiction which renders his monetary theory indeterminate as to price level. A reconstruction of the theory has been made to correct this fault and is presented for criticism.[3]

But were the paper to be an exegetical treatment of an episode in the history of doctrine it would be sterile indeed. A third goal of the paper is to present the theory, interpreted and reconstructed, as a model of a monetary economy quite different from those in use today. It offers the possibility of separating the 'asset' function of money from the medium-of-exchange function to the extent of distinguishing two different markets and two different prices for such functions. By keeping these two markets distinct, we may attack such

* Published originally in *Metroeconomica*, 13 (1961), pp. 95–105, and reproduced with permission. I am indebted to colleagues W. J. Baumol, S. T. Beza, E. O. Edwards, and R. E. Quandt for reading an earlier version of this paper and giving valuable guidance. Errors, however, must be charged solely to the author.

problems as the relationship between money-as-flow and money-as-stock more easily, we can obtain a different outlook upon the restrictiveness, and even the meaning, of Say's Law, and we can discern more easily the true role of money in a stationary system. It is to be hoped, then, that the model will be useful analytically and pedagogically.

I WALRASIAN MONETARY THEORY

Walras's definitive presentation of his monetary theory can be interpreted in several ways, each of which can find support from passages in the *Elements*. The present interpretation, therefore, is not offered with claims of exclusiveness. What is asserted is this: the Walrasian system, taken as a whole, emerged from a 'vision' of economic processes which its author employed with consistency when he thought in terms of equations. From this 'vision' and its rigorous formulation emerges an internal logic which interpretations of the processes must not contradict. It is believed that the one presented here is both consistent with this logic and is, in some 'maximum likelihood' sense, the working framework of reference of Walras himself. But, more importantly, it is asserted that the interpretation is analytically useful.

The basic method of the integration of money into the definitive model is simple: it is conceived as a 'good' yielding no utility in itself, but which, by furnishing 'services' of an inventory nature, allows its possessor to command goods in the absence of receipts from current flows. Consumers equate the weighted marginal utility of this service to that of all other goods in order to determine their own demand for cash balances. Consumers' excess supply is taken to a market where it is lent *for one time period* at a price, p_u', per unit of money, to entrepreneurs (primarily), whose demands for real balances are determined rigorously as constant coefficients per unit of output.

To breathe an economic soul into these mathematical ribs is another matter. The primary tenet of the present writer's interpretation is that Walras was an *economic primitive*, in that his models began *ab ovo*, with consumers and entrepreneurs acting within a complex of interrelated markets where the former *literally* supplied *all* services *in kind* to entrepreneurs (primarily). Walras's vision of his economy never left for long nor strayed far from that of the literal exchange economy which was the chrysalis of the more complex mutations.

Walras consistently considers three states in his analysis: (*1*) the *tâtonnement* phase during which equilibrium is established during a 'market day'; (*2*) the undisturbed execution of the contracts made at the end of the market day during the succeeding 'fulfillment week'; and (*3*) the 'dynamic phase', by which he meant merely a process of equilibrium displacements via successive changes in the data of the economy.[4] By far the greater part of his analysis, and all of his contributions with the exception of his marginal productivity analysis, were concerned with the first two of these states.

Consumers are endowed with a stock of assets yielding a 'weekly' flow of saleable (and, with the exception of raw materials, consumable) services. These stocks comprise land, the laborer himself, fixed capital goods, raw materials, finished manufactured goods of the circulating capital variety, and money. Note the inclusion of money as an asset yielding saleable services; since one sells the *services* of money for one 'week' by lending, Walras's money economy is of necessity one with borrowing and lending, and the coexistence of money and a loan market in it a requisite.

The evidence points to the conclusion that Walras consistently perceived the consumer *literally* taking each of these types of capital assets to a *literal* market in which contracts were made at the end of the market-day for the delivery of the assets against (*1*) payment during the week for the use of the asset, gross of depreciation and insurance charges, and (*2*) return of the asset *in kind* at the end of the week. The word *literally* is stressed. Sometimes Walras attempts to clothe his analysis in borrowed finery by bringing these market processes more in line with reality, but he quickly reverts in these instances to his more primitive constructs.[5]

Nowhere is Walras more difficult to understand than in his frequent attempts to indicate how money capital, or, in his pre-money economy, *numéraire* capital, is visualized in terms of his model. Section 255 best exemplifies this confusion and incomprehensibility. Again and again one is forced back to the 'primitive' interpretation of the model.

All delivery dates in the fulfillment week following the market-day are contract-certain, and payments for all goods and services are made at the time of delivery. Two types of goods can be looked upon as having delivery dates at the beginning of phase (*2*), the fulfillment week: money and real inventories. Since neither can be lent for shorter periods than a week, since they exist as assets at the beginning of the week, and since the purchaser of their services gets maximum use from them if delivery is immediate, it seems a logical construction in accordance with the function they perform in the model.[6] However, Walras does not adopt this assumption explicitly. It is so valuable in pointing up the peculiar qualities of money in the theory, however, that it becomes difficult to understand why he did not make the point explicit.

Now, in this model, the demand to hold cash balances at the *start* of the fulfillment week (which is a demand for the services of money during the execution of contracts) becomes clearly different from the acquisition of money at the *end* of the week. The latter demand is a demand for an *asset* (not a *service*); money in this sense is a source of saleable services in future periods, and as such is desired as any physical capital good would be. This symmetry of treatment in a Walrasian model of the physical and monetary asset, with entrepreneurs (primarily) demanding services from an existing stock for *use* during the fulfillment week and consumers seeking individually to build up or reduce their stocks at the *end* of the period, allows us to distinguish clearly between demand for cash balances (in the sense of money services) and demand for money assets.[7]

The Say's Identity assumption of Walras's model now becomes a more sophisticated concept than that ordinarily associated with it: it implies that the individual does not seek to change his holding of money assets at the end of the week, *not* that he does not seek to adjust the cash balances he hold during the week. The demander of cash balances for use during the week can increase them by borrowing them. Only if he is willing to accept the accretion of money at the end of the period, as he would any other new capital accretion, in the form of assets, not services, can he obtain cash in this way.

I distinguish this market where cash balances are literally hired at a price, p'_u, per unit per period on the promise of their return at the end of the period from the market for securities (discussed below) we shall call the 'promissory note' market. In it the price of money services is determined; in the securities market their price, p_e, is determined.[8]

In still another market, circulating capital (*existing* inventories of raw materials and finished goods) is lent at prices p'_b – a price for each type of good – against the promise of return of capital goods at the end of the period, as well as payment for the service during the period.

The securities market is a mart where those who would purchase new capital goods of the fixed or circulating type go to buy perpetuities guaranteeing the riskless receipt of one unit of *numéraire* at a price p_e. These securities are redeemed at the end of the period in new fixed and circulating capital goods to an equivalent value, which goods are added to similar stocks already in consumers' possessions. If we neglect the existence of consumers' preferences for the direct consumption of the services of these goods, it becomes immaterial in a stationary flow which of these goods (or money) is actually delivered. This remains true when these preferences are taken into account, since the services of delivered assets may be exchanged for those of desired assets at no penalty. Lastly, while new capital goods are delivered in kind at the end of the period, deliveries of the securities occur throughout the period, and payments for them must be made throughout the period.[9]

Securities and promissory notes are not substitutes on either side of the market. If an entrepreneur who desires working capital sells a security and 'covers' his obligation by purchasing a new capital good for delivery at the end of the period, he will not be supplied with cash balances at the beginning of the week. For the consumer, purchase of a promissory note yields a current income, purchase of a security does not. For him also they serve separate functions.

In the securities market it seems clear that Walras thought in terms of p_e, the price of one unit of *numéraire* received in riskless perpetuity, as actively determined by buyers and sellers, rather than the prices of fixed and circulating capital goods being determined, as would be true if securities were truly fictitious. That is, the price p_e and its reciprocal, i_k, the rate of return on 'capital goods', were determined, and this rate of return was transmitted in some unspecified way to the submarkets for individual types of fixed and circulating capital goods.[10]

At this point there is an unwritten chapter in Walrasian capital theory: he never specifies the manner in which changes in the value of i_k are transmitted to the separate fixed and circulating capital goods markets. He *defines* the selling prices of the these goods as the net price of their services capitalized at i_k, so that these selling prices can never depart from this value; yet he does not furnish the links to indicate how he arrives at this result or what mechanism exists to assure it.[11] All of his reasoning about the achievement of equilibrium by *tâtonnements* in the capital goods markets starts from the assumption that selling prices are rigorously determined when i_k is given. In equilibrium this price must equal average costs of production in each capital good market, and the sum of the values of goods produced at this price structure must equal desired savings.

While we may condemn Walras for a good deal of obscurity in his analysis of capital goods in these respects (and others), the basic elements of price and quantity determination are present in the model. For example, if we create a class of 'arbitrageurs' in the stock market who seek to profit by selling securities at p_e and buying capital goods at prices reflecting a higher i_k, but who, like Walrasian entrepeneurs, make no profits in the end, the mechanism can be provided. When this is done, there is no need to assume that the selling prices of all types of capital goods are at all times equal to their service prices capitalized at the rate of return implicit in p_e.

II A RECONSTRUCTION

It is otherwise with Walras's theory of money. In this sector, p'_u is actively determined in the promissory note market, but either i, the rate of return on money assets (the ratio of p'_u to p_u), or p_u, the price of money in *numéraire*, remains to be determined.[12] Walras, as in the case of the circulating and fixed capital analysis, had a choice: given p'_u, how is either i or p_u determined? One other market must be specified.

The whole logic of the model and its symmetry up to this point argue for one solution: *a market for money as an asset.* Consumers who desire to increase their stock of wealth can 'invest' in principle in new fixed capital, new circulating capital, or 'new' additions to their stocks of money. Yet nowhere does Walras indicate this alternative of buying money as an asset: Say's Law, in the Lange sense of a demand for increases in money stocks being identically zero for every consumer, is implicit.[13] This inability of Walras to shake loose the 'neutral money' dictate of his contemporaries led him to sacrifice the major contribution to monetary theory implicit in his model.

In the writer's view this contribution was not the bringing of money into a utility framework for this had always been implicit in the treatment of money by earlier economists; moreover, the extension of marginal apparatus was inevitable and, at least questionable for stationary conditions.[14] A greater achievement was the establishment of money as an asset in a stationary flow and therefore as a logical object of purchase alternative to real goods. No need for uncertainty about interest rate changes was required: merely the preconditions for the existence of money in such a static stationary system.[15]

To savor fully the innovatory character of this achievement one must see that the hoary functions of money – standard of value, medium of exchange, store of value – did not truly describe the nature of the asset demand for money. The accretion of money stocks by individuals at the end of the week is no more a method of storing value than is the purchase of a fixed or circulating capital good. In every sense that is true of these latter types of goods, money is an asset in the Walrasian model.

But Walras did not follow this path. Rather, he takes an incorrect route: although he recognizes that i is determined in a different market from i_k, and is equal to i_k only in equilibrium he falls into the mistake of treating the two as identically equal.[16] Just as he obtains the price of capital goods by capitalizing the prices of their services at i_k so he obtains the price of money, p_u, by capitalizing p'_u at $i_k \equiv i$. That is, he implicitly reverts to the assumption that the rate of return on money and on capital goods is identically the same, contradicting not only his explicit assumption but the logic of his markets.

But Walras further indicts himself. Earlier he had taken great pains to explain that assets would have no prices in an economy where persons were not seeking to increase or decrease their individual holdings of a constant stock. The same must hold for money: if some persons are not seeking to decrease their holdings of money *as an asset*, and others to increase them, there can be no price of money, since all will seek to borrow or lend its *services*. Economically, p_u is indeterminate, even if mathematically it is determinate. But since, unlike physical asset services, money services can be useful only if the asset yielding them has a price, the utility of money vanishes and it cannot exist.[17] Say's Identity is a bar to determining p_u in any asset market by capitalizing the value of its services.

The writer has pointed out elsewhere that an additional difficulty arises because had Walras retained the assumption that $i = i_k$ only in equilibrium, his equations for the demand for nominal balances by consumers would be correct only when the economy was in equilibrium. They must be identical to escape this criticism.[18]

The method by which the theory must be reconstructed is clear: consumers must be given the opportunity of alloting savings to the purchase of additional money *as an asset* for delivery at the end of the period, and i must be allowed to depart from i_k except in equilibrium.[19] The value of p_u is then determined by the necessity of the total demand for increased cash in the asset sense to be zero in equilibrium. The rate of interest, i, then follows identically as the of p_u' to p_u. Let us introduce these new conditions of the model rigorously. Since space limitations forbid the reproduction of the entire Walrasian model, the reader is referred to Walras's *Elements* if the integration becomes unclear at any point.

For the Individual, *i*:

1. In Walras's model, a good, 'securities', exists, whose unit is the riskless receipt in perpetuity of a unit of *numéraire*. At the beginning of the market-day the consumer has a stock of these (in the form of fixed and circulating capital goods), Q_{ie}, and he will desire to change his stock by the amount X_{ie} (positive or negative in sign) in the coming period at the price, p_e, per unit;
2. We create a new good, 'money assets', whose unit is the nominal amount of money necessary to obtain a unit of *numéraire* in perpetuity through purchase of promissory notes. A stock of these assets is held at the start of the market-day, Q_{iq}, and demand (positive or negative) will be X_{iq} at the price p_q per unit;
3. Therefore, desired (gross) saving in units of *numéraire* will be:

$$X_{ie}\, p_e + X_{iq}\, p_q$$

and letting S be savings measured as units of perpetual *numéraire* obtained,

$$X_{ie} + X_{iq} = X_{is},\ Q_{ie} + Q_{iq} = Q_{is}$$

4. With given price expectations a marginal utility function, r_i, may be defined for S in Walrasian fashion:

$$r_{is} = f_{is}\,(Q_{is} + X_{is})$$

5. Letting $\varnothing$ denote the marginal utility of income, we obtain equilibrium when

$$\frac{r_{is}}{p_e} = \varnothing,\ i_k \geqq i$$

$$\frac{r_{is}}{p_q} = \varnothing,\ i_k < i$$

6. From these we obtain

$$X_{is} = f_{is}\ (P)$$

where P symbolizes all prices in the model (including those we have excluded from consideration).

For the Economy, Summing Over All Individuals:

1. For Walras' demand function for securities, we substitute

$$X_e = \sum_i X_{ie} \tag{20.1}$$

and add three new relations:

$$X_s \equiv \sum_i X_{is} = X_e + X_q \tag{20.2}$$

$$p_q \equiv \frac{I}{i} \tag{20.3}$$

$$X_q \equiv \sum_i X_{iq} \tag{20.4}$$

2. For Walras's equilibrium condition that the price of money services over the price of money equal a common rate of return, we substitute the definition:

$$i \equiv \frac{p_u{}'}{p_u} \tag{20.5}$$

3. Since the aggregate demand for money asset *additions* must be zero in equilibrium, we have

$$X_e\, p_e = \sum_i X_c\, p_c \tag{20.6}$$

where X_c is the amount of newly-produced capital good c and p_c its equilibrium price.

Also,

$$i = i_k \tag{20.7}$$

$$i_k \equiv \frac{I}{pe} \tag{20.8}$$

This analysis introduces four new variables into Walras's model – X_q, X_s, p_q, and i – and a net addition of four equations. Lastly, in all the equations of the system, the price vector must now include two of the three quantities p'_u, p_u, and i, while in Walras's system, only p'_u and i_k were needed.[20]

III CONCLUSION

With this reconstruction, the writer believes, Walras's monetary theory is determinate and completely integrated into the body of 'real goods' equations. In so doing we have been forced to give an economic interpretation of the ambiguous Walrasian analysis of savings, investment, capital goods, and money. Because the Walrasian system remains one of the fundamental frameworks of economic reasoning, these two purposes alone would merit our efforts.

But frameworks exist for usefulness – analytical and/or pedagogical. The writer again urges the usefulness on both scores of this reconstruction of the Walrasian theory. First, it is a different model from the classical, Marshallian, Wicksellian and Keynesian models now in use, and as such affords us the rare opportunity of approaching old problems with 'new' tools. Second, it allows money's functions as cash balance and as asset to be split apart for separate consideration. Analysis of 'money-as-flow' and 'money-as-stock' is rendered easier, as is the teaching of this distinction. Third, Say's Law acquires a new sophistication by virtue of the model. Lastly, as the writer has shown elsewhere[21] the model escapes Patinkin's criticism of the existence of the Invalid Dichotomy even though it lacks a 'cash-balance effect' in Patinkin's sense.

Notes

1. This paper will focus upon the definitive treatment of money by Walras within the framework of his general equilibrium system as found in Lessons 29, 30 and 31 of [17], translated by William Jaffe, London, 1954. This is a rendition of the posthumous 1926 edition of the original.
2. See Jaffé [3], pp. 600–2; A. Marget [11], [12]; and D. Patinkin [13], pp. 388–90.

3. It is to be emphasized that the nature of this indeterminacy is *not* the, existence of Patinkin's Invalid Dichotomy. Cf. Patinkin [13], pp. 107–13 and 455–9. In a separate paper, [8], I have discussed Walrasian monetary theory in the light of Patinkin's criticisms, but some of the differences in interpretation between us will be pointed out below.
4. See Walras [17], p. 319, for an explicit presentation of this schema. For a more detailed explanation of Walras's conception of 'dynamics' see pp. 307–8, 318, and 380. In the last reference, note the statement that the discrete market-day–fulfillment week device is dropped in favor of continuous processes *only* when a dynamic framework is adopted. Cf. also G. Leduc [10], pp. 450–3.
5. Critics of this paper have been convincing in arguing that the long documentation supporting this contention had little interest to the non-specialist and was best omitted. It will be included at length in a future book on general equilibrium theory. The crux of the matter is this: in Walras's model, are capital goods lent in kind to entrepreneurs (as we have asserted they are) or is inventory and fixed capital lent in some 'money capital market as money and the real capital goods retained period after period by entrepreneurs?

   Upon the answer hinges the role of cash balances in the model as well as the manner in which the interest rate is determined. Unlike Wicksell, whose adoption of the second interpretation was explicit (see [18], pp. xxvi, 102, 108), Walras is ambiguous. In only two references can the writer find evidence of a 'money capital market': in section 235 where he explains why abstraction from it is possible, but later states that it is distinct from the capital goods market as an annex to the productive services market. Only if this market can be conceived of as our 'securities' market will it fit out interpretation.

   The second reference is the definition of capital on p. 317 as goods hired in money, not in kind; in the next paragraph, however, he moves explicitly to the interpretation adopted here. On the other hand, references to the lending of assets in kind are more numerous: pp. 43, 243, 318, 325, and 329, for example. These considerations do not do full justice to the reasons for adopting the present interpretation, for agreeing to exclude which the writer may be pardoned a grimace.
6. These points are further developed in my [7], pp. 1–10, and [6], pp. 241–9.
7. That this symmetry did not exist for Walras because of his adoption of Say's Identity will be developed below. Also, at this point, the framework becomes valuable as a pedagogical device to distinguish between money stocks and flows.
8. The Keynesian and Walrasian outlooks are similar in that changes in the supply of money are reflected initially in financial market movements, not in goods markets. When Patinkin writes that this case was never advanced in the (neo-classical) literature, he neglects the fact that this is a feature of the Walrasian model. See Patinkin [13], p. 109.

   Also at this point Patinkin's interpretation of the excess supply of money in consumers' hands being used to purchase real capital goods cannot be a correct reading of the Walrasian system, since new capital goods do not begin paying a return until the period after their construction, and no current income is earned on securities. The term that does represent the gross demand for new fixed and circulating capital in Walras, $D_e p_e$, where D_e is the demand for securities, does not introduce any interest payments into the budget equation of the consumer. The term, $O_u p'_u$, in that equation, where O_u is the amount supplied of money services – the term Patinkin seeks to interpret – must be seen as the amount earned in current income by purchasing promissory notes as a means of lending entrepreneurs (primarily) their working balances in a primitive economy in the sense above. See Patinkin [13], pp. 399–400.

9. I would therefore agree with Patinkin and disagree with Jaffé in believing that the good 'securities' did actually exist in Walras's thinking, although my more general tendency is to move in the direction of the primitivistic interpretation used by Jaffé in this instance. Two reasons lead to the conclusion: (*1*) it provides an explanation for the *monnaie d'épargne* that Walras includes in the consumer's demand for money, since cash held against the delivery of securities is necessary, while it would not be necessary if consumers contracted for capital goods to be delivered at the end of the period. More importantly, (*2*) it allows the net rate of return, i_k, on capital goods, identically equal to the reciprocal of p_e, to be the actively determined quantity in the capital market, rather than the prices of the several capital goods themselves. See Jaffé's remarks in [17], pp. 542–3 (in which also my different interpretation of circulating capital services in the budget equation from that of Jaffé will become apparent) and Patinkin [13], pp. 397.

10. This is most clearly seen in Walras [17], but see also pp. 268, 282 and 329. It is therefore misleading to say, as Stigler does, that i_k is *defined* as the ratio of the price of the service of a capital good to its price as an asset, if this be taken to imply that i_k follows passively from two actively determined prices, or that it can differ even momentarily for two or more goods. Cf. G. Stigler [15], p. 242.

11. See [4], p. 331. Schumpeter discusses these markets briefly, but indicates that Walras abandons them for the stock market. He treats securities as fictitious, but is not clear as to whether p_e is determined by capital goods prices or *vice versa*. At one point he says: '. . . it is not the fall in the rate of interest (i_k in our terminology) which plays any direct causal role, but it is the rise in the values of capital goods which reduces (tautologically) the rate of interest . . .' [14], p. 1019. See also pp. 1016–17. But earlier he speaks of the market that determines the price of 'the capitals, hence also the rate of new (net?) revenue.' (p. 1001). This could be interpreted as our securities market.

12. Patinkin is correct, the writer believes, in asserting that it is p_u', not i, that is actively determined in the promissory note market. In the second edition of the *Elements*, before the concept of p_u' and a promissory note market had been developed, Walras speaks in terms of the direct relationship of the demand for money and i. (See [16], p. 383.) But the logic of the definitive model led to the determination in this market of p_u'.

    Patinkin asserts that Schumpeter believed it to be i that was so determined, but on the page where Schumpeter discusses this ([14], p. 1024) he explicitly speaks of i instead of p_u' only under the assumption that money is *numéraire*, in which case they are identical. Patinkin then asserts that the present writer followed Schumpeter in viewing i as being determined in the money market. In proof he cites a footnote in which I pointed out an error in his analysis and loosely referred to p_u' as the rate of interest in an effort to contrast it with p_u'. The body of the article spoke consistently of p_u'. See Patinkin [13], 401–10.

13. O. Lange [9], pp. 49–60. But we must change the specific sense in which Lange refers to the demand for money balances from that for money *services* to that for money as an *asset*. As shown above, desires to change amounts of transactions balances are not identically zero and are effected in the promissory note market.
14. See [4], pp. 333–4, and [5], pp. 631–6.
15. This assertion is open to the challenge of those who have asserted the need for 'dynamic' elements before money can exist. My own position on this point is given in [5] and [7].
16. See Walras [17], pp. 289, 327, and 332 for his realization that i and i_k can be equal only in equilibrium.
17. Walras [17], p. 269.
18. See [6], pp. 246–7.
19. The writer has sought diligently in the rather extensive literature built around Walras's model for an interpretation similar to his own. Most generally, the treatment is of the type of Antonelli [1], pp. 171–89, which merely repeats the Walrasian equations with no attempt to interpret them economically.

    An exception was A. Aupetit, in his [2]. In his analysis, i is consistently distinguished from i_k and p_u' is determined in a clear-cut promissory note market. Having made these two decisions, Aupetit cannot escape the need to determine either p_u or i. Where would he get his needed equation? Would be he forced to find a money asset market and explicitly overthrow Say's Identity?

    It was not to be. He assumes that money is a reproducible good under the gold standard and determines p_u by its cost of production, allowing new production until the rate of return upon it is equal to that of other assets. '. . . Thus is determined for money, under the same conditions as for other capital goods, the quantity of money added to the existing stock and the interest rate . . .' pp. 150–1, and see also pp. 123–8 and 148.
20. The set of equations above fits into my presentation of Walras's model in [5], as follows: equation 20.1 replaces equation 1.1.2; equation 20.5 replaces the third equation of system 1.8; equation 20.6 is a modification of equation 1.71; equation 20.8 replaces the last equation of system 1.8; and equations 20.2, 20.3, 20.4 and 20.7 are net additions.
21. See [8].

References

[1] Antonelli, E., *Principes d'Economie Pure*, Riviere, Paris (1914).

[2] Aupetit, A., *Essai sur la Théorie Générale de la Monnaie*, Guillaumin et Cie., Paris (1901).

[3] Jaffé, W., 'Collation of Editions', in [17].

[4] Kuenne, R., 'Walras, Leontief and the Interdependence of Economic Activities', *Quarterly Journal of Economics*, 68 (1954), pp. 232–54. Reprinted as Chapter 1 in this volume.

[5] Kuenne, R., 'Walras, Leontief, and the Interdependence of Economic Activities: Comment', *Quarterly Journal of Economics*, 69 (1955), pp. 631–6.

[6] Kuenne, R., 'The Architectonics of Léon Walras', *Kyklos*, 10 (1956), pp. 241–9. Reprinted as Chapter 2 in this volume.

[7] Kuenne, R., 'On the Existence and Role of Money in a Stationary State', *Southern Economic Journal*, 25 (1958), pp. 1–10. Reprinted as Chapter 17 of this volume.

[8] Kuenne, R., 'Patinkin on Neo-Classical Monetary Theory: A Critique in Walrasian Specifics', *Southern Economic Journal*, 26 (1959), pp. 309–15. Reprinted as Chapter 18 of this volume.

[9] Lange, O., 'Say's Law: A Restatement and Criticism', *Studies in Mathematical Economics and Econometrics*, University of Chicago Press, Chicago (1942), pp. 49–60.

[10] Leduc, G., 'Schumpeter, Disciple de Walras', *Economie Appliquée*, 3 (1950), pp. 450–3.

[11] Marget, A., 'Leon Walras and The "Cash-Balance" Approach to the Problem of the Value of Money', *Journal of Political Economy*, 39 (1931), pp. 569–600.

[12] Marget, A., 'The Monetary Aspects of the Walrasian System', *Journal of Political Economy*, 43 (1935), pp. 145–86.

[13] Patinkin, D., *Money, Interest and Prices*, Row Peterson, Evanston, Ill., 1st edn (1956).

[14] Schumpeter, J., *History of Economic Analysis*, Oxford University Press, New York (1954).

[15] Stigler, G., *Production and Distribution Theories*, Macmillan, New York (1941).

[16] Walras, L., *Elements d'Economie Politique Pure*, F. Rouge, Lausanne, 2nd edn (1889).

[17] Walras, L., *Elements of Pure Economics*, Allen & Unwin, London (1954).

[18] Wicksell, K., *Interest and Prices*, A. Kelley, New York (1965).

21 Say's Law and Walras's Law Once More: Comment*

In his recent re-evaluation of the role of identities in economic models and his specific interpretations of Walras's and Say's Laws, E. J. Mishan has raised some points which warrant further discussion.[1] My comments will follow the sequence of arguments Mishan employed.

I ON THE ROLE OF IDENTITIES

Although the issue is a definitional, and therefore a secondary, one, it is difficult for me to understand the usefulness of the distinction Mishan urges between an identity 'in the conventional (tautological) definition of the term' and an identity reflecting behavioral assumptions (a 'law' in his usage). An identity of whatever nature – definitional, teleological in some sense, or behavioral – gains meaning only from the analytical system of which it is a part. Say's Law or Identity, as treated by Lange's classic article, asserted that in the neo-classical model within which it was embedded the demand for and supply of money could be viewed as never departing from equality.

There is, therefore, neither logical inconsistency nor a misleading incongruity in Schumpeter's treatment of Say's Law as both an identity and a behavioral hypothesis.[2] Mishan's criticism implies that all behavioral postulates in economic models must find expression as equilibrium conditions, not identities. This is extremely restrictive for no apparent purpose. Moreover, such general prohibitions serve to obscure the relativity of identical relations to the models in which they find employ, and terming them 'laws' as Mishan desires to do furthers their separable existences. Surely the experiences of the profession with Say's Law itself is a sufficient warning of the dangers

* Published originally in the *Quarterly Journal of Economics*, 78 (1964), pp. 478–483, and reproduced with permission.

of lifting a relation which is useful as a postulate in certain models for the study of long-term movements, giving the postulate the semblance of a theorem of proved survivability by christening it a 'law', and then spending years of effort to return it to proper usage in a subset of economic models.

Mishan's criticism of Lange for speaking of the necessary and sufficient conditions for Say's law to exist reflects a confusion on his part, not Lange's. The definition of a phenomenon may be viewed as the statement of the necessary and sufficient conditions for its existence, and this task of defining Say's Law is clearly what Lange is about in the passage cited by Mishan.[3]

II ON WALRAS'S LAW

Mishan's presentation is confusing because at crucial turns of his argument (but not everywhere as we shall see) he loses sight of the essential nature of Walras's Law. As stated by Lange for the simple exchange economy he chose to develop his points, it asserts that the value of all goods (including money) demanded by consumers must be identically equal to the value of all goods (including money) supplied by consumers. Mishan correctly renders this. The basis of this identity – which holds *ex ante facto*, in the sense of consumers' *desires* to purchase or sell, as well as *ex post facto*, and which is true whether or not markets are equilibrated – is simply the assumption that every exchanger will spend all but no more than all of the wherewithal he possesses on goods, including money. Since every actor in the market is constrained by a 'budget restraint', all actors must be, and therefore the value of aggregate demand for all goods must equal the value of aggregate supply for all goods, including money, and identically.

Now it is one of the difficulties with Mishan's strictures concerning the use of identities in models that it may be difficult to decide whether such a relation is in fact a tautology 'in the nature of things' or definitionally or whether it is a behavioral restriction. Surely some such 'wherewithal restriction' on the consumer's planning and actions is necessary if economic problems are to exist, and thus can be interpreted as reflecting first causes. On the other hand, the restriction is behavioral in the sense that it assumes that the individual is not satiated in all goods and that he does not fraudulently seek to spend more than the market gives him. In either sense, within the general equilibrium models we treat it certainly seems to be an admissible and plausible postulate in identity form. Yet, seemingly, Mishan, if it be interpreted as a behavioral restriction, would insist that it be included as an equilibrium condition.

Because Walras's Law reflects the basic assumption that every consumer maximizes his satisfactions subject to a budget constraint, we must view with suspicion Mishan's assertion at this point that Walras did not propose this identity. The budget constraint, at least, appears in each of Walras's systems, for each individual and for all individuals taken together. Indeed, the individuals' and aggregate demand functions for the *numéraire* good are defined as residuals through explicit use of the budget constraints. Also suspect is the sweeping hypothesis 'that identities are superfluous in economic models',[4] given the motivation for Walras' Identity. And, finally, given this motivation, it is misleading for Mishan to assert that Lange derived Walras' Law 'in the way that all identities are derived, by selecting appropriate definitions so as to ensure an identity.'[5]

From this identity of the values of aggregate demand and supply *and* the conditions of equilibrium that demand equal supply in each of $n - 1$ markets follows the necessary equilibrium of the nth market. This deduction is not Walras's Law in Lange's definition – i.e. the identity of the values of aggregate demand and supply –but an implication of it *and* the equilibrium in $n - 1$ markets. Or, alternatively, if when $n - 1$ markets are in equilibrium the model requires the nth market to be in equilibrium also, the value of aggregate demand must equal the value of aggregate supply, and therefore Walras' Law is implied. If the n goods include money, as Lange clearly intends, this last statement is not equivalent to Say's Law as Mishan asserts.[6]

The danger of divorcing the identity from the system of which it is a part is demonstrated by Mishan in asserting:

> this proposition, that if demand equals supply for $n - 1$ commodities it does so also for the nth commodity, can hardly be regarded as an identity. It can certainly be refuted.[7]

In a system with budget constraints operative, and therefore with Walras's Law in effect, the proposition cannot be refuted. It is a theorem of the model in whose derivation Walras's Identity is instrumental.

Dorfman, Samuelson, and Solow (Dorsalow), whom Mishan next

treats, render Walras's Law algebraically and in literary fashion in strict accordance with Lange's definition. However, since they are dealing with a production economy, they state that the aggregate value of all goods purchased by consumers must equal identically the aggregate value of factor services sold. It is assumed, of course, that all factor services are owned by consumers and that all cost payments by entrepreneurs accrue to consumers. Further, Dorsalow explicitly state that the motivation of the identity is the obedience of each consumer to a budget constraint.[8] Once more, Mishan correctly renders the argument.

However, he fails to note that their system does not include a paper money, all of the *n* goods in it being producible and under the regime of price equaling average cost in equilibrium. To me this indicates that they did not wish to deal with the complications of a transactions medium, but, if it does exist, the demand for the money commodity to be used for transactions balances is included in the overall demand for it and therefore the demand for money is found in Walras's Identity. The first is the more sensible interpretation, and if adopted Walras's Law does indeed become trivially Say's Law. But Mishan asserts that Dorsalow's rendition of Walras's Law is the same as Lange's definition of Say's Law in the belief that Dorsalow's system contains money, and therefore in the general rather than the special case. If the second interpretation is adopted, Dorsalow's statement of Walras's Law contains the demand for money and Mishan would be incorrect again.

Although his subsequent criticism of Dorsalow is not relevant because their system does not contain a paper money, Mishan goes on to assert that money was omitted from Walras's statement of the value-of-aggregate-demand-equals-value-of-aggregate-supply-of-factor-services condition, and therefore that it could not have been meant to be an identity. I shall return to the consideration of this question in Section III. But let us remain for a time with Mishan's criticism of Dorsalow and his question of how they 'contrive' an identity if consumers can purchase money as well as real goods.

In such a system Walras's Law continues to hold with money included. If real goods are denoted by *j*, factor sevices by *i*, and money by *m*, then Walras's Law must be written

$$\sum_{i} X_j P_j + \triangle X_m \equiv \sum_{i} \bar{X}_i P_i \tag{21.1}$$

where X symbolizes demand, $\bar{X}$ denotes supply, and $\triangle X_m$ is the *desire* to increase money holdings and the *ex post facto* increase in money holdings. Indeed, it would be helpful if Mishan had written one of his functions in (6) and (7) of his paper explicitly as a residual expression derive from 21.1 above.

If real goods and factor services, taken together, are n in number, we should have n equilibrium expressions of the equality of demands and supplies in each of their markets. Now, however, these n equalities are independent, since Walras's Law includes desires to increase money holdings. If we add an $n + 1$ th equilibrium condition,

$$\triangle X_m = 0 \tag{21.2}$$

we have the same state of affairs that we have in the Dorsalow money-less model; that is, one of the $n + 2$ equations can be eliminated as redundant. Equations 21.1 and 21.2 imply that when $n - 1$ goods or factor markets are equilibrated, the nth is as well; or 21.1 and the n equilibrium conditions for real goods and factor service markets imply 21.2.

What Mishan asserts is that Walras eliminated one equation and one unknown from this system by substituting 21.2 into 21.1 to obtain:

$$\sum_j X_j P_j = \sum_i \bar{X}_i P_i \tag{21.3}$$

Let us suppose for the moment that Walras did in fact do this. Is Walras's Law (Identity) eliminated from the system? No, since the budget constraints give rise to it and allow the substitution which yields 21.3. Relations 21.1 and 21.2 form one subsystem in the model, and relation 21.3 forms another, and both subsystems are equivalent. Nothing fundamental is involved in choosing between them. Therefore, when Mishan writes that Walras, in choosing to write his system with 21.3 instead of 21.2 and 21.2, did away with the need for identities, he is being misled by appearances.

III THE WALRASIAN MODEL

The final question, therefore, is what Walras meant to convey by writing the aggregate budget constraint without including money assets. This is not so important a question as the others raised by Mishan, but may possess sufficient interest to justify a few brief comments of my own. I do not believe that Mishan begins to do justice to the complexities of Walrasian monetary theory. My own interpretations of that theory have been given elsewhere[9] in great detail and I shall not repeat them here. I shall merely assert that my own interpretation of Walras' exclusion of money assets from the aggregate budget constraint is simply that Say's Identity was an assumption of his models. When it is removed, Walras' Law contains money assets, a determinate price level is forthcoming from his models, and the monetary indeterminacy of his models is eliminated.

Several arguments seem to me to tell against Mishan's assumption that Walras's statement was an equilibrium condition to guarantee monetary equilibrium. First, Walras's models were not 'long-run' models in the Marshallian sense, but rather were short-run models in which extreme mobility of factor services allowed long-run characteristics such as the equality of price and average cost to emerge. Walras was not, therefore, concerned primarily with long-term equilibria, as is revealed in his discussions of the models as well as the structure of his models, contrary to Mishan's assertion.

Second, if Walras did accept the possibility of altering money asset holdings, the *individual consumer's* budget constraint should contain net demands for such money increments, and Walras's presentation of the consumer submodel should contain the apparatus of decision for such money stock changes. In neither case does Walras provide for the possibility. (The reader is cautioned that provision for *renting* cash balances is included in Walras's individual and aggregate budget constraints, but they are not relevant to the present argument because of the peculiarities of the Walrasian system.)

Third, Walras was far too interested generally in the manner in which the economy groped by *tâtonnements* from one equilibrium to another after the system was shocked to lead us to suspect that he would be uninterested in the attainment of a new price level with changes in money stock.

For these reasons, I believe Walras did in fact embrace Say's Law so that in writing 21.3 he meant it in fact to be an identity.

IV CONCLUSIONS

I must, therefore, disagree with all of Mishan's major points. I see no advantage to refusing the model-builder the use of behavioral identities. Given the fundamental limitations on consumers' searches for satisfaction, I believe it misleading to assert that Walras's Law as an identity is unnecessary to economic theory merely on the basis of the elimination of a variable and an equation. And I do not believe Walras stated his aggregate budget constraint as an equality, but rather, with the implicit adoption of Say's Law, as the identity Lange noted.

Notes

1. [3].
2. See [3], p. 619, fn. 1.
3. See [3], p. 618.
4. See [3], p. 617.
5. See [3], p. 620.
6. See [3].
7. See [3].
8. Dorfman, Samuelson, and Solow [1], p. 353.
9. See [2], pp. 311–25, 335–9.

References

[1] Dorfman, R., P. Samuelson and R. Solow, *Linear Programming and Economic Analysis*, McGraw-Hill, New York (1958).
[2] Kuenne, R., *The Theory of General Economic Equilibrium*, Princeton University Press, Princeton (1963).
[3] Mishan, E. M., 'Say's Law and Walras' Law Once More', *Quarterly Journal of Economics*, 77 (1963), pp. 617–25.

22 Money, Capital and Interest in Intertemporal General Equilibrium Theory*

I THE PROBLEMS WITH MONEY AND GENERAL EQUILIBRIUM MODELS

Like the classical theory of hydrodynamics which deduced that airplanes cannot fly, general economic equilibrium frameworks are ill equipped to confront economic reality. As a notable example, the existential reality of money as an institution is a jarring reminder of the analytical limitations of the most sophisticated general equilibrium models.

Debate on the issue of money has concerned the most fundamental problems. Is the existence of money compatible with a minimal-risk static model and *tâtonnement* or recontract trading (Hicks [6], Patinkin [11], Kuenne [7]) or do such rationalizations require costs and frictions that are unacceptable to such models' postulates (Hahn [3], Negishi [10])? Given the now admitted problems the Keynesian theory has in integrating money when interpreted strictly within the confines of its author's static model, was his framework of reference essentially neo-classical static equilibrium in nature (Patinkin [11], Kuenne [7]) or dynamic disequilibrium (Clower [1], Leijonhufvud [8])? Was its intended policy message that of the inefficacy of monetary policy in deep depression (see e.g. Hansen [5], ch. 9) or of the primacy of the tool (Leijonhufvud [8])?

Unfortunately, the concern with the existence of equilibrium in a static system with money is part of the rather sterile mathematical casuistry which has characterized general equilibrium modelling in

* Published originally in *Economie Appliquée*, 30 (1977), pp. 617–37, and reproduced with permission.

the last twenty-five years. It has become fashionable to worry about whether lapses from upper semi-continuous excess demand correspondences occur with the introduction of money (Hahn [3]), or whether the ability to pay taxes in money is sufficient to prevent an equilibrium with a zero price of money (Starr [12]). Currently rather less attention is being paid these recondite irrelevancies – whether in the strictest sense the model predicts the aircraft can fly – and more work is being devoted to understanding the relation of money to capital and interest in less formal equivalents of economic wind tunnels where money's existence is taken as a precondition (Malinvaud [9]).

In this paper I shall use some simple model constructs to obtain insights into and novel perspectives on the nature of interest in a productive economy with money, the relation of money to real capital stocks, and the relation of money to the interest rate. The work of von Neumann [13] and Malinvaud [9] will be particularly useful in this task, but I shall depart from their methods to study the insights yielded by disequilibrium states into the nature of money, capital, and interest phenomena, although no tightly determinate general economic system will be constructed. My aim is that of employing simple models to give novel and interesting viewpoints on these matters that point to methods of integrating money in more detailed and specialized models.

My view of large-scale microeconomic systems without numerical specification remains that which I expressed fourteen years ago (Kuenne [7]). They are metaphors or paradigms permitting partial glimpses into a complex economic reality we shall never fully understand, but whose greater understanding relies upon more imaginative and informal theorizing with the model as a guiding vision. Such models can never be expected to give us important operational theorems unless simplifying hypotheses are inserted that usually enhance the metaphoric quality to the mythical. As imperfect and nonperfectible constructs, these models at once threaten to constrain our imagination but free us from a dilettantish concern with 'realistic' assumptions. For example, practically speaking, we are effectively confined to (intertemporal) static depictions of a hugely dynamic economic reality, and my choice of models will reflect that limitation.

A model, then, is a paradigm that serves to guide the necessarily less formal theoretical interpretation of an extremely complex reality. Indeed, much of the sterile mathematical modelling of the past twenty-five years springs from the attempt to place too much of the theoretical burden upon the paradigm and its formality rather than using simpler models more imaginately if informally to attack problems of practical importance. My own feeling in viewing the history of economic thought is that the great theoretical performances were consistently of the latter form.

II THE CLOSED, LINEAR, SLACKLESS MODEL

We start with a closed von Neumann-type model in which time preference is infinitely negative so that no consumption occurs. Unlike the von Neumann model, however, we shall assume the society has a horror of 'waste' and insists upon using all its inputs.[1] All factors are producible goods that perish in one period so that wealth must be reinvested to be preserved and, thus, inventory speculation is not possible. In this economy we have the capital equivalent of the Robinson Crusoe type of pure labor economy in that capital is the only resource.

We assume that in each period t owners of goods sell them in terms of an external *numéraire* (say, the ruble) at present prices P_t to entrepreneurs in purely competitive markets where entrepreneurial skills are free goods.[2] They then use these proceeds to purchase 'bonds' at price $p_{b,t}$ which are denominated as the receipt of one ruble's worth of goods in $t + 1$ at prices P_{t+1}. The bonds are riskless credit instruments issued by entrepreneurs to get the means of production in t, and they are redeemed at the start of $t + 1$.

We have, then, for real goods, the dual system economy

$$A^{-1}X_t = X_{t+1}$$
$$P_tA = P_{t+1}, \qquad (22.1)$$

where A is an input matrix delineated in natural units (i.e. gallons, tons, etc.), X is a column vector of outputs, and P a row vector of prices. Prices are measured in the external *numéraire* (rubles) not priced by the system, and that *numéraire* price is based upon the value of the ruble in $t = 0$. Hence our price system is a constant external *numéraire* system. The *numéraire* serves only as a unit of account, no medium of exchange being used in markets.

System 22.1 implies that the economy is value-conserving over time, for

$$P_tA = P_{t+1}$$

$$P_tAX_{t+1} = P_{t+1}X_{t+1}$$

$$P_tX_t = P_{t+1}X_{t+1} \tag{22.2}$$

We assume that A is *productive* in the sense that there exists some X_t such that

$$A[I + G]X_t = X_t, \tag{22.3}$$

where G is a diagonal matrix with the scalar $g > 0$ on the diagonal. Thus, the economy is capable of sustained physical growth in all commodities.

Since all output in the system is paid to capital, and since competition will insure equal payments to holders of equivalent values of capital stocks, it is clear that capital is of necessity defined in units of value: a ruble's worth of any good in t buys a bond which is redeemed by the equivalent value of goods in $t + 1$ goods. And, from 22.2 the total equivalent value in $t + 1$ is *equal* to the value in t.

The *value* rate of net return in our economy, therefore, is zero. Interest is a return in physical units of goods designed to render equivalent value for equivalent value intertemporally in the same manner as a purely competitive system does intratemporally between markets. Of necessity, in a multigood system, interest must be defined in physical units of the *numéraire* good, or in *value*, and hence, like capital, must transcend the actual physical goods received in redemption of the bonds. Therefore, from system 22.2 and the assumption of purely competitive markets, interest as a net value return disappears, the capitalist in this sense being only as well off in $t + 1$ as he was in t, although in numbers of real goods he may be better or worse off, or indeterminately better or worse off.

In the face of this seeming paradox, theorists have been led to seek measures of interest in physical units of the subject goods themselves rather than *numéraire*, or *own-rates of return*, $r_{i,t}$, which are the incremental units of goods i that must be delivered in $t + 1$ to render equivalent value to a capitalist who lends a unit of the good in kind in $t + 1$. The total quantity of good that must be delivered in $t + 1$, or $q_{i,t+1}$, is

$$q_{i,t+1} = p_{i,t}/p_{i,t+1}, \tag{22.4}$$

and, therefore,

$$r_{i,t} = q_{i,t+1} - 1. \tag{22.5}$$

But the implicit attempt to convert interest into a non-*numéraire* physically-based measure of return is misleading. In general, own-rates will be different among goods and will often be negative (when $p_{i,t+1} > p_{i,t}$). It is tempting in these circumstances for theorists to flee to the stationary or balanced-growth state of 22.3 where all own-rates are equal because all prices in constant external *numéraire* are falling at the same rate. In effect, this stratagem reduces the multi-product economy to a Hicksian-aggregate single-product economy with 'a' rate of interest that *seems* to be defined in own-units. Such economies are extensions of Knight's Crusonia plant world with the potential difficulties Knight's analysis experiences in trying to explain the value dimension in capital and interest (Kuenne [7], ch. 4). And, of course, the appeal that the stationary state had to Wicksell was born of the desire to find a *natural* rate of interest with grounding (or *seeming* grounding) wholly in a universal physical rate of return.

But a general paradigm for capital and interest cannot restrict itself to single-good or balanced-growth multigood states. The value nature of capital and interest is misleadingly distorted by such exceptional states to arrive at a unique 'rate of interest'. That derivation must lie in other directions that fully accommodate the value nature of the phenomena and must not be purchased at the price of distorting the nature of capital and interest. We shall see in Section V that money is the proper instrument for that derivation.

Let us demonstrate that *cross-rates* of return in our economy will result in no advantages over own-rates. That is, suppose a holder of good i in t sells it for good j, buys bonds with the proceeds, receives good j in $t+1$, and sells good j for good i. He will receive no net advantage over selling good i in t for bonds and receiving good i in $t+1$ in redemption. Let $r_{i,j,t}$ be the cross-rate of good i for good j. Then

$$\begin{aligned} r_{i,j,t} &= \frac{q_{j,i,t}(1 + r_{j,t})p_{j,t+1}p_{i,t+1}^{-1} - q_{j,i,t}p_{i,t}^{-1}p_{j,t}}{q_{j,i,t}p_{i,t}^{-1}p_{j,t}} \\ &= (1 + r_{j,t})p_{j,t+1}p_{i,t+1}^{-1}p_{j,t}^{-1} - 1 \\ &= p_{i,t+1}^{-1}p_{i,t} - 1 \\ &= r_{i,t}, \end{aligned} \tag{22.6}$$

Table 22.1 Price, output and own-rate results for t = 1

Goods	$p_{i,0}$	$x_{i,0}$	$p_{i,1}$	$x_{i,1}$	$r_{i,0}$
1	2	1	3.80	.83	–.47
2	5	1	2.30	1.37	1.17
3	3	1	1.60	2.32	.88

where $q_{j,i,t}$ is the quantity of good j received per unit of good i in t (or $p_{i,t}/p_{j,t}$).

Before we move to incorporate more useful complications into our model, let us illustrate system 22.1 with a simple three-good example. Suppose

$$A = \begin{bmatrix} .1 & .5 & .1 \\ .6 & .2 & .1 \\ .2 & .1 & .3 \end{bmatrix}, \qquad A^{-1} = \begin{bmatrix} -.69 & 1.91 & -.40 \\ 2.19 & -.14 & -.69 \\ -.27 & -1.23 & 3.83 \end{bmatrix}$$

and $P_0 = [2, 5, 3]$, $X_0 = [1, 1, 1]$.

Then, for $t = 1$, the solutions are given in Table 22.1.

Profits are maximized at zero, which implies that each capital good's marginal value product equals its price in $t = 0$. Column i of A^{-1} reveals the marginal output of each good that will accrue to an additional unit of good i, all other inputs held constant. Multiplying these column elements by relevant $t + 1$ prices and summing gives the marginal value product of good i, which will equal $p_{i,t}$. For example, from the specification of A^{-1} and Table 22.1 above, for good 1

$$MVP_1 = [(-.69 \times 3.80) + (2.19 \times 2.30) + (-.27 \times 1.60)]$$
$$= 2.00 = p_{1,0}.$$

Finally, let us demonstrate that cross-rates equal own-rates so that arbitrage between goods is ineffective. Suppose a holder of good 1 in $t = 0$ sells it and buys good 2, of which he commands $q_{2,1,0} = p_{1,0}/p_{2,0} = 0.4$ units per unit of good 1 sold. He lends this in kind (eliminating the bond market) and receives $0.4(1 + r_{2,0}) = 0.87$ units of good 2 per original unit of good 1, which he sells for 0.53 units of good 1. But this is exactly the amount he would have received had he sold good 1 and bought bonds which were redeemed in good 1 (or any other good sold at $t + 1$ prices to obtain good 1).

This simple pure capital economy must now be made more realistic, but it serves to give some simple lessons about the nature of capital and interest. Capital in the simple Walrasian sense of an inventory of physical good stocks will not suffice: it can only be meaningfully defined in a value dimension. In this regard the Austrian goal of measuring it in terms of the period of production can be sympathized with, if not accepted. Similarly, interest also must be defined in value terms, and attempts to convert it into purely physical terms – a weakness of productivity theorists of the interest rate – are contrived and misleading. And, finally, in this sense, interest does not exist at all! Payments to capitalists constitute rather a means of preserving the equivalence of value intertemporally – a result of the competitive market mechanism which in its intratemporal setting is usually accepted as equitable.

III AN ECONOMY OPEN TO CONSUMPTION AND PRIMARY FACTORS

Let us assume first that capitalists consume a proportion k_i of good i each period, so that system 22.1 becomes

$$\begin{aligned} &1.\ A^{-1}[I-K]X_t = X_{t+1} \\ &2.\ P_tA = P_{t+1}, \end{aligned} \qquad (22.7)$$

where K is a diagonal matrix with the k_i on the diagonal. Prices are not affected, of course, as no new costs are introduced into the system, as they would be if the stock reduction were caused by spoilage or other depreciation. But system 22.2 becomes

$$P_t[I-K]X_t = P_{t+1}X_{t+1}. \qquad (22.8)$$

The value of capital stocks is preserved, since the value of output each period is equalized to the value of the capital stocks used in production. Of course, the value of output falls from its potential size by the value of consumed goods converted to lost output. And the own-rate of return on a unit of good actually invested remains the same as in system 22.1.

Let us also open up the economy to the existence of primary, nonproduced factors with payments W_t for their use. These payments will be assumed for convenience to be wholly consumed and to account for all consumption. Hence, the consumption proportions k_i are adjusted each period to yield the amounts of goods so consumed.

System 22.1 must therefore be rewritten as

$$\begin{aligned} A^{-1}[I - K]X_t &= X_{t+1} \\ P_t A + W_t B &= P_{t+1}, \end{aligned} \tag{22.9}$$

where B is the input matrix for factor services. It follows from 22.9 that

$$P_t[I - K]X_t = P_{t+1}X_{t+1} - W_t B X_{t+1}. \tag{22.10}$$

From the assumption that all primary factor earnings are consumed and only such earnings are consumed,

$$W_t B X_{t+1} = P_t K X_t, \tag{22.11}$$

so that 22.10 can be written

$$P_t X_t = P_{t+1} X_{t+1}. \tag{22.12}$$

The introduction of the source of consumption payments into the cost and therefore the price structure returns the system to the value conservation of our closed model. The value-added components increase prices through costs, but consumption subtracts out an equal amount of value through reduction in the physical availability of produced inputs. In general, prices will be higher and quantities smaller, therefore, than in the closed economy, but competition will ensure the value return to capital will be zero, returns to capital equalizing values intertemporally.

Let us illustrate once more with another simple example. Suppose A remains that given in our previous example and that P_0 and X_0 are those given in Table 22.1. Assume further that B, W_0, and K_0 are

$$B = \begin{bmatrix} .15 & .11 & .30 \\ .12 & .24 & .02 \end{bmatrix}, W_0 = [1.50, 3.25],$$

$$K_0 = \begin{bmatrix} .184 & 0.000 & 0.000 \\ 0.00 & .229 & 0.000 \\ 0.00 & 0.000 & .273 \end{bmatrix}$$

Table 22.2 Price, output, and own-rate results for $t = 1$

Good	$p_{i,0}$	$x_{i,0}$	$p_{i,1}$	$x_{i,1}$	$r_{i,0}$	*Consumption* ($c_{i,0}$)	*Investment* ($n_{i,0}$)
1	2	1	4.42	.62	–.55	.184	.816
2	5	1	3.25	1.18	.54	.229	.771
3	3	1	2.12	1.61	.42	.273	.727
	$P_0X_0 = 10$		$P_1X_1 = 10$			$P_0C_0 = 2.33$	$P_0N_0 = 7.67$

Value Added = $W_0BX_1 = 2.33$

$P_0X_0 - P_0C_0 = P_0N_0 = 7.67$ = Capital goods input

$P_1X_1 - W_0BX_1 = = 7.67$ = Depreciation plus interest

The solution values for $t = 1$ are those given in Table 22.2.

Interest in a Ricardian constant external *numéraire* economy where wages and rents are fully spent and all capital reinvested must be zero for the preservation of value through time. If capitalists consume some portion of their stocks the same principle holds for a lessened capital input. If wage and rent earners save a portion of their incomes they too would, as new capitalists, receive an own-rate or cross-rates on the redeemed investments in the bond market which preserved the value of the investment over time. Hence, in the open economy, the fundamental principles we established for the closed economy remain intact.

IV THE ECONOMY WITH POSITIVE TIME PREFERENCE

We have been dealing up to this point with an economy whose capital sector reveals an infinite preference for future over present goods. Let us suppose now that to induce the reinvestment of total product in the closed economy requires a premium of r units of *numéraire*, so that the current price of bonds determined in the market will be $(1 + r)^{-1}$ rubles rather than 1 ruble. For simplicity we will return to the closed economy because we have seen that the complications of the open economy do not change the basic principles of the simpler system.

This innovation affects the price system of 22.1:

$$A^{-1}X_t = X_{t+1}$$
$$P_t[I + R]A = P_{t+1}, \qquad (22.13)$$

where R is a diagonal matrix with r along the diagonal. Then, from 22.13,

$$P_t[I + R]AX_{t+1} = P_{t+1}X_{t+1}$$

$$P_t[I + R]X_t = P_{t+1}X_{t+1}$$

$$P_tX_t = (1 + r)^{-1}P_{t+1}X_{t+1}. \qquad (22.14)$$

The real outputs of the system are unaffected, so that in a real sense the society is exactly as well off as before the introduction of positive time preference, because our simple system is one in which no allocation effects depend upon prices. The values of all goods in $t + 1$ merely rise by $r \times 100$ per cent over their nominal value-preserving levels. But, in a real-value sense, interest once more is a zero-value net return to keep value equivalencies. Own-rates of return are the same as those in the economy of Section I where $r = 0$, and hence are negative or positive as prices rise or fall in $t + 1$.

Von Neumann's model, of course, is the special case of the economy that has attained balanced growth in which all own-rates are the same and positive (for scarce goods):

$$\begin{aligned} A[I + G]X_t &= X_t \\ P_t[I + R]A &= P_{t+1}, \end{aligned} \qquad (22.15)$$

where $g = r$ in matrices G and R which are diagonal with g and r on the diagonals respectively. Von Neumann identified the only nonproduced factor of the economy as 'waiting' and hence denoted the physical return g on each type of good as the reward for waiting. But this use of a physical own-rate as the rate of interest in an economy and a net return is misleading, as our analysis of Section I indicates. Our present analysis reinforces this by showing that the time preference factor in this economy merely affects prices. Payment to the owners of the capital stock renders the intertemporal real values of the goods equal to yield a zero real rate of value return. That this special case implies that such equivalence requires an equiproportional physical return on each good is a peculiarity of the state, not its essence. In the general case, the only 'reward' for waiting is to receive an equivalent value in $t + 1$ goods, which may imply receiving less of specific goods, not more.

Once more, the existence of positive time preference and the consequent need for bonds to sell at an r per cent discount in period t to achieve the same degree of saving as was achieved at a 0 per cent discount in our Section III open economy will merely boost all prices by r per cent and leave all real magnitudes unchanged. System 22.9 becomes

$$A^{-1}[I - K]X_t = X_{t+1} \quad (22.16.1)$$

$$P_t[I + R]A + W_t[I + R]B = P_{t+1}, \quad (22.16.2)$$

where value added payments are advanced in period t from capital stocks accumulated through the sale of bonds. By assumption, value added is consumed and constitutes all consumption, so 22.11 continues to hold and 22.16.2 can be rewritten

$$P_t[I + R]AX_{t+1} + W_t[I + R]BX_{t+1} = P_{t+1}X_{t+1}$$

$$P_t[I + R][I - K]X_t = P_{t+1}X_{t+1} - W_tBX_{t+1} - W_tRBX_{t+1}$$

$$P_t[I + R]X_t - P_t[I + R]KX_t + W_tRBX_{t+1} = P_{t+1}X_{t+1} - W_tBX_{t+1}, \quad (22.17)$$

which, from 22.11 may be written

$$P_tX_t + P_tRX_t - P_tRKX_t + W_tRBX_{t+1} = P_{t+1}X_{t+1}. \quad (22.18)$$

Hence

$$P_tX_t + P_tR[I - K]X_t + W_tRBX_{t+1} = P_{t+1}X_{t+1}, \quad (22.19)$$

and prices will increase proportionally by the interest paid on both forms of capital since all $x_{i,t}$ and $x_{i,t+1}$ will be the same as in the economy of Section III.

Therefore, whether equivalent amounts of investment must be induced by more or less discount on bonds due to time preference, and whether the economy is open or closed to consumption and primary factor payments, does not interfere with the basic lesson of these simple constant external *numéraire* economies:

1. Capital can only be defined meaningfully in value terms, that is, in physical equivalents of the *numéraire* good, and to seek exceptional models that permit the theorist to argue as if it were physically measurable in own-units is to mislead.
2. Interest can only be defined meaningfully as a net value return to a value capital investment, own-physical returns being derivative from such fundamental values.

3. The rate of interest in the sense of 2 is zero in real terms, being determined by the value of capital in $t + 1$ that exactly equals the value invested in t in real terms.
4. The holder of capital goods receives a return in nowise different over time from that received by an exchanger of goods within a time period – equivalent value.
5. It is impossible to understand the general phenomenon of interest in a general equilibrium context without taking into account the changes in the prices of goods that occur in general from period to period in a productive economy. In this dimension, too, to find refuge in stationary or balanced-growth economies where relative prices remain unchanged in order to net out price changes is to hide from view an important aspect of the problem.

V THE CURRENT-*NUMÉRAIRE* ECONOMY

From the narrow standpoint of understanding and defending interest, it is unfortunate that the constant-*numéraire* economy is unfamiliar to us. We are used to a current-*numéraire* economy in which prices in each period are related to a standard commodity whose value is artificially fixed from one t to the next despite its fluctuation in constant-*numéraire* value. This realistic and useful device perverts the notion of equivalent valuation over time, but its adoption explains the derivation of a single interest rate for all goods based on value and different from zero.

To illustrate the current-*numéraire* economy let us arbitrarily choose good 3 in Table 22.2 to fulfill that role by assuming its value is identically 1 in each period. This conversion of the system from a constant external *numéraire* economy to a current internal *numéraire* economy is accomplished formally by multiplying the prices of $t = 0$ and $t = 1$ in Table 22.2 by the factor $\beta_t = 1/p_{3,t}$ derived from that table. The prices on the new basis are those given in Table 22.3, and we place bars over them to distinguish them from constant-*numéraire* prices.

Of course, the transformation does not disturb the as yet moneyless economy because such economic systems are homogeneous of degree zero in all constant-*numéraire* prices, which is to say that only relative prices are determinate. We are free to anchor such relative prices absolutely as we have done in Table 22.3. Therefore, the current-*numéraire* prices yield the same *intratemporal* information as

Table 22.3 Open consumption economy prices on a current-*numéraire* basis

Good	$\bar{P}_0$	$\bar{P}_1$	r_0
1	.67	2.08	–.54
2	1.67	1.53	.54
3	1.00	1.00	.42

the constant-*numéraire* prices. But this is true of intertemporal prices only if we know the β_t, which we have derived from the constant-*numéraire* system, and which are ordinarily unattainable in a current-*numéraire* economy.

If we translate the definition of own-rates into the new pricing system, we obtain

$$r_{i,t} = (\beta_{t+1}/\beta_t)(\bar{p}_{i,t}/\bar{p}_{i,t+1}) - 1. \tag{22.20}$$

If we know the ratio of the current-*numéraire* prices in two periods *and* the ratio of the conversion factors for those periods we may derive the own-rates for the goods. In the special case of the *numéraire* good,

$$r_{3,t} = \frac{\beta_{t+1} - \beta_t}{\beta_t}. \tag{22.21}$$

If we can derive the own-rate for the *numéraire* good we have solved the problem of reclaiming the vital intertemporal pricing information necessary for the functioning of the economy over time.

The difficulty, of course, occurs because the price of good 3 in a constant external-*numéraire* sense, like the price of every other good, is subject to change each period. Were it to remain constant in terms of an external *numéraire* from period to period $\beta_{t+1}/\beta_t = 1$, and direct computation of the $r_{i,t}$ using $\bar{P}_t$ and $\bar{P}_{t+1}$ would be possible. In the absence of this ideal *numéraire*, the problem of determining intertemporal value linkages can be solved only if a futures market exists in t for the *numéraire* good. In this market a unit of good 3 in week t will be exchanged for a given amount of good 3 in $t + 1$, yielding direct observation of $1 + r_{3,t} = \beta_{t+1}/\beta_t$. Since $\bar{p}_3$ cannot change from its value of 1, quantitative differences in such contracts must reflect value changes in good 3 in a constant-*numéraire* sense. Hence, instead of the value rate of return being zero – which we have seen it to truly be in a constant external-*numéraire* economy – we have shifted the price system to make the current-*numéraire* value rate of return on good 3 equal to $r_{3,t}$. Herein lies the heart of the intertemporal transformation and also a fundamentally deceptive illusion: what is in a truly intertemporal sense a zero rate of return is made to seem (usually) a positive net return.

And, of course, the bond market provides the necessary futures market for the *numéraire* good. In it, a bond promising one unit of *numéraire* in $t + 1$ is priced in units of t *numéraire*. Therefore, $p_{b,t} = (1 + r_{3,t})^{-1} = p_{3,t+1}/p_{3,t} = (\beta_{t+1}/\beta_t)^{-1}$, the last being the inverse of the ratio we seek. To earn a zero net value return, holders of bonds must earn $(1 - p_{b,t})/p_{b,t} \times 100$ per cent return in *numéraire*.

Thus, the *numéraire* good acquires an intertemporal as well as an intratemporal function: its own-rate of return is converted to the economy's value rate of return, now no longer zero in the nominal sense. A different choice of *numéraire* good, therefore, results in a different value rate, since own-rates are in general different. Own-rates from 22.20 are now defined as

$$r_{i,t} = (1 + r_{3,t})(\overline{p}_{i,t}/\overline{p}_{i,t+1}) - 1. \tag{22.22}$$

The ratio of prices in 22.22 indicates the number of units of good i in t that are the equivalent of a unit in $t + 1$ on the assumption that a unit of value is equivalent in both periods. The good 3 term corrects this ratio by introducing the true rate of exchange between *numéraire* units between the periods. This term becomes a common discount factor relating *numéraire* values over time. Since own-rates are unchanged from those in the constant-*numéraire* economy, it follows that cross-rates will be equal to own-rates once more.

We now begin to see the emergence of an 'interest rate' for the economy even when all own-rates are unequal and the system is not in a state of stationarity or balanced growth. For, from 22.22

$$1 + r_{3,t} = \frac{\overline{p}_{i,t+1}}{\overline{p}_{i,t}}(1 + r_{i,t}), \tag{22.23}$$

with $r_{3,t}$ the nominal net value rate of return on all goods. We may demonstrate this by defining $v_{i,t}$ as the nominal value rate of return on good i in t in a current-*numéraire* economy:

$$v_i = \frac{\bar{p}_{i,t+1}q_{i,t+1}}{\bar{p}_{i,t}q_{i,t}} - 1. \tag{22.24}$$

We know that in a constant-*numéraire* economy $v_{i,t} = 0$, or:

$$0 = \frac{p_{i,t+1}q_{i,t+1}}{p_{i,t}q_{i,t}} - 1. \tag{22.25}$$

Thus, if we subtract 22.25 from 22.24

$$v_{i,t} = \frac{q_{i,t+1}}{q_{i,t}}\frac{\bar{p}_{i,t+1}}{\bar{p}_{i,t}} - \frac{p_{i,t+1}}{p_{i,t}}, \tag{22.26}$$

or, from the derivation of current-*numéraire* prices,

$$\begin{aligned} v_{i,t} &= \frac{q_{i,t+1}p_{i,t+1}}{q_{i,t}p_{i,t}}\frac{\beta_{t+1}}{\beta_t} - 1 \\ &= \frac{\beta_{t+1}}{\beta_t} - 1 \qquad (22.27) \\ &= r_{3,t}, \end{aligned}$$

the last substitution coming from 22.21. Thus, the value rate of return on every capital good becomes nominally the own-rate of the *numéraire* good.

If, in addition, the cross-rates of the *numéraire* good are of more practical concern than the own-rates of other goods, this distinction of the *numéraire* good's own-rate will be enhanced. For instance, if good i is typically not lent in kind in t for a future receipt of good i in $t + 1$, but is rather lent against the receipt of a specific amount of the *numéraire* good in $t + 1$, and this receipt may be converted into good i or another good or held as *numéraire* good, then own-rates other than the *numéraire*'s become uninteresting. The *numéraire*'s own-rate emerges in full distinction as 'the interest rate' of the economy – the only intertemporal measure of equivalence that has fundamental importance, all others being truly derivative.

We have then arrived at the intertemporal function of the traditional concept of money. It provides, first, the *numéraire* function of measuring intratemporal and intertemporal values in physical amounts of itself; second, its *services* each period as a medium of exchange, forming the obverse of every spot and futures market in the physical sense of being the 'other good' traded; and third, its *stock* as that of any other capital good which yields scarce services and a means of storing value intertemporally (indeed, in our simple economy, where all goods are perishable, the only means).

VI THE MONEY ECONOMY

Money combines these features of *numéraire*, medium of exchange and store of value admirably in reality; but as we have noted in Section I modern general equilibrium theory still struggles to understand the conditions for existence of such an instrument and the roles that transaction costs, disequilibrium trading and price uncertainty play in defining its relations to other goods.

One cause of this failure is the bona fide complexity of the forces of uncertainty and disequilibrium trading that converge upon the money good in a modern economy. To a certain extent the sense of failure is generated by a desire for mathematical purity that has no practical significance and the implicit refusal to accommodate more informal and nonrigorously specified preconditions for money's existence. But, as I indicated in [7], ch. 5, I continue to believe that an important contribution to the causes of the imperfect integration of money into general systems has been the dominance in Anglo-Saxon economics of the Marshallian framework and vision. I include in this category of analysis (*pace* Clower and Leijonhufvud) the Keynesian variant, and I interpret Patinkin's contribution as largely that of instilling a Walrasian correction to these shortcomings.

Those limitations are essentially two. The first is that highlighted by Patinkin's work: the Marshallian framework incorporated money in a manner that did not permit its linkages via substitutability and complementarity relations with other goods to manifest themselves fully. Partial analysis with *ceteris paribus* was crippling in these dimensions for goods other than money, but this general limitation was worsened by the isolation of the latter good in the cash-balance equation. The second limitation was one that my own work has striven to eliminate: the Marshallian theory did not permit a valid separation of the *services* of money as a transactions medium from its *stock* role as an alternative asset. We shall discuss each of these limitations separately.

The Linkage of Money to Other Goods

Most germanely for present purposes, Keynes strove desperately to overcome this first theoretical handicap of his Marshallian heritage, but in my reading of the *General Theory* was constrained by his overriding desire to explain underemployment equilibrium and in the end failed to correct it (Kuenne [7], pp. 346–61). Leijonhufvud [8] has disputed my understanding and that of other 'neo-classical' interpreters of Keynes with an ably argued alternative presentation of the Keynesian theory, dynamic and disequilibrium in nature, as opposed to the Keynesian model, which was static and equilibrium in form, where the terms theory and model are distinguished in the manner of the discussion in Section I. That is, Leijonhufvud argues that Keynes used the static, formal paradigm he created largely in the Marshallian image to present a more informal and novel explanation of the persistence of unemployment in a money economy.

In an important sense, Leijonhufvud's argument makes important concessions to those of us who attempted explicitly to probe the nature of Keynes's *model*. In Leijonhufvud's view it is static and best approached through a general equilibrium excess demand interpretation. Four aggregate goods exist: (*1*) consumer goods, (*2*) labor services, (*3*) nonmoney assets, containing a Hicksian aggregate of bonds and capital goods whose prices move in lockstep, and (*4*) money. A general equilibrium with unemployment in this economy is impossible since the interest elasticity of investment in Keynes's system is assumed to be high because nonmoney assets are typically taken to be long-lived. The short-run consumption function does possess a significant real-wealth effect via interest rate changes. There is, therefore, at least an implicit agreement that to the extent Keynes's theory was tied to his static model it was a failure.

Leijonhufvud ridicules as 'the same old compromise' ([8], p. 334) my assertion ([7], pp. 360–1) that 'involuntary unemployment is most readily, perhaps only, explainable in terms of the dynamics of adjustment of the market economy. To the extent that Keynes focussed attention upon the slowness, or weakness, of these adjustments his analysis contributes fruitful insights.' Yet it is the conclusion that he reaches, using Clower's interpretation of Walras's Law in a dynamic disequilibrium process as the clue to understanding Keynes's theory.

Hence, Leijonhufvud's effort to interpret Keynes's theoretical departure from his model is an important one. Whether or not the theory was actually Keynes's – that is, whether Keynes was so successful in escaping the Marshallian money tunnel-vision – is not of great importance. The insights that interpretation yields into the functioning of a money economy are, and here Clower and Leijonhufvud have made important contributions. Hence, a brief review of the dynamic disequilibrium theory Leijonhufvud presents as Keynes's is important in the more general task of judging the role of money and capital in modern general systems.

The Keynesian theory, Leijonhufvud argues, is a disequilibrium theory employing a realistically constrained market adjustment mechanism that cumulates initial disturbances and prevents re-establishment of equilibrium for long periods. Three distinctive Keynesian features characterize this crippled dynamic adjustment. First, there exists the familiar Clower [1] failure of the excess demands for goods to mirror an excess supply of labor in markets where the second good is money. Second, this inability of entrepreneurs to sense the latent demand for goods leads them to reduce quantities, not prices, in the short run as they search for better information. Hence, an initial unemployment disturbance cumulates via the multiplier mechanism. Lastly, given the sensitivity of investment to interest rate changes and therefore to price changes in Keynes's bonds-capital goods aggregate, the economy could be righted with relatively small declines in the interest rate. The barrier to such an outcome – even if sought with an aggressive monetary policy – is not the liquidity trap but the difficulty in the short run of altering the public's longer-run notion of 'normal' interest rate levels. Liquidity preference theory, therefore, was not Keynes's theory of interest rate determination but rather of the *changes* in interest rates in the short run.

Therefore, although Keynes's *model* was solidly in the Walrasian general equilibrium tradition, with the money sector fully integrated via real-balance, real-wealth substitution and income effects with the real sector, to the degree that money was a crucial variable in the system and monetary policy all-important in a static world of *tâtonnement* adjustment, the real and largely unread message lay in the imperfect dynamics of the *theory*. Hence, Keynes had escaped – and grandly – from the first defect of the Marshallian money construction. Indeed, it was the malfunctionning of the price structure that ultimately caused the disequilibrium in that the overly high prices of capital goods compared with consumer goods and wages constituted the core of the disequilibrium problem and the inability of interest rate changes to correct it the source of capitalism's depression problem. Disturbingly, the logic of Leijonhufvud's analysis leads one to conclude that Keynes should have placed great stress upon the reduction of money wages as the efficient way out of unemployment in order to get the cumulative mechanism operating in the desired direction!

The Distinction Between Money Services and Money Assets

The second, and not completely independent, defect of the Marshallian money framework, is that it never permitted the cash-balance or service role of money to be adequately distinguished from the asset or stock function. The desire to hold money for transactions purposes to employ its services as a universal inventory and to pay the price of such services – the interest rate – was not clearly distinguished from the desire to accumulate stocks of money as capital goods yielding such saleable services. Keynes strove valiantly within his Marshallian heritage to emphasize the asset role in his liquidity preference theory, but intermingled the demand for cash-balance services somewhat inadequately in the same market as that in which asset demand is active. I have argued elsewhere ([7], pp. 341–61) that both the needs of Keynes's model and his Marshallian background led him to eliminate the links between paper and real-good markets, so that the excess demand for money for either purpose was reflected wholly in the value of excess supply of bonds – Keynes's Identity. Hence money is not truly integrated in a framework that remained stubbornly Marshallian, most particularly in the sense of linking it to real capital goods and 'the interest rate'.

My model, inspired by the Walrasian framework, does permit a more successful interweaving of these phenomena. We simply assume a degree of transactions friction – be it credit risk, transactions cost, etc. – that leads economic agents to hold cash balances at positive cost.[3] The demand for cash balances takes the form of selling promissory notes guaranteeing the delivery of $1 + i$ units of money in $t + 1$ for every unit borrowed in t. Because the money unit is the standard of a current-*numéraire* price system, having the value of unity in both periods, i becomes a physical own-rate of return for money *and* the *value* rate of return for the economy. These cash balance services, m, are provided by holders of money stocks, M, which are stocks of earning assets like any other capital stocks.

Cash balances, which must be returned in $t + 1$, can be acquired only in the promissory note market (including the fictitious issuance of such notes to onesself at the implied opportunity cost if one uses one's own money assets). The excess demand for cash balances, therefore, is directly reflected in a value of excess supply of such notes. On the other hand, the decision to 'purchase' money as a capital asset is to add it 'permanently' to one's capital stocks. It may be 'purchased' by reducing one's demands in goods markets. Hence, excess demand for money as an asset is reflected in excess supplies of all goods and service markets – not simply a bond market, as in the Keynesian interpretation.[4]

Let us illustrate this integration by starting with the simple three-good constant-*numéraire* economy of Section I, with a money sector added. The investor now has two options: to purchase real capital goods in $t + 1$ through the purchase of bonds at $p_{b,t}$, or to buy money stocks at the price $p_{M,t}$ in *numéraire* by reducing his demand for bonds in our simple economy (and for all goods and services as well in less simplistic systems). The new money holdings are added to wealth stocks in $t + 1$. In an economy in which transactions motives are dominant, the desire to hold money as an asset inheres in the ability to sell its cash-balance services each period through the purchase of promissory notes that must be redeemed next period in the money borrowed plus net payment. Where money is 'productive', or useful in overcoming transactions frictions, the premium will be positive, hence its constant-*numéraire* price will tend to decline over time.

The own-rate of return on money in a constant-*numéraire* economy will be positive if the *numéraire* price of money falls. But, of course, there is no reason to expect this induced fall in price to occur. In a regime with a fixed amount of paper currency, the price of money is free to change without the need to conform to a cost of production. If $p_{m,t+1}$ is the *numéraire* redemption price of a promissory note lending 1 unit of money in t, then

$$p_{m,t+1} = q_{M,t+1}\,p_{M,t+1} = (1 + r_{M,t})p_{M,t+1},$$

and

$$r_{M,t} = \frac{p_{M,t} - p_{M,t+1}}{p_{M,t+1}}.$$

The prices of money in t and $t + 1$ must adjust to yield the same zero net value return as yielded on bonds and capital goods. In so doing, the exchange value of goods for money – the absolute price level – is determined. This implies a positive own-rate on money if money is depreciating and a negative own-rate if it is appreciating.

When money is made the internal *numéraire* good, so that prices are measured in it and all goods and securities traded for it, own-rates for other goods become derivative and uninteresting, as noted in Section V. The own-rate of return on money – the ratio of terms in 22.21 – is directly established in the promissory note market and becomes the value rate of return for the current-money-*numéraire* economy, as we have shown there. Even though in our simple economy it is the money market's return that must adjust to the return on real capital (by virtue of purely competitive production), the choice of money as a *numéraire* seems to impose the reverse discipline. We have finally arrived at an 'interest rate' common to all assets in the simple economy with which we are dealing, even in its intertemporal disequilibrium phases. Money takes its place as an earning asset on full footing with physical capital goods and adjusts its own rate of return to yield a zero real value return over time.

VII CONCLUSION

The key to the successful integration of money with the real sectors of general equilibrium theory lies in the development of models in which money's saleable services can be split away from its role as an asset, and markets can be identified for the pricing of each. In our simplest of models the pricing of money as an asset is accomplished by excess demand adjustment for bonds (i.e. capital goods). The price of its services (the interest rate) is determined in a market in which they are sold. The mutual adjustment of these two prices, under the general requirement that the rate of return on all assets be equal provides the means of integrating the capital and money markets.

The common rate of return on assets is dictated by the ratio of the value of money services to the value of the money *numéraire* in the promissory note market. This is determined by the demand for and supply of money services and money assets in full competition with other paper and real goods. But, more deeply and less deceptively, the phenomenon of interest is interpretable only within a framework where prices of all goods and services are changing and where the fundamental principle is that of replacing equivalent value for equivalent value intertemporally. The true value rate of return is necessarily zero.

Notes

1. This causes us certain technical problems. Inability of the model to proceed directly to the 'balanced growth' input configuration by 'wasting' some resources means that it will be unstable over time and ultimately destroy itself unless initial stocks happen to be in the optimal configuration. For our purposes this will not be of concern, since the nature of our disequilibrium solutions is that of a von Neumann model when it is off the turnpike.
2. We will follow the convention of signifying vectors by capital letters and scalars by lower case letters. No distinction will be made between row and column vectors, this characteristic being clear from context.
3. For an extended attempt to justify the existence of money in an intertemporal static general equilibrium, see Kuenne [7], pp. 291–305. More recently, Grandmont and Younès [2] have made an assumption similar to the one made by me.
4. See, for example, the discussion by Negishi [10].

References

[1] Clower, Robert, 'The Keynesian Counterrevolution: A Theoretical Appraisal', ch. 5 in [4].

[2] Grandmont, Jean-Michel, and Yves Younès, 'On the Role of Money and the Existence of a Monetary Equilibrium', *Review of Economic Studies*, 39 (1972), pp. 355–71.

[3] Hahn, Frank H., 'On Some Problems of Proving the Existence of Equilibrium in a Monetary Economy', ch. 6 in [4].

[4] Hahn, Frank H., and F. P. R. Brechling, (eds.), *The Theory of Interest Rates*, Macmillan London (1965).

[5] Hansen, Alvin, *Monetary Theory and Fiscal Policy*, McGraw-Hill, New York (1949).

[6] Hicks, John, 'A Suggestion for Simplifying the Theory of Money', *Economica*, NS 2 (1935), pp. 1–19.

[7] Kuenne, Robert E., *The Theory of General Economic Equilibrium*, Princeton University Press, Princeton (1963).

[8] Leijonhufvud, Axel, *On Keynesian Economics and the Economics of Keynes*, Oxford University Press, New York (1968).

[9] Malinvaud, Edmund, 'Capital Accumulation and Efficient Allocation of Resources', *Econometrica*, 21 (1953), pp. 233–68.

[10] Negishi, Takashi, 'Market Clearing Processes in a Monetary Economy', ch. 8 in [4].

[11] Patinkin, Don, *Money, Interest, and Prices*, Row Peterson New York, 2nd edn (1965).
[12] Starr, Ross M., 'The Price of Money in a Pure Exchange Monetary Economy with Taxation', *Econometrica*, 42 (1974), pp. 45–54.
[13] Von Neumann, John, 'A Model of General Equilibrium', *Review of Economic Studies*, 13 (1945–46), pp. 1–9.

Index